O
TD   Fuel and the environment.
195
.M5
F83
1973

Fuel and the environment

Congress Theatre, Eastbourne. 26–29 November 1973

Fuel and the environment

Conference proceedings

ann arbor science PUBLISHERS INC.

P.O. BOX 1425 ● ANN ARBOR, MICHIGAN 48106

ANN ARBOR SCIENCE PUBLISHERS, INC.
P.O. BOX 1425
ANN ARBOR, MICHIGAN 48106 USA

Library of Congress Catalog Card No. 74-75704

First published in England by
The Institute of Fuel.

First published   1973
© The Institute of Fuel

ISBN 250-40058-8

This edition is published by special arrangement with the British
Institute of Fuel. The content is the same except in the edition
published by The Institute of Fuel, original figures were published in color as follows: in Chapter II—figures 1, 3, 6, 7, 8;
Chaper IV—figures 1-5; Chapters V—figures 1, 2, 3, 4, 5, 6,
8, 12, 14, 15, 17, 18, 21, 22, 23, 24, 26, 27, 28; Chapter VI—
figures 12, 13, 15, 16, 17.

Papers

Organizing Committee

Chairman
HRH The Duke of Edinburgh, KG, KT, FRS, HonMInstF

Deputy Chairman
F. E. Ireland, BSc, CEng, FInstF

D. Armstrong, CEng, MInstF
A. N. Britten, BSc, CEng, MIMechE
D. H. Broadbent, BScEng, CEng, FICE, FIMechE, FIEE
F. Broomhead, CEng, MIChemE, MInstF
S. Cayton, MBE, FRSH, MAPHI, MInstPC, MInstF
J. H. Chesters, OBE, DScTech, PhD, FRS, FInstF
D. R. R. Fair, OBE, BSc, CEng, AKC, AInstP, FInstF
J. H. Flux, BSc, CEng, FIGasE, FInstF
B. R. Fuller, OBE, BSc, ARCS
J. W. Furness, MIMechE
J. E. Garside, BSc, PhD, CEng, FInstF
H. J. Hay, CEng, MIChemE, MInstF
R. F. Hayman, BSc, CEng, FRIC, FIGasE, FIChemE, FInstF
G. E. Mercer, OBE
G. A. Payne, BSc, CEng, MIGasE, MInstF
L. E. Reed, MSc, PhD, CEng, MIMechE, MInstF
C. A. Roast, CEng, FIMechE, MIEE, FInstF
G. G. Thurlow, BSc(Eng), MSc, PhD, CEng
G. Whittingham, MA, PhD, FRIC, FInstF

Secretary, The Institute of Fuel
R. Jackson, MSc, PhD, CEng, FInstP, FIM, FIEE, MInstMC, FInstF

Editor, The Institute of Fuel
Christopher Payne

Publicity Sub-Committee
D. Armstrong, CEng, MInstF (*Chairman*)
R. F. Hayman, BSc, CEng, FRIC, FIGasE, FIChemE, FInstF
F. E. Ireland, BSc, CEng, FInstF
G. G. Thurlow, BSc(Eng), MSc, PhD, CEng

Public Relations Officer
J. Hamson

Lord ZUCKERMAN, OM, KCB, MA, MD, DSc, MRCS, FRCP, FRS*

1 Fuel, pollution and the environment

SUMMARY

Each source of energy has its special environmental problem. The cost of dealing with them and of assuring an acceptable environment must be regarded as part of the price that has to be paid for providing ourselves with heat, light and power. Equally, the various problems that are entailed have to be tackled with a proper sense of political realities and priorities.

While there is broad agreement that environmental amenities must be safeguarded, the diverse and often subjective interests of different amenity groups makes this an area in which decisions have to be reached through democratic debate. Other aspects of environmental protection, especially those associated with the generation of waste products which can be detrimental to human well-being, are less controversial because they can be approached scientifically.

On the other hand, since environmental problems are usually relative and not absolute in nature, any judgment of risk, whether to amenity or health, needs to be balanced against the assessment of the benefit associated with the factor responsible for the corresponding hazard. The more widespread an understanding of facts, the more likely will we be able to approach the problems that relate to energy and the environment in a rational way, and the more likely to get our investment priorities in these fields into sensible perspective.

★　★　★

The title of this opening session, 'Exposition of the Problem,' is deceptively simple. The scope of our Conference extends far beyond the confines of a single problem which could be formulated in any restrictive fashion. Moreover, if one were to ask whether anything significantly new could be said today about pollution and the environment, or about fuel and environmental pollution, the answer would surely be that in the light of all that has been written or said about the environment in recent years, it is hardly likely that a new slant could be imparted to this particular term in a Conference which is essentially concerned with the matter of energy supplies. One might even go further and ask, why bring in the word environment at all. The problem of assuring the

supplies of energy, without which our species cannot continue, is surely difficult enough without linking it to a word, which, in its various connotations, has over recent years revealed itself as a complex of an almost infinite number of interacting and often conflicting issues and problems.

There used to be a sour joke in the troubled days of the 'thirties' about what was then called the Polish question—a matter which at the time was always in the minds of people who were concerned with the precarious state of Europe. As I remember it, the story went that an Englishman, a Frenchman and a Pole were each asked to write an article about an elephant. The Englishman's essay was a learned treatise about the way one hunted the beast, and about the most vulnerable points at which to aim a bullet. The Frenchman, in what was then assumed to be his corresponding preoccupation, wrote about the elephant in its relation to the art of love. And the Pole, true to his concern, provided a disquisition about the relation of the elephant to the Polish problem, and to the political dismemberment of his country.

Today, one might well say that if there were a parallel story, one would use the word 'environment' for elephant. It imparts a twist to almost all discussion. To some it means the preservation of our physical amenities; but there is certainly no single view about which amenities matter most. To others it means protecting our physical environment from any noxious pollution which could be a hazard to health. The concept of the conservation or exploitation of our primary resources is foremost in other people's minds, while protecting the balance of nature—however that concept is defined—is what others again think of when they talk about the environment. The term has even more connotations, and almost all of them conflict with each other. That is why 'the environment' has become more a political than a scientific issue. And that is why those who are concerned to advance the subject of ecology as a scientific discipline— a discipline which deals with the interaction of the fauna and flora of particular habitats—now almost inevitably find themselves at odds with the propagandists who, in the name of ecology, are concerned to influence political decisions about matters that impinge on environmental amenities, and the use of resources. It is not at all surprising that the

*Former Chief Scientific Adviser to HM Government, now consultant to HM Government.

issue of safeguarding our physical environment has joined the problems of the improvement of our social environment and of assuring our energy supplies at the centre of the political stage.

I make no apology for introducing the opening paper of this Conference on a political note. The protection of our physical environment is an immensely important and continuing responsibility. But it is a responsibility that needs to be approached with a proper regard for political realities and priorities. Energy and the environment—or pollution and the environment, which is the specific title to which I have been asked to address myself—are particular aspects of a general problem which has other and more immediate aspects that are every bit as critical to human survival. There are still enormous social and political problems that derive from the unequal distribution of wealth and resources. In many countries population continues to grow at a rate which far exceeds that at which resources can be mobilized in order to improve the conditions of life. Starvation and famine are still a condition of life in many parts of the world. In some countries malaria control has broken down. Illiteracy is widespread. Nations are still at war with each other. Those of us who have participated in international discussions on any of these matters will appreciate just how urgent, how disturbing, and how acute they are, and why it is so difficult to make real progress in dealing with them. In comparison with some of these other problems, the protection of the physical environment should surely be a relatively simple problem. We all want to live in clean surroundings and enjoy not only good working conditions but also a beautiful countryside. Why then, is there so much heated argument in the debate about the protection of the environment?

The human race has evolved as part of the animal kingdom. Like every other species of animal or plant, we are subject to the selective evolutionary forces that have determined which of the myriad living forms, which in their time have represented life on earth, should survive. Species that have failed to adapt to climatic change in their habitat, or representatives of those which have not had the wit or capacity to move from adverse conditions to those in which they could survive, have disappeared from the earth. Of all the species of living matter that have ever existed, we alone can deliberately change the physical environment within which we have to guarantee our continuity. Only we have been endowed with the intelligence necessary to survive in almost any kind of environment. So proud are we of our achievements that we are now ready to talk—even if it is still in terms of science fiction—about surviving in outer space.

We determine our environment for better or for worse. From the caves of the Palaeolithic we descended to village life in the valleys, and from the village we moved to the city. We have developed many forms of building material, and have liberated ourselves from the domination of the seasons. We neither freeze in the winter nor succumb to the heat of the summer. We make clothes out of natural and synthetic fibres. We cultivate and distribute food supplies from all parts of the world. We have developed a science of medicine, and we have made ourselves independent of the direct light and heat of the sun by using other immediate sources of energy. If we had

not done these things, *Homo sapiens* would still be a species of at most a few million nomadic food-gatherers and hunters, as opposed to one consisting of thousands of millions of individuals, who either belong to, or strive to belong to urbanized societies which have become totally dependent on the fruits of manufacturing industry, and on international trade in goods, including primary commodities such as fuel and food.

It is only recently that we have seriously started to count the environmental cost of providing ourselves with heat and light and energy. The cave dweller slept around his hearth, and his ashes and midden heaps became the floor on which his successors lived. Deserts or open moor replaced the forests which were destroyed to make farmland, or to provide fuel for the hearth and for primitive industry. A large part of the land which we in this country now classify as derelict derives from colliery tips and disused mine workings, buildings and machinery. When the mines were active, neither their owners nor those to whom they gave employment, nor indeed anyone else, minded or probably were even conscious of the fact that as a result of their actions an environmental debt was being piled up. Industrialists of the 19th and even of the first half of this century were proud of the plumes of smoke that belched from their factory chimneys and many of the wealthy pioneers of modern industry lived within the confines of their factories, in an environment that would today be condemned by the average person. There was also a day when the citizens of Southern California did not mind oil wells and pumps in their back gardens. No doubt some still do not mind.

Contemporary attitudes about environmental spoliation always reflect the tastes, knowledge, and political awareness of the times to which they relate. With the spread of education and with the growth of resources, and with the change of social attitudes deriving essentially from transformations in the relation of employers to employed, or of capital to labour, political pressures to improve living conditions and the physical environment have gone on mounting steadily. Today people are acutely aware that there is a need to protect the physical environment and, so far as possible, to conserve the remaining fauna and flora of the world. This awareness is bound to increase over the foreseeable future, not only because more and more people are becoming both educated and exposed to the 'media,' but also because of the increase in disposable wealth, which adds to an ever-growing demand by a multiplying population not only for more goods, but also for greater mobility in a better, even if shrinking, physical environment. Not one of these issues should be seen in isolation. Increase in population, increase in demand for resources, increase in wealth, a dislike of pollution, a demand for better amenities—all interact within a framework of contradiction and conflict.

We demand more fuel and energy, but this adds to pollution and often to the spoliation of our surroundings. We demand more wealth, but this conflicts with the conservation of resources. More wealth means more motor cars and more travel, and so more pressure for roads and amenities outside the cities. Increase in population goes on adding to the momentum of demand. Nothing stands still. If only world population could be stabilized at its present level for say, 20 years, if only we all shared the same views about the priority of the problems which beset the future of humanity, all might be well.

But nothing stands still. We cannot expect world population to be stabilized in the near future. National interests clash, and interests clash within nations. We shall never be able to satisfy all the divergent views that exist about amenities and environment. When one talks about fuel and the environment, we must therefore do so in the realization that compromises will have to be made on all sides. This, unfortunately, is not always evident.

For example, it has been reported recently that Mr Nader and his environmentalist colleagues in the USA have begun a law-suit aimed at the shut-down of nearly all the nuclear power stations operating in the United States, on the grounds that they are unsafe, and a hazard to the health of millions. Mr Nader might well succeed in his doubtlessly well-intentioned action; but in equity, and in fairness to his fellow citizens, he ought to fail unless he has practical suggestions to put forward about how to economize significantly in the use of fuel, and about alternative measures whereby the United States can satisfy its immediate and future demands for energy. If he and his colleagues want more coal to be used, but are opposed to strip-mining they will have to make themselves responsible for reversing a powerful social trend, and show that underground mining is an attractive, safe and worthy occupation. If they cannot do this, they should be able to tell their fellow-citizens where the USA can find new deposits of oil within its own territory, or how it can now use tar sands or oil shale. If this turns out to be beyond their capacity, they should be able to demonstrate how the United States can increase its imports of Middle East oil, without either imperilling their country's trading position or putting too much political and financial power into the hands of the Middle Eastern countries. If this, too, is beyond them, they will have to instruct engineering and scientific research workers in measures that can bring about major economies in the use of energy; in the mobilization of solar or geothermal power; or at the very least prove themselves better nuclear engineers than the professional ones now employed in the design of safe reactors, and in devising measures for disposing of nuclear waste. And if they cannot do any of these things, they will have to be ready to take over the Government of their country and control the social upheavals which would inevitably follow from a catastrophic decline in America's standard of living if she were denied all plausible means for providing for the energy demands of her industries, public services, sky-scrapers and apartment blocks.

I have read that 5% to 6% of the gross national product of the United States is accounted for by energy generation and distribution. If this percentage were to rise even fractionally. the cost of all goods and services would rise, and the tasks of improving the living conditions of millions of poor, and of 'cleaning up the environment' would also prove much more difficult than they already do. If it does nothing else, Mr Nader's demand that existing nuclear power stations be closed down thus reveals the enormous complexity of the problem of fuel and the environment—if in the term 'environment' one is ready to include man as well as hills and rivers and the natural fauna and flora.* Most people concerned with environmental problems already recognize that any deliberate and sudden change in any single element in an 'ecological system', for example eliminating some weed or insect, could have secondary results which are neither expected nor welcome. We need always to remember that socio-economic systems are no less sensitive to sudden change than are those called ecological.

While this Conference is mainly concerned with fuel and the environment within the context of the United Kingdom, our approach to the problem is undoubtedly constrained by its international complexities. Fortunately we have been immensely lucky in the discovery of North Sea gas and oil. Estimates suggest that we may be able to satisfy at least half of our national demand for oil by the middle 'eighties.' But we are still in the early seventies, and I need hardly remind the Conference of the immediate and enormous economic and political power of the OPEC countries, and of the fact that for some years ahead we face the prospects of rising costs for imported oil. I have read some interesting articles about these matters, mainly by American writers, and it is not for me to speculate about them. But what is already clear is that our own economic prospects, and our hopes for political peace in our own country, would be far worse than they are had hydrocarbons in the bed of the North Sea not been discovered and exploited. Without this new source of fuel, our political and economic situation would be gloomy indeed.

On the other hand, we dare not forget that the protection of our physical environment is as real an issue as is the need to assure our supplies of fuel. Neither is probably as urgent a problem as is, say, housing; and our policies about, or rather our attitudes to both, can and should be flexible both in themselves and in relation to each other. For example, we should not be too bemused by the figures which are inserted into current economic equations about energy. I do not know how many fuel balance sheets have been drawn up since the end of the second world war, or how many are being spelt out now. We have been proved wrong in the past, and if experience is any guide, we are likely to be proved wrong in the future. We ran down our coal mines, but coal mining now needs to be encouraged. This is why the Government introduced the 1973 Coal Industry Act which keeps open the option of a large industry for the future, and which provides the Government and the industries concerned (the CEGB as well as the NCB) an opportunity to make a thorough appraisal of future prospects. We were over-optimistic in our estimates of both the potential and the costs of nuclear power; today we are cautious, perhaps overcautious. No significant allowance for North Sea oil was made in the calculations which led up to the 1967 White Paper on Fuel Policy, which at the time appeared to provide a sound basis for the policies we were then following but which, as it turned out, and because of

---

*I am fully aware that Mr Russell Train, the Chairman of the President's Council on Environmental Quality, rejects the common belief that the delay in starting new nuclear power plants, as well as the lack of refining capacity in the United States, are due to the 'environmentalist movement'. Other considerations, such as the price policy for oil and gas, were obviously involved.

over-optimism in the nuclear component of the sums, expedited the run-down of coal. I doubt whether the figures in current equations are going to be any more appropriate to an unforeseeable future than were those in the equations which were written in the late 'forties,' 'fifties' and 'sixties.' That is why we need to be flexible in our policies for fuel. And that, equally, is why we need to be flexible, or more important, open-minded in our approach to the environmental problems which are related to the provision of the energy we need, whether we are talking about the so-called fossil fuels or about nuclear power.

Environmental problems are relative, not absolute matters, and because dealing with them costs money, they add to the unit cost of energy. The environmental cost needs to be regarded as part of the price we have to pay for our fuel. I hope I shall not be thought unduly cynical if I were to suggest now that in the final analysis—that is to say, were we ever up 'against the stops' over our supplies of fuel—finding the resources to pay for energy supplies would be accorded a higher priority than would, let us say, reducing the sulphur content of oil to some arbitrarily low level.

In dealing with the very broad issue of fuel and the environment, this Conference will be considering the potential contribution of various sources of energy to estimated total demand over the foreseeable future, and with the environmental problems—using the term very broadly to include the social as well as the physical environment—each source of energy generates or could generate. Each phase of the total energy cycle, from the exploration for primary resources, whether of coal, gas, oil or uranium, to the final step of dealing with the waste products of combustion—including here radio-active waste—has its special environmental problems, which differ for each source of energy.

The problems of coal, oil, gas, electricity, and nuclear generation àre going to be dealt with in the following session, and that of measuring pollution and assessing its effects in the third. Both the international scene and 'the future' will then be discussed in the light of such pointers as are available. All these topics are live issues in other countries, in some of which they are being debated even more urgently than they are here. In the field of extraction our environmental problems are the same in kind as those in the United States, but their scale is different, as is also our experience. For example, open-cast mining in this country does not provoke the public protest which strip mining now does in the United States, for the good reason that since 1952, statutory regulations by which the National Coal Board abides has made it necessary for rehabilitation of the land to go hand in hand with coal extraction. Since the end of the war, more than 110 000 acres have been released by the Board after working and restoration, and nowhere does one any longer see permanent dereliction. Indeed, some of the mined land which has been released after restitution has turned out to be better farming land than it was before it was 'stripped'. Whether underground gassi-fication is still regarded as a practical possibility I do not know; but were it to be, it might help reduce some of the pollution that results from the extraction, transport and surface utilization of coal.

There are the obvious environmental risks associated with the extraction of oil from the North Sea. Recently I heard of one which was new to me, but which nonetheless may be common knowledge, namely that of land subsidence resulting from the extraction of gas. This, I understand, has already occurred in Holland. When I was told that this had happened, I found myself asking whether the amount of energy that would be required to reverse the change in order to prevent the possible incursion of the seas, would be less than, equal to, or more than the value of the energy that had been extracted in the form of gas. Those who will speak in the third session will no doubt be telling us about the changes which have taken place in the air of our cities since the introduction of the Clean Air Act of 1956, and about the costs. We shall also be brought up-to-date about the subject of car emissions and about the hazards to health associated with carbon monoxide, hydrocarbons, and oxides of nitrogen. In the case of nuclear power, we shall no doubt be hearing about reactor safety, and about the disposal of radioactive waste.

Clearly there are enormous problems which we, like all other industrialized countries, have to face in assuring that sufficient energy will be available to permit of life, as we know it, continuing into the future. The formidable costs of solving these problems, in which has to be included the price we shall need to pay for protecting our physical environment, clearly cannot be assessed—even for the immediate future—except in the broadest terms. Whatever R & D resources are applied to their solution, we cannot even be certain that the theoretically ideal solution will necessarily be found to certain problems—for example the disposal of radio-active waste, or the 100% assurance of reactor safety, or the large-scale exploitation of solar energy. In my view, absolute goals, wherever they may be set in the field of energy and the environment, are useful only as directional guides and as a spur to action. We shall not have failed if we do not reach them in every case.

This brings me to the two issues of standards and risks. The criteria we use in order to set environmental standards are still mainly of an arbitrary nature, and few have a sufficient scientific foundation. For example, we know that the salts of the heavy metals are toxic when taken in sufficient amount, but to take one example, we do not know what threshold for lead in the blood constitutes the average danger line, or indeed how to prevent lead getting into our bodies by way of the air we breathe, and the food and water we take in. Arbitrary standards for this hazard have been set on the basis of expert judgment, both by WHO and national authorities, but there are those who declare that they have been made too exacting. Others claim that with lead levels in the environment at what they are, lead constitutes a growing risk to health. But who is right, and who is to decide? It is almost certain that the cry that we are being poisoned by lead is far more shrill today than it ever was in the days when lead-colic was something which the average medical practitioner might have been expected to recognize. Surely the risk of lead poisoning is far lower today than it was, say, at the beginning of the century? And surely, too, the average expectation of life for all age groups of the population is far higher than it was when lead was an accepted hazard.

With the increasing public awareness of the dangers to health of some kinds of pollution, an awareness which is always sharpened by the attention given the matter by the press and television, coupled with the power of modern instruments to detect and measure chemicals in sub-microscopic amounts, there is a danger that standards for pollution may be made unnecessarily rigorous. I hope that we shall be hearing more about this matter later in the Conference. If my fears turn out to be unwarranted, so much the better—or so much the worse, depending on the way one regards this particular problem.

Defining a possible specific environmental hazard to health is an immensely difficult task for the scientist even when he knows that some suspected substance or compound is toxic at certain concentrations. It is even more difficult to establish the cause of a known clinical condition by means of statistical correlation. The relationship between cigarette-smoking and lung cancer is the most recent success story of this kind of enquiry—an earlier one was the establishment of the link between goitre and iodine deficiency in water. But there have not been many others in recent years.

The 1972 Report of a WHO Expert Committee on 'Air Quality Criteria and Guides for Urban Air Pollutants' is careful to point out that an association between a pollutant and illness or death may be accidental rather than causal. This cautionary note is exemplified by the current suggestion that the incidence and prognosis of coronary thrombosis and other cardio-vascular diseases is positively correlated with the softness of water in the areas concerned. There certainly seems to be evidence of a correlation, but whether this reflects a causal relationship, and if so whether the causative factor is the softness *per se*, or some as yet undetected and unmeasured constituent of the water, remains to be determined. The lung cancer story was not established just by a process of simple correlation. In the end it was all but an experimental enquiry.

There are some who say that from the point of view of health, our 1956 Clean Air Act was a gamble that paid off. The Committee on Air Pollution which reported in 1954—it is better known by the name of its Chairman, Sir Hugh Beaver—was concerned with all aspects of the uncontrolled emission of combustion products, and its report was the basis of the 1956 Act. The paragraphs which it devoted to the question of health referred to the 4 000 people, mostly elderly people who were suffering from bronchitis and other respiratory disorders, who died during the famous four-day London 'smog' of December 1952. Since then there has been no similar disaster. At the time the Clean Air law was enacted, it was recognized that it was the combined effects of coal smoke and SO_2 that had accelerated the deaths of those who died during the 'smog'. The Act proscribed the emission of smoke but not of SO_2, which it was realized would be much more difficult to control. As it turned out, the gradual decline of particulate matter in the air after the burning of coal in open grates was banned, and the disappearance of fogs due to coal smoke, is what is now believed to be the explanation of the fact that we have been spared any similar catastrophe.

This brings me to the question of risk. What is an acceptable environmental risk? Mr Nader wishes to condemn nuclear reactors, not because they have been proved by experience to be unsafe, but because one day there might be an accident which could affect if not millions—his figure—then certainly thousands of people. Yet if ever a technological development was started and pursued with an urgent appreciation of the hazards against which it is necessary to guard, it is nuclear technology. Is it ever possible to guarantee that an accident cannot take place? When, one has to enquire, would a social environmentalist like Mr Nader be satisfied, and through what kind of technical interpreter would he derive his facts?

The judgment of risk in environmental matters always has to be balanced against our assessment of the benefit which may be associated with the factor responsible for the corresponding hazard. There are, for example, circumstances in which people are exposed to the risk of ingesting harmful quantities of lead. At the same time it is believed that lead in petrol is responsible for some of the lead which gets into the tissues of most people. But is the benefit to health that we might derive from the abolition of the leaded petrol commensurate with the additional use of resources—including fuel—which might be required to replace the small high-compression engines the leaded petrol makes possible? Are we confident that our knowledge about the toxic effects of trace amounts of lead is adequate as a basis of decision? Or consider DDT (even though it has little direct relevance to the energy problem); is the prevention of any of the environmental harm it does commensurate with the world-wide disaster to agriculture that would result from its withdrawal, or with the denial of the benefits it provides in helping to keep malaria at bay? These are the kind of risk-benefit questions that need to be asked in the assessment of possible environmental hazards to health. Obviously the fact that they can be asked is a spur to finding less harmful devices or compounds to replace those which are now suspect. The USA 1976 car emission standards not only stimulated much work on catalytic devices which would help consume noxious exhaust gases, but also provided a spur for more basic work on the redesign of the conventional motor car engine. This is all to the good. The demand to reduce pollution can be as powerful a factor in technological innovation as consumer demand is in general.

There is yet another aspect to the assessment of environmental risks to life and health which I feel should be mentioned. Risk is associated with most things we do. Statistics about motor accidents, which are broadly regarded as something we have to accept, or are prepared to accept, are published widely. Do we publicize sufficient statistics for accidents in, let us say, the building industry or chemical plants, or in the home? Are we up-to-date about figures for occupational diseases generally? When people speculate about reactor accidents, or about SO_2 in the atmosphere, or about the dangers of lead or mercury poisoning as a result of environmental pollution, would it not help if we were able to compare numerical estimates of the gravity of risks they conjure up with the known risks we run in our everyday lives?

This is one of several steps that might be taken in order to help get the matters with which this Conference is concerned into sensible perspective. We are living in today's world, and we cannot pretend to see the energy picture as it will appear, say 30 years ahead—for the simple reason that certain elements which we know will eventually have their place in the total composition do not yet exist—for example, the exploitation on a significant scale of nuclear fusion, or that of geothermal power. Conferences such as the present one serve a very useful purpose in providing a balanced view of a problem which is beginning to generate not a little public unease. I was much struck recently by a Report by a Special Committee of the Association of the Bar of the City of New York, published towards the end of 1972.* It pointed to eight critical defects in the present system for regulating electricity and the environment, all of which are relevant—so I believe—to our own problems. I make no apology for quoting the main ones here, though neither in the order they appear in the Report, nor with the intention of suggesting that the conclusion to which they lead the authors of the Report—namely that the USA needs an overall Energy Commission or Agency— necessarily has any relevance to our own situation.

What the Committee said is, first, that the way in which the USA goes about its business fails to keep in the forefront such key issues as the changing scale of demand for energy, and the proper allocation of resources for research; and, second, that major issues which recur time and time again are dealt with on a piecemeal basis rather than what the Committee called 'generically.' The Committee also emphasized that in some cases it takes far too long to reach decisions about the construction of necessary facilities. This, the Committee felt, is often due to the fact that there is inadequate public trust in the objectivity of the decisions that are taken.

I myself believe that the issue of public participation is all-important, and that we, like the Americans, have not devised a means whereby the man-in-the-street can obtain an authoritative and balanced view of the problems we face. When it comes to the environment we are far more inclined to listen to the voice of the amateur than to that of the professional. As a recent article† put it, we hearken to the declarations of lawyers who write as automotive experts, of nuclear physicists who preach about nutrition, or sociology professors who speak out on chemical engineering, and of life-scientists who suddenly discover they are ecologists. Nor do we in the UK lack those scientific journalists who, without any professional expertise at all, are ready to engage in argument with nuclear physicists and engineers about the dangers of nuclear reactors. Sometimes I find myself dreaming that a new brand of quack doctors, Doctors of the Environment, has been created, and that their 'degree' has become an obligatory qualification for members of the many 'environmental companies' which are now mushrooming so fast. Not so long ago I passed a window in the centre of London which was taken up by a threatening poster suggesting that if my home was not properly 'ionized' I should immediately step inside. I

propose not to do so until I have heard the matter of environmental ionization, and its relation to human well-being, properly discussed in some authoritative scientific forum.

The better and more widespread an understanding of the facts, and the less we allow ourselves to be bemused by unrealistic statements about the environment, the more likely are we to get our investment priorities, both for energy and the environment, into sensible perspective in relation to whatever writing one can already read on the wall. It has also become obvious—and if our own situation does not make the picture clear, then the current American one certainly does—that, however difficult it is not to abide by conventional 'discounted cash flow' sums, we cannot approach the problem of investment in energy and the environment by accountancy principles which relate to today's costs and values. If one is permitted to use a customary figure of speech, the environment becomes increasingly 'scarce' as a commodity as every year passes; and so, too, do our known supplies of energy.

Since almost everything we do depends upon adequate supplies of energy, it is obvious that the judgments that we make today about investment in energy will determine almost everything we do tomorrow. At the same time, we have to recognize that resources are always limited, and we must resolutely avoid the temptation of making the best the enemy of the good. There never will be enough resources to satisfy all claims about environmental amenities. There never will be enough resources to clear up all the ravages of the past, or to assure that the changes that necessarily have to be made in our physical environment as our numbers increase, and as our standard of living rises, are to everybody's satisfaction. Above all, the concept of zero pollution levels has no practical reality. One can reduce the sulphur content in the fuel we consume to low levels, but it is always necessary to remember that each step which is taken to improve on what has gone before will cost proportionately more.

I have been told of a smelter project in the United States which was abandoned because the power cost that was demanded in order to reach a very low specified emission level represented the energy requirements of something like ten thousand homes. Above all, we cannot legislate against the laws of thermo-dynamics. Whatever we do, waste products will be generated and will have to be dealt with. For example, if we wish to replace the conventional internal combustion engine with an electical power unit, we should be discharging increased amounts of CO_2 and SO_2 from power stations; or adding to the amounts of radio-active waste which would have to be disposed of. In such a change-over there would always be offsets to the gains. But in the final analysis, and however much we economize in the use of fuel through more efficient burning devices and through better insulation, we must have supplies of energy. It is pointless to design nuclear power plants, in order to economize in investment, for operational ratings which carry the risk of the failure of materials and of corrosion. If we have to have nuclear power we shall have to pay the price for safe nuclear power. For example it has been suggested that nuclear stations could

*Electricity and the Environment. The Reform of Legal Institutions, Report of the Association of the Bar of the City of New York Special Committee on Electric Power and the Environment, 1972, St. Paul, Minn., West Publishing Co.
†Walker, E. A., Engineers and the Environment: Being heard and being right, Journal of Metals, April 1973.

be built underground, or even at sea. Were this to prove necessary, we should have to pay the price for solving whatever novel and potential technical problems such a departure from conventional practice would entail. We cannot live without fuel and energy, and there is no point in designing power plants which would destroy us.

As I have said, our species is all-important in determining the nature of the physical environment in which we live. But we are not so dominant as to be superior to the natural processes which also transform our physical environment; and it should not be necessary to repeat King Canute's demonstration that it was not within his power to control the tides. I understand, for example, that in north west Europe we add more oxides of sulphur to the atmosphere than get there by way of natural processes. But in the world as a whole, biological decay is said to contribute three times as much sulphur dioxide to the environment as we do, and that the ratio is ten times in the case of carbon monoxide. We cannot stop these natural processes. We may regret the destruction of coral reefs due to the spread of the Crown of Thorns starfish, and we can try to slow down these natural happenings. We should also be careful not to precipitate or to reinforce natural processes which are to our disadvantage. But it would be a rash man who would claim that it is within our power to arrest all natural change. Who knows, the arctic ice cap might one day encroach yet again over northern Europe.

But having said this, I want to repeat, in conclusion, that we live not in a state of nature, but in the state of man. We are able to approach the problems which relate to energy and the environment in a rational way, in the confident knowledge that with more research, all the goals which we are trying to achieve in the energy field, such as nuclear fusion or the release of geothermal energy, in the final analysis will reduce air pollution and, we may hope, other kinds of pollution as well.

That is the note on which I would like to end. If some of the fears that we entertain about our future, and in particular about the environment and about our energy supplies, had prevailed in the days when our forbears lived in caves, if the kind of message we are accustomed to hear from the media had been ringing out then, our ancestors might never have left their primitive shelters. But that they did, and we are now committed to an urban civilization. As we map our future, let rational forethought, therefore, take the place of the kind of despair which today so frequently obstructs change which is critical to our survival.

D J EZRA, MBE, MA*

2 The coal industry

SUMMARY

The energy industries are vital to the economic well-being of the Nation. The National Coal Board, amongst others, believes that, if living standards are to continue to increase then more fossil fuel will be used in the decades ahead than is used today. In addition, the Board believes that the British coal industry operates now, and will continue to operate in the future, without significant adverse impact on the environment.

Since nationalization, the Board has allocated resources to environmental control on a steadily increasing scale. Significant technological advances have been made in coal preparation and processing which, not only improve efficiency, but also reduce pollution. Automatic and semi-automatic loading and transport systems have been developed reducing both noise and dust. Techniques for stabilizing, contouring and cultivating spoil heaps have been developed in parallel with improved open-cast reclamation techniques and in addition to a pioneering effort which has established the use of unburned spoil for bulk fill. Finally, but most importantly, the Board has ensured that the impact of its operations on the environment, both now and in the future, are considered at all stages of its planning procedures.

In the future, increasing shortages of energy as well as environmental considerations should encourage significant changes in the processing and utilization of energy. The Board, in parallel with other agencies, is already researching into various energy conversion processes, as well as assessing the potential for integrated energy conversion facilities for fossil fuels, and believes that there is scope for an increasing tempo of collaborative research in this field.

1. INTRODUCTION

The energy industries of the United Kingdom are a major economic social and environmental force in the life of the nation. The National Coal Board believes, and

it is not alone, that if the standard of living of the nation is to go on increasing then more fossil fuel will be burned for many decades ahead than is burned today and that, on present trends, a significant proportion of that fuel will have to be coal. The purpose of this paper is to demonstrate that a large United Kingdom coal industry can operate now, and in the future, without significant adverse impact on the environment, and can in some ways improve environmental conditions.

At the present time the National Coal Board spends more than £16 million on environmental control each year and in the future we expect this amount to grow in real terms. There is not a single aspect of our wide-ranging activities, from the siting and underground development of a new colliery through to clearing away the last vestiges of a closed one, where environmental aspects are not considered at the earliest stage of planning. Unfortunately our concern for the environment was not shared by our predecessors before nationalization.

We are all familiar with the traditional image of the coal industry, the gaunt winding tower and its attendant chimney belching black smoke, the drab clutter of railways and pit head buildings and, beyond this, the even drabber huddle of terraced cottages all set against the sombre backdrop of a sprawling and sometimes smouldering spoil heap. This popular image refers to the attitudes and activities of a coal industry which has passed into history.

One of our problems has always been that we inherited the dereliction and the widespread neglect of the environment left by 200 years of indifference, and this is still taken as representative of the modern coal industry. It is true that, in spite of vigorous action taken in the last 25 years, much of which will be described in the paper, there are too many signs of past neglect still visible; but coal can take its place alongside all other sources of energy as a fuel that can, and should, be used in the future without any fear of adding disproportionately to the pollution of our environment. It is submitted that the firm commitment given by the Board to protect and, if possible, to improve the environment at all stages of coal mining, preparation, transport and use can, and is, being implemented, and that as a result there is no risk to the

*Chairman, National Coal Board.

FIG. 1   Killoch Colliery, Ayrshire. Started production 1960

country's environment arising from the nation's continuing dependence on coal.

The organizers of this Conference, conscious that 'environment' can embrace an almost infinite range of topics, have rightly sought to concentrate our thoughts by limiting the subject matter of papers to the effect of the fuel industries on the environment beyond the boundaries of their own activities. These terms of reference preclude me from discussing the vast amount of effort the Board has put into improving the working environment of the miner. The advanced methods of roof support, better ventilation standards, comprehensive dust and noise suppression, improved transport and permanent underground lighting even on the face; all these have been pioneered by the Board in a successful effort, not only to ameliorate the miners' working environment, but also to make mining a much safer occupation than it once was.

The experience so gained has pointed the way to the methods that are being used to tackle responsibilities to the community at large and which are the issues to be raised in this paper.

2. COAL WINNING
2.1. Exploration and mining
As everyone who has followed the recent controversies on mining in the National Parks will be well aware, the harbinger of any mining operation is the survey to assess and prove the resources of the required materials in the ground. In the case of the coal industry, this means sinking bore-holes.

The National Coal Board has always considered it essential to act, from the outset, in open consultation with local and Government authorities concerned with the

environment. In the case of the sinking of test borings, this is carried out in consultation with the Institute of Geological Sciences, and after consent has been obtained from the local Planning Authorities. Sites are usually occupied for only two to four months and, on completion of the borehole, the hole is filled with cement from the bottom to the top and the site reinstated and returned to its original use. As it is important to ensure that no pollution to the local water resources occurs, the River Authority and local water undertakings are kept informed, while the Institute of Geological Sciences, by examining core samples, provides a check that free flow-through aquifers are maintained.

Where the Board sinks new mines, these are, of course, subject to the whole gamut of planning and environment-related statutes. Meetings are held between the Board and the local Authority at which the Board's plans are fully discussed. Every effort is made to ensure that the works are designed to harmonize with the surrounding environment and to minimize any unsightliness visible from main roads or other principal viewpoints. A modern colliery is now a much more attractive complex of buildings than in the historic past, as may be seen from the example of Killoch colliery illustrated in Fig. 1. This is very different from the traditional silhouette of a colliery alluded to earlier.

The great majority of today's mines were, however, planned prior to the coming into force of effective planning control which really started with the Town and Country Planning Act 1947 and is now exercised under the T. and C.P. Acts of 1968 and 1971. Local authorities have to show on the first County or Borough Development Plan the extent of colliery and related surface activities for which ministerial approval is required, and quinquennial reviews are undertaken. This is augmented by less formal, continuous contacts by which surface development is co-ordinated with active or projected mining.

A good example of the Board working closely with local and Government Authorities in the planning of mining and associated surface activities is Peterlee New Town. Following concern at the possible conflict between the developments by the NCB and those proposed by the Peterlee Development Corporation, a joint working party was set up in 1949. Its successor comprising representatives of the New Town Corporation, the Board and Statutory undertakings still meets today. The result of this co-operation has been that the township has become established during the extraction of a total of $8\frac{1}{2}$ million tons of coal from beneath its designated area, of which $3\frac{1}{2}$ million tons has been worked under built-up areas, both residential and industrial.

Modern mining techniques and, in particular, the deeper mines that it is now possible to work, have greatly reduced the incidence of subsidence and damage at the surface. The Board has a legal responsibility in general to repair structural damage to property, while statutory undertakings such as the railways and water services have considerable rights of support. Some subsidence has had useful results, for example by improving the irrigation, but in general the effects, whether it be on structures such as roads, railways or reservoirs, on property or on agricultural land, are harmful, and the Board has spent much research effort to minimize damage from this cause, resulting in marked improvements.

The collaboration of the NCB and the local authorities does not end at the colliery or open-cast working. It must be remembered that the influence of a coal mine on its neighbourhood extends through the work force to those who provide consumer goods and services for the miners and their families. This is not the forum to discuss the provisions made by the NCB for the housing and welfare of their work force—but it is of interest to note that the siting and lay-out of housing and associated buildings is another environmental issue with which the Board staff is often deeply involved. An example is at Hetton-le-Hole in County Durham where 118 houses have been substantially improved in collaboration with the local Council, as part of a general scheme for clearing dereliction in the area.

A colliery, once operating, is liable to emit much the same pollutants, and is subject to the same controls, as any other industrial plant. The main potential problems are air pollution, effluent discharges and the disposal of minestone in its various forms.

2.2. Air pollution from collieries

Atmospheric pollution from the boiler plant installed at collieries has been reduced over the last decade to such an extent that it is no longer a significant problem. This has been due to two factors. First, the power required for colliery purposes is now mainly taken from the grid: over 80% of collieries being designated as 'all electric.' Secondly, the Board has modernized the steam plant still retained using the developments which, as discussed more fully in Section 4, have made it possible to operate coal-fired industrial plant in such a way that the levels of smoke, grit and dust emitted are well below the statutory limits, and do not cause a nuisance. Owing to the measures taken to reduce the levels of dust below

ground, run-of-mine coal is now sufficiently wetted during production as to reduce excessive dust during much of the subsequent processing including preparation, blending, stocking or loading on to transport for despatch. Where dust is produced—for example during crushing—it is possible to contain this by installing dust extraction hoods and by the use of water sprays.

It is obvious that dust formation cannot be entirely prevented; ground deposits which have dried off and are then dispersed by the wind are an obvious source of trouble. It is our experience, however, that having provided adequate equipment and facilities for dealing with dust (such as mobile sweepers), the Board can guarantee that the good 'housekeeping' necessary is carried out by providing adequate supervision, and by ensuring that, through the management structure and training, the necessary care and commonsense is applied, for example in locating plant.

2.3. Effluent discharges

The liquid wastes to be dealt with from a mining and washery complex may comprise mine water pumped from the workings or overlying strata, washery and general yard drainage: and supernatant water from effluent settling lagoons, tip seepage and surface water run off.

In the past, it was often acceptable to discharge unaltered mine water from the mine to water courses under the provision of the various River Acts, but it has now been suggested by the DoE that these mine waters should be under the control of the River Authorities; this could lead in time to a tightening of discharge conditions and the need to treat increasing quantities of the more highly mineralized waters. The Board regularly meters the suspended solids, the temperature, acidity, biochemical oxygen demand, toxic metals, etc., of the effluents, and is making an increasing use of settling tanks and treatment plants to bring the properties of the liquids it discharges well within the limits where they will not mar the wholesomeness of rivers and streams. Another approach, being adopted at Tilmanstone Colliery where the mine water is very saline, is the construction, at a cost of £450 000, and in co-operation with the Thanet Water Board, of an outfall to discharge mine water to the sea, so avoiding any risk of polluting the chalk aquifer, used for the public water supply.

Water from the overlying strata of a mine is often of very good quality and is used at the mine; any surplus may be used to augment the town supply.

General washery and yard drainage is usually highly contaminated with suspended solids. Here the only practicable treatment is to pass the water through settling ponds and oil traps until it is clean enough for discharge. Much can be done to make this treatment effective. Ponds need to be of the right shape for settlement and dredging, and of the right size to give a steady discharge. Clean water must not be passed unnecessarily through dirty water systems to avoid disturbing material already settled; and, above all, control of coal preparation operations must be correct. The Board insists that up-to-date control instrumentation is installed at all washeries.

The practice now is to draw off supernatant water from lagoons under controlled conditions. The discharge is usually fairly clean but further treatment is often added to meet the most stringent conditions. Fine strainers may be used, but these tend to block easily and restrict the flow. The more usual practice is to provide a final settlement pond before the discharge point to reduce the suspended solids to an acceptable amount. The quality of the final effluent depends on particle size and floculating agent; where the flocs are colloidal complete settlement may be impracticable. One method, now under development, of separating the water and solids in washery tailings by evaporation is discussed later.

Seepage from tips is not, at present, the cause of a major effluent problem but, as the Board tightens up its control on all potential environmental hazards, it will no doubt become more common to collect and treat these waters as well. The main points from which the water escapes are located, often using old maps which show the natural water courses of the underlying ground before the tip was built. It is then a fairly easy business to meter the quantity and nature of the flow and to treat the water at these points when necessary.

2.4. Disposal of solid wastes
Probably the two mental pictures most people get when pollution by the coal industry is mentioned are first, black smoke and, second, large conical colliery tips. The 'black smoke' image will be dealt with later, but I would like here to describe the large effort we are now putting into making the disposal of solid waste, wherever we can, of positive benefit to the community rather than the reverse.

Mineral matter is unavoidably extracted with the coal. It may be associated shale, mudstone, siltstone, sandstone or seatearth, and in a few areas limestone or igneous rock. In total about 55 million tons of coarse material has to be discarded each year to which must be added about a further 5 million tons of fine material, either as a slurry or as a filter cake discarded from coal washery plants.

The coal industry has also been made responsible for

FIG. 2   *Proposed pilot-scale fluidized-bed tailings combustor*

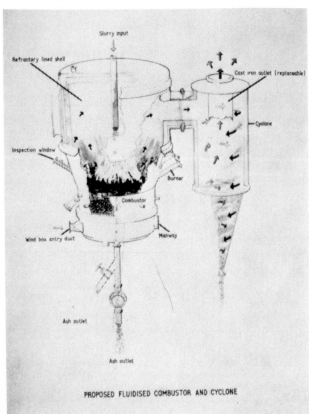

PROPOSED FLUIDISED COMBUSTOR AND CYCLONE

the legacy of 2 000 million tons deposited in the past, and contained in existing tips. Much of this and all the new mineral waste or minestone (sometimes referred to as unburned spoil) contains some coal though a proportion of the older heaps have lost the carbon they once contained by internal burning.

Our intention is to use as much of the material as we possibly can and our activities in this direction will be described a little later on. Before doing so, however, it is proposed to deal briefly with the ways we are now using to ensure that this material, where it has to be tipped on the open landscape if only temporarily, does not present a hazard to the environment. Our purpose is to ensure that there is no danger or nuisance caused by instability, burning, effluent pollution, erosion, blockage or flooding.

As a result of much research, we have now developed detailed and careful regulations and procedures[1] for laying and consolidating tips which ensure adequate stability. The incidence of fires on, or in tips has been much reduced, not only by a conscious effort to prevent spontaneous (much less induced) combustion but also by improved coal preparation procedures which have reduced the coal content of the tipping material. The pollution of water courses and nuisance from erosion have also been much reduced by flatter slopes and improved drainage but still exist as minor problems requiring continuing attention, especially in high rainfall areas. The lower, flatter shaped tips now built, expose more area to erosion and leaching but this disadvantage is more than offset by the degree of compaction now being achieved. The soiling, seeding and planting of tips, now common, goes far towards reducing these nuisances. Experiments have shown that it is possible to cultivate tips made of material which, due to, for example, a high sulphur or alkalinity level, would appear most unpromising. Once vegetation is established, probably by the use of a layer of imported soil, nature has a wonderful habit of adapting the environment to produce a variety of plant life which soon belies the sub-soil that it is hiding.

The 'tailings', that is the material left at the far end of the washing processes of coal, consists of very fine dirt particles and a little coal suspended in water. In most cases, it is sufficient to dry this material to about 20% moisture when it can be added to the other dumped material as a stable, firm material or cake. This drying can be carried out by filter-presses or by 'deepcone' separators developed by the Board. In some cases, however, where the mineral matter makes a particularly persistent slurry with the water, drying by the established means to give a satisfactory 'cake' is difficult and the Board is building a 1 ton/h pilot-scale plant which will dry this material completely in a fluidized bed, heated by the combustion of the fine coal present in the slurry. The dried material has the big advantage that it has been altered chemically in the heating process such that it will not 're-slime' when water is re-added to it. It is intended to mix the fully dried material with partially dried material from a 'deep cone' separator to give a fully satisfactory cake for dumping; though the dried material may also have some use for upgrading into blocks, etc. Fig. 2 shows the proposed pilot-scale fluidized-bed tailings combustor.

There are obviously nuisances caused while a tip is under construction: such as noise and dust, and the possibly objectional sight of the workings themselves. Here also, the Board makes every endeavour to mitigate the nuisance by careful operation in consultation with the local authorities.

It has been implicitly assumed up to now that the best that can be aimed at in tipping the waste mineral products from mining is to minimize the environmental problems that might arise. However, it is becoming increasingly evident that, in many cases, it is possible to use tipping to improve an area: and the Board now endeavours to arrive at schemes not only acceptable to, but also of positive benefit to the community in consultation and co-operation with Planning and other Local Authorities, local societies and, where appropriate, individuals.

At New Stubbin Colliery a sloping valley site of comparatively poor land has been levelled by ordinary tipping, and the West Riding County Council have prepared and laid out, the surface for playing fields for a local school. At Rickershard, in Lancashire, the Board are tipping two million tons of dirt to an agreed pattern to be used to make a golf course. At Leigh, artificial hills are being created to improve an otherwise flat and monotonous sky-line. Other landscaping schemes carried out in connection with open-cast mining are described in Section 2.6.

While the Board accepts the duty of keeping old tips within its ownership safe it has not the resources in general to do much more. One way in which the Board is justifying its spending money on the reshaping of the old, pre-nationalization tips is by rewashing those with a recoverable coal content. The residue can be then placed in new, more acceptable tips. Another way in which the Board is getting rid of old tips is by so arranging open-cast mining that adjacent old tips can be lifted and reformed with the overburden:[2] examples of this are given in the next section.

Mention must be made of the shaping and rehabilitation of other old tips being undertaken by County Council Land-Reclamation units operating within the government grants system which is not directly available to the Coal Board. About 4 000 acres of previously derelict land has now been transferred to local authorities. It is hoped to more than double this in the next five years.

2.5. Utilization of solid wastes

In the last section was described the efforts made by the Board, and others, to dispose of the mineral waste products extracted during mining, to the best possible advantage of the environment. How much better, however, to make positive use of this material. The Board has for some years pursued a vigorous policy of expanding the amount of mineral waste put to useful purposes to the extent of setting up a special marketing group, the Minestone Executive, to deal with this. As a result, we are now ensuring that the equivalent of some 10–15% of the minestone we produce is being put to useful purposes.[3] We have every intention over the next few years of increasing the quantity of spoil so utilized.

An obviously important use for this material is as a fill for civil engineering projects and both burned and unburned spoil has been used on a number of sites, such as in the construction of the Hoverport at Pegwell Bay which is built on reclaimed land, and which used more than 300 000 tons of colliery shale. Spoil has also been used for a number of road works up and down the country. For example, an embankment 23 ft high is being built for the Wrexham by-pass which will consist eventually of

over 500 000 tons of burned spoil and minestone. Many other similar examples may be quoted. The use of minestone (which is much more freely available than burned spoil) now has the approval of the Ministry of Transport following vigorous technical representations by the Board which demonstrated that the two reasons for which unburned spoil was originally suspect as a road material are invalid. It has been made clear that this shale does not exhibit poor wet-weather workability: in fact civil engineering works have proved possible on unburned shale in weather conditions where work on other materials was impossible. It has also been shown to be safe from the possibility of spontaneous heating if it is laid and compacted to the normal Ministry of Transport's Specification. As a final demonstration, the NCB built an embankment, where 180 thermocouples were embedded to monitor any temperature changes and confirmed the absence of heating on the full-scale. Temperatures recorded daily since December 1969 have given no indications of spontaneous heating. As local authorities gain confidence in the use of minestone for roadworks, and other uses are developed, it is hoped that more and more of this material can be put to useful purposes.

This will, it is worth pointing out, have a major beneficial effect on the environment other than that of reducing the amount of minestone that has to be accumulated in tips. It will replace the natural gravel and other materials which are now so often dug out with little or no regard to the loss of good agricultural ground, or to the unsightly scars on the landscape that result. It would seem so much more sensible to use a material which is extracted as the necessary by-product of our industry wherever possible, rather than to compound the environmental problem of disposing of this by-product material by the digging, at the same time, of equivalent fill-material; and it would be urged that there is scope here for incentives or controls. These should discourage the contractor from raping what is often good agricultural land for back fill, often with only marginal financial gains, and persuade him to perform a useful community function by substituting pit-heap spoil, which is often only a short haul away. Any regulations should also ensure that the use of colliery spoil is considered as a material for all big civil engineering projects. Economics have to be carefully considered, but how much better would it be on purely environmental grounds to use spoil as a base for the new airport at Maplin Sands, rather than to run the risk of the ecological disturbances caused by the major dredging of sand which will otherwise be needed. In this context the Board welcomes the recent willingness of the DoE to institute dual tendering on motorway contracts whereby the total costs, including environmental costs, are taken into account before deciding between opening new borrow pits, or importing fill from existing tips.

Colliery spoil can also be 'up graded' into an increasing number of products. If a small amount of cement is mixed with suitable spoils and the water content of the mix correctly adjusted, after consolidation, it will set to a strong enough material for use as a sub-base for a roadway or as, for example, the ground for a stocking area, or car park. It has been found experimentally that many unburned spoils are not susceptible to frost heave, according to the DoE regulations, even without cement stabilization. The Board is providing technical advice on the suitability of various spoils and on the strength of the compacts which can be produced, the number of possible industrial applications is growing.

Another growing use for colliery spoil is in the manufacture of light weight aggregates. The demand for suitable materials, mostly for making into blocks for building purposes, is growing, and because the supply of clinker is dwindling consequent upon the closure of the older chain-grate-fired power stations, the demand for minestone as a suitable substitute in aggregate manufacture is also growing. The material, sometimes blended with an inert material such as pulverized fuel ash, is fired on a sinter-strand where the coal present in the unburned spoil provides the heat for the process. The resultant lightweight material (lightweight because of the porosity introduced by the agglomeration, calcination and combustion of the material) is crushed and used for making precast concrete blocks. The manufacture of this material is being undertaken by a growing number of private companies but the NCB, as suppliers of the raw material, supply also a technical service, evaluating both the raw materials and finished products, and also advising on the sintering process.

A final use of minestone is its conversion into artificial roadstone. The initial interest in this development was to provide a manufactured substitute for the somewhat scarce, and therefore expensive stones whose special properties make them suitable for areas of high-skid resistance on busy traffic lanes, road junctions, and similar. It was found that a suitable substitute could be made by careful heat treatment of a compact of finely ground burned spoil which, while bettering the required physical properties of the best natural stone, was competitive as regards cost. The development of this synthetic stone may in fact prove even more vital due to a general shortage of stone for use in road building and maintenance; indeed a general shortage of aggregates has been predicted for the not-too-distant future. Thus it is possible that manufactured roadstone from minestone may not only provide a useful contribution to road safety but also, in the longer term, find wider markets in the construction industry. Such uses of minestone will be of positive benefit to the environment as opposed to the negative effects of quarrying and gravel digging.

2.6. Open-cast mining

Open-cast mining started in this country during World War 2. The Board assumed responsibility for it in 1952 and current production is about 10 million tons of valuable coal p.a. The mining operation itself is carried out with the minimum of intrusion on the lives of the surrounding community. Before coaling starts, top soil and sub soil are stripped and preserved for return after the extractive process. These temporary mounds are used to form embankments which, sown with grass, help to screen machinery noise and dust from the community. Silencers are fitted on mechanical equipment, water sprays are used to lay the dust on site roads and the wheels of vehicles leaving the site are washed to avoid soiling public roads. Many draglines that remove the overburden lying above the shallow coal seams are electrically-powered and virtually silent. Work at night or on Sundays is controlled and blasting is carried out to independently determined safety standards, and at specified times.

Once the coal has been extracted, the land is restored to use, and very often is left in a much better state than it was before. This is particularly so where, as is often the case, the open-cast workings are covering ground where deep mining has been carried out in the past or where there were, before the open-cast mining, other signs of neglect stemming from the industrial expansion over the last century. In these cases, open-cast mining has often proved the incentive and the means for an extensive restoration which has brought back into use many acres of derelict land.

All open-cast mining is planned in close contact with the local authorities and the relevant Government departments with the eventual reclamation of land in mind: not only that directly in the area of the open-cast workings themselves, but on occasions incorporating neighbouring pit-heaps and other derelict areas. Much of the 115 000 acres so far returned to use is now being farmed as if the mining had never been; and often with improved yields. For example, the first crop on a West Cumberland site yielded 12 to 20 tons of grass an acre, surprising the owners with its quality. A long-closed colliery at Bonds Yard in Derbyshire together with four coke and coal stocking grounds and two large tips were transformed into grazing land (winning incidentally a European Conservation Year Award[2]). In Fife, after extracting 30 million tons of coal from poor quality, barren land, a new loch is being built and 50 million tons of earth will be moved to reclaim a 350 acre derelict site for good farmland. In Northumberland, after mining 7 million tons of coal, the site left has been shaped to the specification of an international golf course architect. The results will be a championship standard course which may attract national tournaments to the area.

There are many other examples from which might be chosen one more to illustrate how restoration can lead to important new leisure amenities.

The Haigh Colliery Site lies just east of the M1, between Barnsley and Wakefield in the West Riding of Yorkshire. Before working started nearly the whole of the 57-acre site was covered by buildings and tips containing about 500 000 yd³ colliery waste belonging to the disused Haigh Colliery.

The area was an eyesore which could not be missed by anyone travelling on the motorway. When the site was worked the colliery buildings were demolished and the rubble deposited in the excavation. The tips, parts of which proved to be quite hot when opened up, were incorporated into a gently-graded area, blending in naturally with the surrounding countryside. Over 110 000 tons of coal were produced from the site, most of which is now under the management of the Ministry of Agriculture being re-established to agricultural use. The south-west corner of the site has been designated as an area where planning consent will be forthcoming for the building of a hotel to serve motorway users, and this has been sown with grass to check erosion and to give it a pleasing appearance until development takes place. An area of derelict land has therefore been made productive again following open-cast working, and a blot on the landscape has been removed. Fig. 3 illustrates the site as it was before opencasting and as it is now.

In the tidy-up operations following open-cast mining, the opportunity is taken, where possible, to remove the remnants of spoil heaps, abandoned colliery buildings and any other dereliction left over from our 200-year industrial heritage.

FIG. 3   The Haigh colliery site before opencasting and after reclamation

The positive co-operation between county council or local authority and the NCB required to make maximum use of these large restoration programmes has already been referred to. In the south of Durham, for example, where the County is planning to spend £1 million a year throughout this decade on reclaiming derelict land, the Board has already handed over 900 acres and has agreed to relinquish further substantial areas over the next five years.

3. COAL PROCESSING

The National Coal Board operates a number of substantial coal processing complexes based, in the main, on coke ovens. It must be admitted that many of these processes are inherently difficult to operate without creating environmental problems from time to time. However, since nationalization the Board has mobilized considerable technical and financial resources directed towards minimizing pollution from these plants. Many notable improvements have been made. But the NCB is far from satisfied and expects over the next few years to increase still further the level of effort devoted to outstanding problems.

Historically considerable effort has been devoted to locating and identifying the sources of any pollution. This could be a combination of a number of things, such as dust, smoke, toxic chemicals, unpleasant odours, each emanating from a different point in the complex. Once the specific problems had been identified the Board was able to finance the installation of a whole range of pollution control equipment, electrostatic precipitators, cyclones, filters (both gas and liquid), gas washers and

liquid coolers. Wherever feasible closed circuit washers and coolers are installed thereby controlling both air and water pollution simultaneously. In parallel the Board has encouraged, in its coal processing plants, the same 'good housekeeping' policies as those that have done so much to alleviate the environmental impact of colliery surface operations.

At the same time the Board's engineers have pioneered a number of technical developments aimed at significantly reducing the remaining pollution associated with coal processing. The chief among these have resulted in a substantial reduction in quantities of dust, smoke and obnoxious liquids emitted during the operation of coke ovens.

The formation of black smoke as the coke is pushed out of the oven is much reduced by a stringent control of oven temperature to avoid under-carbonization. There is, however, still a pollution risk when ovens are opened, either for charging with coal, or for emptying the coke. A system of 'sequential' charging has been developed.[4] In this system the coal is charged by stages: each increment through its own chargehole, which is opened for the least time necessary. The coal heaps in the oven are levelled mechanically, and a clear space always left for the volatile matter to find its way to collecting mains. These latter are connected ultimately to exhausters which pull the volatiles through the successive absorption stages. The system has been demonstrated to be practicable and efficient. It has been accorded favourable comment by experts including the Alkali Inspectorate and its use will be extended considerably in order that charging emissions can be eliminated from an increasing number of coke oven plants. The dramatic improvements

FIG. 4   Non-sequential charging of the Manvers Coke Ovens

FIG. 5   Sequential charging of the Manvers Coke Ovens

affected by sequential charging are illustrated in Figs. 4 and 5.

The liquid effluents produced in these plants could, if discharged untreated, seriously harm the life systems of rivers and streams. Research within the Board showed that a biological treatment system used for the treatment of domestic sewage could be adapted for the purification of coke oven and tar plant effluents. Seven of these plants—costing over £100 000 each—have been installed. The effect to date—together with other cleaning plant—has been to reduce the total amount of polluting matter discharged by nearly 50%, and the quantity of suspended solids by approximately 70%.

The comfort and wellbeing of the operators cannot be overlooked. Some tasks have to be carried out in hot and rather unpleasant conditions. A start has been made on the air-conditioning of the cabins of the coal-charging cars, and will be extended to other control cabins. Where it is not possible to avoid working in unpleasant conditions, plans are advanced for limiting the exposure by automating operations where possible, and by reducing the time spent on the job and providing clean cool air 'sanctuaries.' Automation will in itself, help to reduce pollution by substantially reducing, and eventually eliminating those emissions attributable to operator error.

Apart from these on works efforts, the Board's Scientific Control Department maintains a comprehensive system for monitoring the ground level values of potentially harmful substances around our various works, in addition to monitoring the plants themselves.

The coal industry has made significant contributions to the technology of pollution measurement by the development, and improvement, of special techniques for the sampling and analysis of dust and gases. For example the BCURA grit and dust measuring cyclone is quoted in the Schedule of Government Regulations setting out the procedure to be followed when making grit and dust measurements; and the requisite instrument is now widely accepted, and used, for most measurements of grit and dust emissions from boiler plant and furnaces in the UK.

In the future the Board intends to continue its exchanges of views with the Alkali Inspectorate, Local Authorities and other involved and interested parties,

thereby ensuring that everything practical is being done to reduce pollution due to coal processing. At the same time it is expected to make an increasing investment in still further reducing that pollution as soon as new, and improved pollution-control techniques become available.

4.   COAL MARKETING AND UTILIZATION
4.1.   Storage and transport
The two main environmental problems associated with the large scale handling of solid fuels in the past have been dust and noise. Dust blown by the wind from open stock piles is normally contained by the conventional methods of damping and sweeping of ground deposits, reinforced by special precautions such as the designing of stockpiles to minimize exposure to strong winds, rolling and contouring and covering the coal heaps on particularly exposed parts, as required by local circumstances.

Noise is occasionally put forward as a source of nuisance from mining and stocking operations; but it is a fact that modern practice has in large measure eliminated nuisance from these sources. Large steam-driven winding and pumping machinery has been superseded by electrically driven plant, and the replacement of steam by diesel locomotives for shunting has also reduced noise from wagon-moving operations. Existing coal handling machinery can, in general, be effectively silenced and all in all, we believe that pollution by noise is no longer a general problem in respect to surface operations. The internal problem of noise underground is of course something entirely different which is receiving closest attention.

The transport of coal in modern steel-bodied vehicles and wagons has put into the past, where it belongs, the problem of leakage of fine coal from holes in wooden bodied vehicles. Deliveries to the industrial and domestic consumer can now be made cleanly and quickly using automatic feeding devices attached to the delivery vehicles—either by means of a conveyor or a pneumatic feeder—or by modern bagging methods.

The sheer bulk of solid fuel transported to large power stations, each requiring up to 30 000 tons per day could have created substantial environmental problems, particularly if large volume road transport had been involved

FIG. 6 A liner train in the process of loading coal at a colliery

The bulk rail transport of coal by modern means is reducing environmental pollution. 'Merry-go-round' trains are used in which coal is delivered continuously, and steadily, from pit to power station under modern conditions of loading and unloading, such that a train load of some 1 000 tons can be unloaded in less than half an hour by means that minimize dust pollution of the atmosphere. This facility provides the main 'interface' between this paper and that of Paper No. 5, by Mr Hawkins. It is a facility in which I hope he will agree both our industries, and British Railways, collaborate to minimize the effects on the environment of transporting these very large tonnages of coal from producer to customer. Fig. 6 shows a 'liner' train in operation at a colliery.

4.2. The conversion of coal to heat and power
4.2.1. General considerations
The larger part of the coal produced in this country is burned to create heat, and the combustion process involves the chemical combination of carbon, hydrogen and other constituents of the fuel with the oxygen of the air to form products of combustion which are passed to the atmosphere. Those interested in conserving the environment—and that means all of us—are concerned about the effect of such gases on the air we breathe but it can be claimed that we have come a long way in the coal industry towards ensuring that combustion products are

now released into the reservoir of our atmosphere in such a form as to have negligible effect on health, agriculture and any other factor affecting life.

The flue gases we regard as pollutants also occur naturally in the atmosphere; many play essential roles in ecological cycles. In this respect it is interesting to observe that the amount of naturally-formed sulphur dioxide passing to the atmosphere throughout the world is six times greater than that from man-made sources. The basic pollution question is obviously one of localized concentrations, but there will clearly be continuing argument as to the extent to which man-made amounts of pollutants can be added to the atmosphere without adversely affecting our environment. If it be accepted that man wants to live on this earth at all, much less live in any degree of comfort, it is as exaggerated to talk of banning all man-made pollution as it is to give the industrialist—and the householder—*carte blanche* to burn what he likes, where he likes, how he likes.

The position to aim at is one of minimal pollution yet consistent with the attainment of healthier living conditions. As far as the combustion of coal is concerned, there are now the means to reduce the emission of pollutants to such an extent as to ensure that those pollutants emitted are in such a form, and are so dispersed, as to make a negligible effect on the environment. It will be shown later that, should the evidence for tighter emission standards (than now appear necessary) become apparent, there are means under development by which, looking ahead over the next few decades, these requirements can be met.

In respect of control and reduction of air pollution it is the Board's concern:

(a)   to demonstrate that it is economically feasible and desirable to burn coal or coal products without causing harmful pollution to the atmosphere;

(b)   to provide a service to its customers so that they can meet the requirements of Clean Air Legislation both at present and in the future, with due regard to efficiency and economy.

The methods to be employed to reduce pollution and the part played by the NCB vary with the type, and more particularly the size of the appliance and it is convenient to discuss the domestic, industrial and power generation markets separately.

4.2.2. The domestic market

There is no doubt that many people (particularly, but not exclusively, in the United Kingdom) like the comfort and visual appeal of the open coal-fire. In recognizing this the Board, which has always given its active support to a Clean Air policy for Britain, has adopted a two-pronged approach to the problem of smoke reduction from domestic solid fuel heating, i.e. by making available smokeless fuels and, more recently, bituminous coal burning, smoke consuming, closed appliances.

Lord Robens has graphically described (in his Coal Science Lecture, 1971, 'Science and Coal—the Future') the difficulties that the implementation of the Clean Air Act in the late '50's raised for the coal industry. Even if the NCB had been willing to, it could not have left it to the gas, oil and electricity industries to take over domestic heating in this country. They had not the capacity, nor indeed did many of the users of solid fuel wish to change, particularly those with appliances with a lifetime of anything from 15 years upwards. In these circumstances, the NCB took the decision in the first instance to concentrate on the production of smokeless fuels, suitable for the appliances then available, as well as for the improved appliances and, particularly, for the boilers needed to fire the rapidly increasing number of central heating systems.

We have now reached the stage where smokeless fuels account for approaching one-half of the domestic fuel sold, creating a market with a turnover at retail prices of about £160 million per year.

The two reactive smokeless fuels made by the Board, Homefire and Roomheat, are especially suitable for the open fire, where they give the physical appeal and the heat associated with the coal fire, but without smoke. The technology necessary to produce these fuels is very complex but, thanks to the tenacity of the Board's scientists and engineers, the problems have been largely overcome and we are gratified that the fuel now marketed is finding a satisfied, often enthusiastic, clientele.

Homefire is made from powdered Midlands bituminous coal, de-smoked in a fluidized bed under controlled time and temperature conditions. The heat required to maintain the bed at around 400°C is generated in the reactor by the partial combustion of the coal. The residual material is immediately compressed into briquettes while still at around 400°C, using reciprocating presses.

Roomheat is made by a similar process to Homefire except that the briquettes are pressed in double-roll presses, and input coal having mildly swelling characteristics is used.

To meet the need for a less reactive fuel suitable for certain modern deep open grates, roomheaters and closed boilers the Board markets its long established Sunbrite, which is a domestic coke manufactured mainly in the Midlands, Yorkshire and North East where suitable coals are available.

The Board also produces another long-established briquetted smokeless fuel in South Wales, known as Phurnacite. This fuel is made by the medium temperature carbonization of pitch bound ovoids specially blended from carefully selected naturally smokeless coals. The resultant fuel is very hard and strong and is suitable for solid fuel boilers, cookers and closed appliances.

There is, then, a complete range of approved smokeless fuels available to meet all purposes.

Let us turn now to the alternative approach, that of developing appliances to burn normal, smoky bituminous coal smokelessly. Scientists and engineers, in collaboration with the appliance manufacturers, have developed a range of smoke-consuming appliances that burn selected cheap bituminous coals within the requirements for 'smoke-free' areas. The appliances, known collectively as 'Smoke Eaters' achieve their result by ensuring that the smoke is drawn back through the incandescent fuel bed and burned. Fig. 7 illustrates the principle of the device.

It is obvious that there is still scope for further development of these appliances, particularly as a higher degree of aesthetic and mechanical amenity becomes desirable. Improvements in thermal efficiency automatically mean lower emissions for the same heat output, whilst less clinkering and easier ash removal will reduce the work associated with the appliance. The NCB is also continuing the development of a smoke-reducing open fire.

It is clear, therefore, that coal and coal-based fuels can continue to be used on the domestic market without causing significant pollution due to smoke. There remains the problem of sulphur dioxide emission.

It is unfortunate that the effects are not fully understood of various concentrations of SO_2, or other sulphur

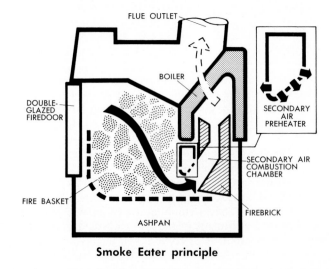

Smoke Eater principle

FIG. 7   A diagram illustrating the principle of the 'Smoke Eater' devices

compounds in the atmosphere, on health, much less on agriculture and materials. Any local problems caused by an emission of sulphur dioxide are much less in the absence of smoke. Elimination of smoke emission has led to an appreciable, probably very substantial, reduction in adverse effects on buildings, parks and health in urban areas even without any change in sulphur dioxide concentration. It is therefore necessary to establish the facts before being stampeded into unnecessary action; and it is submitted that it remains to be proved that the amount of sulphur dioxide emitted from domestic chimneys, even with the density of chimneys reached in some dormitory areas, causes an appreciable effect, assuming the virtual absence of smoke.

There is almost no reliable data on the emission of nitrogen oxides from domestic chimneys. Preliminary measurements that we have made on solid fuel appliances using the methods which, as mentioned later, we have now established with the CEGB and British Gas as reliable, suggest that NO_x emissions from such units are unlikely to be significant.

4.2.3. The industrial market
In the years after the last war the need for more efficient and, therefore, more economic operation coupled with the impetus of the Clean Air Acts has led the NCB to foster the development of automatic firing equipment such that the emission of smoke, grit and dust from coal-fired industrial steam-raising plant is now generally well within acceptable limits: a latter-day Blake could no longer talk of 'dark, Satanic Mills.' Modern plant is available for the arrestment of grit and dust to a high standard of efficiency, and the Marketing Department of the Board is anxious, through its technical and fuel advisory services, to advise existing and new solid fuel users on means of minimizing pollution of any kind from plant and achieving compliance with the appropriate standards.

The discharge of flue gases from industrial plant is influenced, in respect of potential sulphur dioxide discharge, by the Memorandum on Chimney Heights issued by the Ministry of Housing and Local Government; a document which the Board's engineers helped to prepare. Provided the requirements for height of discharge are observed, together with attainment of reasonable gas velocity, the emission from solid fuel-fired industrial plant should not contribute significantly to local pollution.

However, should the pressure to reduce emission of SO_2 gain still further recognition in the form of additional legislation, much of the work already appropriate to large combustion plant for the 'fixation' of sulphur in the fuel bed will be applicable to general industrial coal-fired plant.

As well as pollution from chimneys, local dust nuisance can also arise from coal and ash handling, and can be prevented by mechanical handling systems now available at a reasonable price.

One example of an NCB-sponsored development in this field is the open-screw feeder for feeding coals from flat-bottomed bunkers.

It is not possible to deal individually with the many specialized furnaces and kilns which use coal as a fuel. In the case of the larger users, such as the cement industry, they are fully competent to take the necessary steps to minimize atmospheric pollution. Where coal or coke is used in other industrial processes, as in cupolas, driers, brick works and so on, the NCB technical staff collaborate with the manufacturers of the plant, the user and the relevant local authority, or the Alkali Inspectorate, in arriving at the best practicable means to minimize any pollution. In many of these plants, of course, the main source of the pollutant is often not the fuel but, whatever the cause of the nuisance, means can always be found for minimizing it; the extent to which this is desirable depends upon the cost set against the advantage to be gained.

A Working Party, set up by the DoE, with one of the Board's engineers as chairman, is in the process of reporting its recommendation for limits of grit and dust emission covering various furnaces, irrespective of the fuel used to fire them.

4.2.4. The power generation market
The burning of coal at power stations is backed by a highly competent technical staff employed by the Generating Boards. The role of the NCB in these cases is therefore much less direct, being restricted to that of a collaborator, in addition to our prime responsibility of making available suitable coals. Much important work on the effects of gaseous emissions from power stations has been done by engineers and scientists within the CEGB, but the technical effort within the NCB has played its part especially in the development and testing of measuring equipment. Earlier work on the detection of sulphur trioxide and particulate matter is now being followed by work in conjunction with the CEGB on the measurements of the oxides of nitrogen and sulphur dioxide. The NCB is one of the first large organizations in the UK to put into use a 'chemiluminescence' monitor which has been shown capable of giving reliable measurements of NO_x in combustion equipment. Non-dispersive infra-red photometry is used for the continuous determination of SO_2. Particular attention is now being paid to the often difficult problem of obtaining, and preserving, a truly representative sample of flue gases. In the United States, considerable interest is being shown in emissions of nitrogen oxides from combustion plant, varying from motor cars to large boilers. In the case of coal-fired boilers, the situation in the US appears to be different from that in the UK in so far as, in a recent study by the Esso Research and Engineering Company, the average emission of NO_x from a large US coal-fired boiler was reported to be just less than 1 500 ppm. In contrast, the highest short term emissions measured recently in this country are only about half of this, and the average emission lower still. It would seem important to establish the reason for this difference, since it may point the way to possible improvements.

As may be evident from statements made earlier, the NCB fully supports the 'tall stack' policy of the CEGB and it is felt on the evidence at present available that the effect of the various gases emitted from the modern well-operated coal-fired power stations, whether they be sulphur dioxide, nitrogen oxides or any other flue gas constituent, on the environment in this country is insignificant. This whole matter is, as will be discussed elsewhere in this Conference, being studied by a joint OECD exercise, the British participation being directed by the DoE. This exercise has the NCB's support and the Board looks forward with interest to the outcome.

Should it be shown necessary to remove in large measure the sulphur dioxide and other gaseous pollutants

FIG. 8   One of the Board's pilot-scale fluidized-bed combustors, capable of liberating 1.4 MW of heat

from the flue gas of coal-fired plants, the Board is developing the technology to make this possible. In the meantime, close study has been made into possible means of reducing sulphur content of coals by selective washing, but the technical and economic feasibility of this approach unfortunately appears very limited and, at present real prospects are not seen for meaningful reductions in the sulphur content of coal before combustion.

5. FUTURE DEVELOPMENTS IN COAL UTILIZATION

It now appears certain that coal, in around the same quantities as now used, will be needed in this country, at least for the next few decades. There are, however, bound to be changes in the way coal is used. In fact, it would be extremely short-sighted not to take full advantage of the technology available, and becoming available, to utilize this coal as efficiently and cleanly as possible.

The future combustion of fossil-fuel in the USA is already dependent on the development of methods to make sure the stack gases are substantially sulphur-free. Legislation to restrict the emission of nitrogen oxides is also becoming stronger. While, for reasons earlier listed the NCB does not feel that similar restrictions would be appropriate in this country now, it is accepted that the steady increase in world fuel consumption and the improvements in other pollution sources may make similar restrictions justified in this country in time. It must be reiterated, however, that stronger justification is needed for such a move other than climbing on to a 'bandwagon' of unsubstantiated, often emotional, general claims.

It does not, at present, appear feasible to remove the

sulphur from the coal before it is burned. While methods to give some sulphur reduction are possible, much of the sulphur is so intimately associated with the coal material that separation without completely altering the coal structure is very difficult. One way that it can be done is by solvent extraction, in which the coal substance is dissolved into a suitable solvent and separated from the mineral matter to give a material which contains not only little sulphur, but also very little ash. The Board has actively researched this field, and the technology is well known to us. In the raw state, it can have sufficiently low melting temperatures for it to be handled as a liquid if required, and this could obviously be used as a very attractive fuel should the economics change sufficiently to make it viable. Alternatively, it can be used as a feed-stock for the manufacture of synthetic natural gas or oil or, if preferable for direct combustion, as a low calorific value gas. Recent developments by NCB scientists include one in which the extraction is carried out by gas in the super-critical condition thus eliminating ash altogether, and opening up exciting new possibilities of producing carefully tailored materials to give the most efficient conversion to a range of products.

There are a vast number of systems for removing the sulphur from the flue gases, once combustion has been completed, and more are being developed almost weekly as, in particular, the Americans seek the most economic system. All of these suffer the disadvantage that additional plant, merely for sulphur removal, has to be added to the existing plant at a high capital cost, and a high running cost, which it is difficult to offset against a problematical income from the sale of by-products. For this reason, there is increasing interest in the development of methods in which sulphur recovery is made part of the combustion system and, here, it is suggested that fluidized-bed combustion, which the NCB has done so much to develop, should have a major part to play.

This combustion system was initially developed because of its potential savings in capital cost as compared with conventional pf firing. A number of design studies have shown that these capital cost savings should be from 10 to over 20% depending on the size of the boiler. It is the capacity of fluidized-bed combustion to retain the sulphur, in addition to reducing the capital cost of the boiler, which has proved the greater incentive to its development. The increase in generation cost consequent on sulphur retention is the lowest of any system yet devised.

In a fluidized-bed combustion system, the coal in granular form, is fed into a fluidized bed of inert material (in the case of coal the ash particles themselves) where it burns at a relatively low temperature of about 800-900°C, at which temperature the ash does not soften. If limestone or dolomite particles are added to the bed at the same time as the coal, the sulphur reacts to form calcium, or magnesium sulphate, which is removed from the bed along with the ash.

The amount of sulphur retained is dependent on the amount of limestone or dolomite added, methods of operation and so on, but research work on rigs burning more than 1 000 lb coal/h, operating both at atmospheric pressure, and at pressures up to 6 atmospheres absolute, carried out by the National Coal Board in collaboration with the Environmental Protection Agency of the American Government, has shown decisively that the retention of well over 90% of the sulphur in the coal is both technically and economically feasible.[5] Fig. 8 shows one of the NCB's pilot-scale fluidized-bed combustors.

The National Coal Board has continued to encourage and support (in some cases under contract) the development of this system in the USA where there are plans to have pilot plants up to 30 MW operating by 1975 and a 200 MW prototype well tested by 1980. It has, moreover, joined in a Consortium set up under the aegis of the National Research Development Corporation to exploit in this country (and overseas) the pioneering development of this concept by the National Coal Board: and it is interesting that, besides its use with coal, fluidized-bed combustion is also finding applications as a means of burning, without pollution, other fuels such as oil and domestic refuse.

Another possible way of removing pollutants when burning coal is to gasify the coal first, clean the gas and then burn this in the combustor. It is possible to adapt this approach to existing boiler units, and it can be competitive with the use of flue gas cleaning plant in spite of the loss of part (maybe 10 to 20%) of the energy in the coal in the conversion process, and of the additional capital costs.

More advanced concepts, most of which involve gasification or combustion under pressure with a gas turbine combined with steam generation, may use fluidized-bed combustion, or pre-gasification, or some combination of the two.

If it be looked ahead, to the day when it becomes necessary to convert coal to gas (or oil) to augment or to replace the existing sources of supply, it is possible to incorporate a cleaning stage in the process so that the resultant secondary fuels will be free from the causes of pollution. The technology for converting coal to liquid or gaseous fuels is outside the scope of this paper, although much effort is at present being expended, particularly in the USA, and being monitored by the NCB, since breakthrough there would provide the impetus for moving away from the conceptual stage.

A logical outcome of the present Research and Development trends being pursued, is the use of coal as the major feedstock into an Integrated Energy Conversion complex for the production of electricity and gaseous or oil products.

The number of possible combinations of processes is potentially very large. These large plants to which one can give the generic title of 'Coalplexes' can also be linked to the manufacture of chemical feedstocks, and coking products. Whatever outputs are required, each unit in the combined plant will be required to have not only maximum efficiency and flexibility, but also to make full use of the pollution control systems already discussed in this paper. The Board is actively engaged in economic appraisals of the Coalplex concept, which will become all the more urgent and relevant given the possibility of a severe energy shortage by the mid 1980s.

Present methods of extracting energy from given fossil fuel resources are relatively inefficient in terms of the waste and therefore the pollution created. Unless a fundamental alteration of energy conversion processes occurs we will be faced with a clear choice between faster growth or environmental protection; where the former carries also the seeds of even more rapid resource depletion.

The Coalplex concept provides one means of escaping from this dilemma before it reaches crisis proportions, as indeed President Nixon implicitly recognized earlier this year in his Energy Message, both verbally and in increasing by 27% the central energy fund allocation to be devoted to coal research and development.

Through concentration of energy conversion processes on to a single site, maximum advantage can be taken of common services and energy exchange between the various processes. Efficiency and pollution control within the 'coalplex' concept are two sides of the same coin.

Residual waste streams of e.g. sulphur, coal ash, alumina and iron ore, can ultimately be treated as valuable by-products. These will, of course, have a very low opportunity cost in terms of the further environmental pollution and associated social costs that would otherwise have to be borne; coal ash, for example, could be converted into a range of aggregates, reducing the need for quarrying into the country's landscape.

Pollution is a world-wide problem, and efforts are being made in all industrial nations to find satisfactory solutions. In the coal industry there exists particularly good relations both with colleagues in the USA, as seen in the collaborative work on fluidized-bed combustion, as well as through working in close collaboration with colleagues in the Common Market. Through these contacts, and through those with other nations, the NCB intends both to play a part in developing the most efficient and effective means of avoiding pollution from coal, and also to adapt and introduce into this country the most suitable methods that there are for our particular circumstances, as these become both necessary and available.

6. CONCLUSION

This paper has, of necessity, covered a wide range of topics because pollution appears in many forms, means different things to different people and has to be tackled in many different ways. In addition there is always the danger that, in arresting a pollution from one source, a new source of pollution is initiated somewhere else. For example, when discussing the removal of sulphur from flue gas, as in the last section, it must not be forgotten that the sulphur is only removed from the gas to the cleaning medium such as, with fluidized beds, the ash. It is necessary to consider, as indeed the NCB is doing, the disposal of this ash or, preferably, its conversion into useful products (reflecting back on the issues discussed in Section 3) as parts of the exercise on flue gas cleaning. To give another example, with perhaps wider connotations, some years ago the Board closed the last colliery in a particular coal field. Three years later it became apparent that a large area of domestic and industrial development was reverting to the swamp that it had been before the emergent coal industry of the 18th century initiated pumping.

The NCB feels that its first priority in the coming years must be to ensure that it can supply the demand that is going to be asked from it. How efficiently the NCB does this will depend on how far this country is prepared to accept now the need to maintain a viable and progressive coal industry to ensure that what is still by far the dominant indigenous source of energy in this country will be available when it is needed. It is hoped, however, that it can be seen that the National Coal Board is giving almost equal importance to ensuring that all its activities are carried out with the minimum of harm to the environment.

The task in front of us should not be belittled. There will be many difficult technical problems to be solved and those in a position to do so are urged to encourage as much integrated research and development as possible by the industries concerned, by suppliers and users, and by the Government agencies involved. This is one field where as much collaborative research and development as possible must yield dividends. With this continued effort, and the necessary drive to implement it, the fuel industries as a whole can face their future responsibilities to the community with every confidence.

7. REFERENCES

1. Spoil heaps and lagoons; NCB Technical Handbook, Sept. 1970.
2. ARGUILE, R. T., The reclamation of five industrial sites in the East Midlands; *J. Inst. Municipal Engineers*, **98,** June 1971.
3. TANFIELD, D. A., Construction uses for colliery spoil, *Contract Journal*, 14 and 21 Jan. 1971.
4. MEADES, M. R., and RANDALL, G. E., Smokeless charging; Coke Oven Managers Association Year Book 1961.
5. WRIGHT, S. J., The reduction of emissions of sulphur oxides and nitrogen oxides by additions of limestone or dolomite during the combustion of coal in fluidized beds; Proceedings of the 3rd international Conference on Fluidized Bed Combusion, Hueston Woods, Ohio, 29 Oct. to 1 Nov. 1972, US Environmental Protection Agency.

GEOFFREY CHANDLER, MA*

3   The oil industry

SUMMARY

Oil is now the largest single source of energy in the UK. It has risen rapidly to this position as a result of its competitiveness of price, versatility, relative cleanliness, and ease of handling. This growth—which has been worldwide—has contributed significantly to the welfare of society, but the scale of operations involved has inevitably brought associated environmental problems. The paper discusses the problems arising from exploration and production, ocean transportation, refining, distribution, and product usage, and the measures that the oil industry has taken both for prevention and cure. These measures include plant design, codes of practice, scientific methods for pollution abatement, industry co-operation through voluntary agreements, and also require well-founded legislation, effectively enforced, and public awareness of individual responsibility. Emphasis is laid on the need to avoid pollution, but also to provide immediate remedial measures and compensation in the event of human error or accident. Many problems do not have a definitive solution—short of ceasing an industrial activity altogether: the expectation of continued growth of oil demand and of more stringent standards in the future means a continuing emphasis on the research and development which must be the basis of effective action.

1. INTRODUCTION

Oil is the largest single source of energy in Britain. It now contributes some half of total supply, and the indications are that its share of the energy pattern will continue to grow (Fig. 1).

The reason for this development is not far to seek. Oil has had a relative stability of price (though taxes in Europe have risen to about two-thirds of the selling price of certain products) and has remained highly competitive with other fuels. The versatility of petroleum provides a range of useful products which no other fuel can match—from the lightest gaseous hydrocarbons, to the heavy residual fuel oils. There is so far no practical alternative to oil as a propellant of motor cars, aircraft and ships; industry and agriculture are heavily dependent on

petroleum fuels. Its relative cleanliness and its ease of handling have given it additional advantages.

Against these positive assets must be set the fact that the soaring growth in oil demand has brought certain environmental problems. By the nature of its activities the oil industry operates very large plants whose emissions and effluents, unless strictly controlled, are capable of polluting air, soil, and water. Its raw material, crude oil, is transported across the oceans by tanker fleets which incorporate the largest ships afloat. Storage and distribution of the products to the consumer call for unremitting care both on grounds of safety, and to minimize pollution risks. The industry was well aware of its responsibilities to the communities it serves long before environmental conservation became a sensitive public issue. Intensive research has been translated into positive action. Considerable progress has been made over the last two decades by sustained attention to what William Blake called 'minute particulars'* in every phase

U.K. Inland energy consumption (primary fuel input basis)

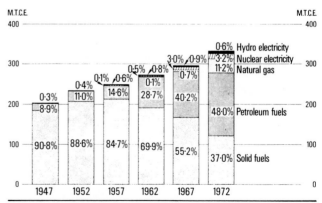

Source: Monthly Digest of Statistics, February 1973

FIG. 1   Oil in the UK energy pattern

*President of the Institute of Petroleum; a Director, Shell International Petroleum Co.

*   He who would do good to another must
        do it in minute particulars:
    General good is the plea of the
        scoundrel, hypocrite and flatterer,
    For art and science cannot exist but in
        minutely organized particulars.
                        (Poems from MSS *c.* 1810).

of operations. This has been achieved by the application of scientific methods for pollution abatement, co-operation between individual units of the industry, voluntary agreements, sustained by well-founded legislation and increased public awareness of individual responsibility.

The measures taken—from the well head to the customer's tank—are discussed in some detail in succeeding sections of this paper, and only certain highlights need be mentioned in this introduction.

The major exploration and production enterprises that have taken shape in the North Sea in recent years have focussed attention on the potential safety and pollution hazards involved in offshore activity, though this has a background of many years' experience elsewhere. Drilling operations are carried out by highly experienced crews working to strictly enforced procedures. The operating companies have the strongest possible incentive to be scrupulous about their observance. A major blow-out could be fatal to drilling crews, and a spillage of oil or gas harmful to bird and marine life.

The search for ways of preventing oil pollution at sea has occupied governments and oil and shipping concerns for many years, particularly since crude oil is now moved across the oceans in such vast quantities. The 'load on top' system, pioneered by the oil industry and introduced in 1964, enables all but a minute part of the oil residues from tank washings to be retained on board the tanker and discharged with the cargo at the terminal port. Without this technique, exponential growth of world tanker traffic would have led by 1972 to a discharge rate into the ocean of some 3 million tons per annum: 80% of the world's tanker fleets have accepted the 'load on top' system. If the remainder fell into line only about 75 000 tons (or 0·006%) of the crude oil transported would be discharged. Their failure so far to comply arises from lack of accession by all governments to an international instrument, and underlines the fact that good practice requires political will expressed in effective legislation to reach the intended goals. Still more stringent standards are now being sought, and these may involve segregated ballast tanks in ships and improved disposal facilities ashore.

Sophisticated navigational aids, segregated traffic lanes in the congested English Channel, and careful crew selection and training all contribute to safe operations, but accidental spillages are bound to occur from time to time. Although the responsibility for cleaning up the sea lies primarily with national and local authorities, the oil industry is geared to swift action in mobilizing facilities which can immediately be brought into action, backed by voluntary joint insurance schemes to meet the high cost involved.

Pipelines have proved to be the safest and most pollution-free means of transporting large volumes of crude oil or products overland, and accidental spillages are very rare events. Good housekeeping is practised at petroleum storage installations and distribution centres, and in the transport and delivery of products to the customer.

In oil manufacturing the amount of oil in the effluents of a modern refinery has been reduced to one-tenth or even one-hundredth per ton of crude oil refined as compared with old-established plants. A cost of up to 10% of the total capital expenditure on the refineries concerned is involved in air and water pollution abatement. Refinery plants are now designed with built-in

anti-pollution devices on the principle that prevention is better than cure. Particular attention is also devoted to noise control.

Sulphurous emissions are a general problem not confined to the oil industry, although there has been progressive improvement due largely to the change-over from coal to other sources of energy.[1] The emergence of natural gas as an important fuel and the increasing use of low sulphur (African, and eventually North Sea) crude oils should make a substantial contribution to helping solve this problem during this decade. A second possibility to reduce sulphurous emissions is the removal of sulphur from oil fractions before they are burned. For distillate fuels there are well-proven processes, but the treatment of residual fuels presents a massive economic and engineering challenge requiring a capital outlay of hundreds of millions of pounds. The removal of sulphur from residual fuels is recognized as a major problem to which environmental requirements demand a solution, and much research and development are being devoted to this end.

A third method is to remove sulphur dioxide from furnace combustion gases. A great deal of work needs to be done before such a solution can be regarded as widely applicable economically, rather than viable only for very large plants, such as major electricity generating stations. Another promising method is to gasify residual fuel oil under conditions that facilitate sulphur removal from the gas, which is then burned to drive a gas turbine and then the residual heat used to raise steam for steam turbines. Such a combination has an additional potential for increasing the efficiency of a power station beyond that attainable with a steam cycle alone.

It is the responsibility of the oil companies to produce and transport crude oil, to manufacture products tailor-made to the needs of the market, and to store and deliver them to the user—all without unacceptably injuring the environment. The use (or misuse) of the products lays some responsibility on the consumer. 'Misuse' implies not only the use of a product for a purpose for which it was not intended, but also use in poorly designed, or inefficient equipment which may give rise to avoidable emissions, smell or noise.

In the UK the industry has supported organizations such as the Institute of Petroleum and the Oil Companies Materials Association to produce codes of safe practice, and design methods and materials specifications to ensure safety in equipment and operations. These are constantly being reviewed and revised, as instanced by the Institute of Petroleum's codes of safe practice. In Europe 20 oil companies have a joint organization (CONCAWE, short for CONservation of Clean Air and Water, W. Europe) to obtain and pool information on matters affecting the environment[2] and to advise governments and inter-governmental organizations on anti-pollution measures. The major oil companies give extensive assistance to their customers to help them burn fuel efficiently, and in a way that safeguards the environment.

One piece of equipment that currently arouses much controversy regarding emissions is the motor car, although its performance in this respect depends more upon the design of the engine and exhaust system than upon the composition of its fuel. In recent years of growing conservation awareness, more than 50% reduction in the objectionable products of combustion that come out of the exhaust has been achieved initially by simple adjustments to the carburettor and ignition-timing system. Lead compounds have long been used as an economic means of raising petrol performance. Although there is no clear evidence that lead from automotive sources is a health hazard, legislation is substantially reducing lead in petrol.

Removal of lead, or too great a restriction of it, would result in lower performance and higher fuel consumption at a time when the need to conserve energy is being increasingly acknowledged. Basic re-design of engines and matching fuels will be necessary in the long term, and this will be a costly process both for the oil and car industries in collaboration. Meanwhile, the fuel needs of current generations of cars must be met.

There is no room for complacency. The industry must continue to try to anticipate problems, take the initiative in seeking solutions, and be seen by all concerned to be doing so. This implies good communications, a readiness to pool appropriate research findings and to make the results available through normal licensing procedures. Because the whole community is involved, the merits of co-operative effort are manifest. There needs in general to be more education, a greater exchange of know-how, and better use of the skills of universities and other academic institutions in co-operation with other energy industries.

It must be recognized that there is usually an inescapable correlation between environmental constraints and cost, and a point where the law of diminishing returns sets in. At the recent Washington Oil Spills Conference it was stated: 'It has to be decided if it is justifiable to impose very strict limits on effluents (with all the expense involved in treatment facilities) to improve

FIG. 2 *A blow-out preventer stack as used in North Sea drilling platforms. The size of these pieces of equipment can be gauged from the man working on the stack*

conditions on, say, 100m of shore. With finite resources there would seem to be greater environmental priorities.'

Soundly-based legislation calls for constructive consultation in advance with the oil industry, which will be primarily responsible for devising and implementing the remedies called for. Legislation may, according to need, be national, regional (e.g. in the European Economic Community), or international, but it should not be so restrictive or inflexible as to waste resources and ignore the regenerative capacity of the environment itself.

Standards affecting products that move in international trade—such as motor cars and petrol—should be harmonized at least on a regional basis. Emission standards from fixed sources (such as power stations and oil refineries) should be allowed to vary between region and region, country and country, and urban and rural areas. This will make maximum use of the self-cleansing ability of the environment and release financial and technological resources for more urgent conservation problems. Rigidity in international legislation could prove the enemy of a healthy environment.

The oil industry believes that prevention is better than cure; that adequate environmental standards must be maintained; that accidental pollution exceeding the standards set must be dealt with quickly, and compensation paid for attributable damage. The costs of meeting legislative standards should be reflected with other production costs in the price of goods and services. This is what the 'polluter must pay' principle really means.

Industry is the servant of society. It has a responsibility to assist in creating and in meeting the standards that society requires. Governments for their part have a responsibility to consult the technological and economic experience of industry, to weigh the costs and benefits of alternative courses of action, to pass legislation, and to ensure that it is equitably and effectively applied.

Absolute guarantees cannot be given: in any activity the only guarantee of zero risk is to cease that activity. Zero pollution is a technical impossibility. What is required is an appropriate philosophy, serious intent, practical measures to minimize risk, standards effectively enforced on all practitioners, the designing-out of pollution at the planning stage, stand-by measures, and compensation in the event of accident.

The debate on the environment has often been marked by high idealism and understandable emotion; but often also by an insufficiency of analysis and intellectual integrity. It is only by an honest exposure and examination of the facts that society will be able to determine the required balance.

2. EXPLORATION AND PRODUCTION
2.1. The North Sea

Now that natural gas is being piped ashore from the UK sector of the North Sea—and oil will soon be produced and landed from beneath Scottish waters—the environmental problems involved have attracted increasing attention because they are now on our very doorstep, and are not remote hazards in faraway places with which we have no direct concern.

However, offshore exploration and production is no new thing. In the Gulf of Mexico 13 500 wells were drilled between 1938 and 1972: in those 34 years only seven incidents posed an environmental risk.[3] Over the period findings and observations along the Louisiana coast have established certain facts about the effects of petroleum activities on the marine ecosystem.

(1) Accidental and short-term oil pollution, while troublesome, expensive, and locally damaging, appears not to have any permanent effect on the ecosystem.

(2) Operations in offshore waters do not involve serious ecological problems, although they may create regulatory problems with respect to navigation and the use of the areas by fishing interests. In many instances, offshore structures have been beneficial to recreational and fishing interests.

The accumulated knowledge and experience gained from under-sea operations, not only offshore North America but elsewhere in the world, are available to the companies engaged in the North Sea search. On the other hand, the water depths are greater, and the meteorological conditions more severe. So far, however, more than 600 wells have been drilled without any accident leading to environmental damage.

The governments of the countries bordering upon the North Sea have taken the problem of oil pollution very seriously. The UK Government has drafted Construction and Survey Regulations and intends to seek agreement with other governments concerned so that uniform standards and practices shall operate throughout the whole area. Recently the Institute of Petroleum published its revised and updated 'Code of Safe Practice for Drilling and Production Operations in Marine Areas', which embodies the most modern practices for safe operations.

When a company with a North Sea concession has planned its drilling and production programme, it has to be approved by the Department of Trade and Industry before operations can begin. All drilling equipment and production facilities must meet approved engineering specifications and incorporate fail-safe devices. Each company must supply weekly progress reports, and periodical visits by DTI inspectors ensure that rules and procedures are complied with. The rules also cover the

FIG. 3   Lay-out of an under-sea oilfield

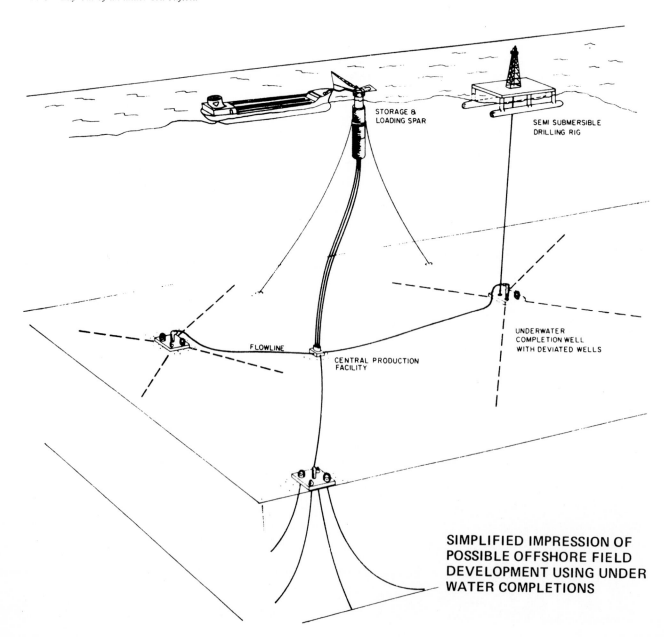

STORAGE & LOADING SPAR

SEMI SUBMERSIBLE DRILLING RIG

FLOWLINE

CENTRAL PRODUCTION FACILITY

UNDERWATER COMPLETION WELL WITH DEVIATED WELLS

SIMPLIFIED IMPRESSION OF POSSIBLE OFFSHORE FIELD DEVELOPMENT USING UNDER WATER COMPLETIONS

lighting and marking of offshore installations, the locations of which are notified to mariners, marked on charts, and a safety zone established around each which is prohibited to shipping.

The directional drilling techniques used enable as many as 30 wells to be drilled from each platform—all surfacing and having their well head equipment inside the platform. Multiple well production from a single platform reduces the risk of ecological damage, and also avoids congestion of sea space—an important factor from navigation and fishing aspects (Fig. 2).

There are other comprehensive safeguards during drilling operations: the use of drilling mud (a blend of specialized chemicals) of sufficient density to hold against pressure surges; casing protection to avoid breakthrough of hydrocarbons from deep formations into shallower ones; and blow-out preventers to close off the well in the event of any sudden pressure breakthrough. Each platform will incorporate shut-down facilities, both manual and automatic, to stop production immediately in the event of an emergency.

When drilling is complete and the rigs become entirely production platforms, there will be several main shut-off points—one a fail-safe valve installed in the flow string and held open by hydraulic pressure from the platform (cut off the pressure and the valve closes). All the flow lines from the wells will be gathered together into a manifold with a valve on the inlet to the separators, again of the fail-safe type. Platforms are so designed that drips and loss of small amounts of oil and chemicals are not discharged into the sea.

Should any oil be accidentally spilled, a mobile system of booms and skimmers will be available to pick it up. Chemical dispersants will be employed which are non-toxic to marine life. Application is by conventional spraying methods using boats equipped with surface-breakers to achieve thorough mixing of oil and dispersant.

The Forties and Auk fields are due to start producing in 1974. From Forties the oil will be piped to the Scottish coast through a 115-mile sea line constructed of 32-in diameter special steel with a wall thickness of three-quarters of an inch overlaid with $2\frac{1}{2}$ in of reinforced concrete; for much of its length it will be buried. Oil from the Auk field will be landed by tanker. Production wells will be drilled from a platform as already described and the oil conveyed to single buoy moorings at which tankers can load. SBMs are well-tried pieces of equipment with a good non-spillage record, and especially robust versions have been designed for exposed locations in the North Sea.

For the initial stages of development of the much larger Brent field a floating oil storage facility is being built, shaped like a huge fisherman's float, and capable of holding 300 000 barrels of oil. This 'spar,' as it is called, will provide operational flexibility to cope with irregular tanker movements. Special skimming and filtering facilities will be incorporated to ensure that ballast water discharged to the sea is oil-free.

Research is in progress which will enable oilfields to be completed on the sea floor and linked by suspended flow lines to floating spar storage or submerged storage tanks. Among other advantages, under-sea fields considerably reduce the risk of ships colliding with surface structures (Fig. 3).

For billion barrel fields further from shore the scale of production may justify the laying of underwater pipelines to land. But the technical problems of laying such pipelines at 500 ft depth will be very challenging and they may take two or three summers to complete. Historically and statistically, pipelines are by far the safest transport system where large volumes have to be handled. Regular radiographic examinations are made to check for splits in welds, and if a leak develops everything shuts down until a repair has been made. Past experience has shown that pipelines when buried have not caused any serious pollution troubles.

2.1.2. Spill clean-up

The UK offshore operators have devized a co-operative emergency oil spill clean-up procedure backed with equipment and supplies, based at Aberdeen, Dundee, and Great Yarmouth. Additional locations are under consideration.

The detailed emergency procedures and contingency plans worked out for Aberdeen (northern sector) and Great Yarmouth (southern sector) ensure the rapid alerting of all the necessary services to give maximum assistance in the shortest possible time to any offshore rig or platform faced with an emergency which could amount to a disaster, the foremost objective being the safety of personnel aboard.

The clean-up facilities are designed to cope speedily with a spill of the order of 10 000 barrels. There is, moreover, a very substantial back-up. Most of the operators—certainly all the big ones—have their own arrangements which can be brought into operation, to support the DTI's comprehensive emergency facilities in case of a major spill. Co-operative possibilities are being examined with other North Sea countries, such as Norway, Denmark, and the Netherlands, and a comprehensive plan for the whole area is envisaged, with measures harmonized and co-ordinated through operators' committees representing all companies in each sector.

The basis of the procedure derives from equipment developed by the Warren Spring Laboratory of the DTI[4] which relies on the use of dispersants applied by simple equipment available at low cost, and which have a very low level of toxicity.

Discussions at national level, as well as between companies, are proceeding to set up a scheme to provide compensation for oil pollution damage. Most of the UK operators already have their own insurance cover for exploration operations. Since currently each company must find the first £1 million of any claim involved, this

TABLE 1 Increasing oil movements accommodated by rapid increases in tanker sizes

Year	Typical size of crude oil carriers
1938	15 000 dwt
1948	25 000 dwt
1956	36 000 dwt
1960	47 000 dwt
1961	100 000 dwt
1964	150 000 dwt
1968	200 000 dwt
1972	200/300 000 dwt

will intensify awareness of the importance of spill avoidance.

3. MARINE TRANSPORTATION

Not all oil pollution at sea is caused by tankers. However, the huge volume of oil transported—1 225 million tons of crude and 275 million tons of refined products in 1972[5]—would, without the adoption of new measures, have created a major environmental problem of disposing of dirty ballast and tank washings without harming the marine environment or damaging coastal amenities.

Something like 100 million tons of oil were imported into the British Isles in 1972, the bulk of it through four major arteries: the Thames estuary, Milford Haven, Fawley and Mersey. Much larger volumes pass through the English Channel to continental oil ports: more than 100 million tons a year to Rotterdam alone, making the Channel the world's most congested waterway.

The dramatic expansion of oil refining capacity in Europe and in other areas of high consumption which has characterized the past 20 years, has resulted in virtually the whole petroleum requirement being imported as crude oil. Oil consumption in Western European countries has doubled roughly every 10 years and shipping statistics indicate that total ocean-going movements of crude oil have trebled every 10 years over the past 30 years. Vastly increasing oil movements have been accommodated by the equally rapid increases in tanker sizes, although supplies to many parts of Europe are conveyed by a combination of sea transportation and cross-country pipelines. The larger the tanker, the fewer needed to do the job (Fig. 4).

Some European crude oil pipelines have their starting point on the French and Italian Mediterranean/Adriatic coasts; to the extent that such lines supply areas in Germany and elsewhere, which might otherwise be

FIG. 4 Comparative tanker sizes

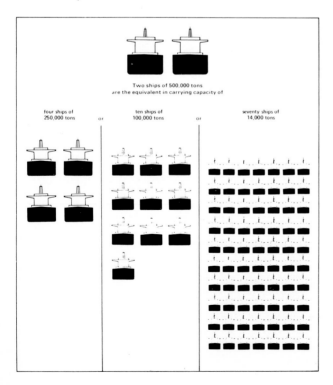

supplied through North Western European ports, the number of seaborne movements in the North Sea area, and consequently the pollution risks, have not increased as rapidly as they might otherwise have done. The use of pipelines for transporting crude oil and products is increasing yearly. Surveys taken over the last few years in Western Europe indicate a very small proportion of spillage in this category: ·002 per cent of oil throughput.

Although the wide use of fuel oil in ship propulsion is a contributory factor, there is no doubt that the bulk of the problem of beach pollution stems from tanker operations, and in particular from the procedure of washing dirty cargo tanks at sea. This fact has long been recognized, and in 1954 an International Convention was formulated which envisaged as the eventual solution eliminating discharges to sea and putting all dirty residues and washings ashore.

It was realized that implementation would be difficult and, as an interim measure, it was initially laid down that no washings should be discharged within 50 miles of land. The limit was further extended to 100 miles from shore by the 1962 amendments.[6] Even from this distance, however, experience showed that there was risk of 'persistent oil' (and particularly crude oil tanker washings) drifting ashore.

In 1964 the oil industry introduced the 'Load on Top' principle,[7] which has the potential for completely solving the problem. By this method the oil residues are kept on board, the new cargo of crude is loaded 'on top,' and the residues put ashore with the crude oil when the tanker reaches her discharge port (Fig. 5). Not all refineries were prepared to accept slop residues from tankers with the additional equipment costs involved in building crude oil de-salters which, themselves, contribute to the refinery effluent load. However, a large proportion of refineries owned by companies of major international oil groups have fallen into line.

The 1969 amendments to the Convention required black oil tankers to retain their washing residues on board until they can be put ashore at loading port, discharge port, or repair yard. These amendments have not so far been implemented by governments except—in the last year—by the UK and Sweden. 'Load on Top' objectives will not be fully realized until all governments accept and enforce the necessary measures. The oil and shipping industries have, for their part, been less energetic than they might have been in promoting skilful operation of these methods. The techniques of load on top have, however, been steadily improved, and a manual[8] to guide masters on the most effective operational procedures has recently been produced and given the widest circulation throughout the international shipping world.

In view of the intense concern over environmental matters which has manifested itself in recent years, the Inter-governmental Maritime Consultative Organization (IMCO)[9] held an international conference in October 1973 to take a fresh look at the problems and formulate a new convention to control all forms of pollution from ships. A radical solution envisaged is that new tankers should be equipped with segregated water ballast tanks in order to minimize the mixing of oil and water. The provision of separate ballast tanks would substantially increase the cost of new tankers, and it would be some years before the new concept could become effective.

In the meantime governments of importing countries whose coasts border upon the Mediterranean (an enclosed sea) want the aims of the original 1954 convention enforced so that all residues and effluents are put ashore at crude-loading ports. This would call for expanded and improved reception facilities and create problems, particularly in the handling of waxy residues; and especially at ports where very large tankers are loaded in deep water; in some cases many miles from shore.

Enormous technological and political difficulties are involved in implementing a clean seas policy. Nonetheless, the oil and shipping industries are convinced that the application of all the means practically possible—'load on top,' improved shore reception facilities where appropriate, and the segregated ballast concept—can eventually solve the problems, given the necessary support by governments in enforcement.

3.1. Accidental spillage

In avoiding accidents, sophisticated navigational aids, segregated traffic lanes in congested waterways such as the English Channel, and careful crew selection and training all have important parts to play. Nonetheless, accidental spillages are inevitable from time to time with such huge volumes of oil being transported by sea.

The progressive introduction of large ships is proving a positive advantage, particularly in the Channel. Large ships are no more accident prone than small ones, provided they are properly handled, and fewer of them are needed to carry the same amount of oil (Fig. 6).

After the Torrey Canyon incident in 1967 the major oil companies formed a technical committee to place their expert knowledge at the service of the Government. The large number of unresolved problems that came to light led the Institute of Petroleum to set up a Co-ordinating Committee on the Prevention of Sea Pollution in November of that year; a number of working parties have since

FIG. 6 *Scale model of a tanker, capable of reproducing the handling characteristics of a 200 000-ton ship, under test on a lake in southern France. (Rainy weather calls for improvised protection for the two-man 'crew')*

FIG. 5 *How the Load on Top system works*

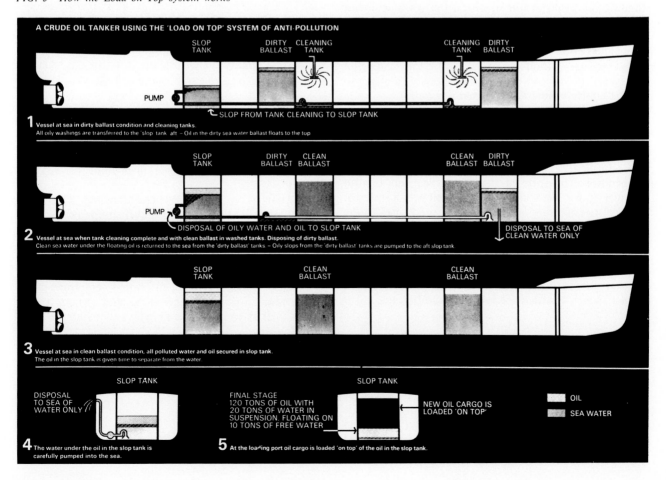

FIG. 7 The 'sand sink' method of dealing with spilt oil at sea

been active in finding solutions.[10] The Committee and its working groups are liaising closely with government and shipping organizations. Oil companies are well represented in the Tanker Section of the UK Chamber of Shipping, and the Chamber in turn is represented on the International Chamber of Shipping, which has consultative status with IMCO, the body responsible for considering international legislative aspects. The more recently established Oil Companies International Marine Forum has members from virtually all oil companies with marine interests, and has consultative status with IMCO.

A lot of work has been done to obtain a better understanding of the behaviour of oil on the sea after it has been spilled, and considerable effort continues into the study of improved methods and materials for dealing with spilled oil;[11] and, as a special case, into effects on sea birds.[12] Several methods are now available for cleaning up oil quickly, in order to minimize the possible damage to beaches, inshore waters, or marine life.

Spilled oil can be contained by booms, or a surface-active agent applied around a spill to 'corral' the oil and keep it from spreading until it can be collected and dealt with by mechanical skimmers. For large-scale oil spills a method called 'sand-sink' has been devized (Fig. 7) and is operational in parts of the North Sea and English Channel. It involves spraying the spilt oil with a slurry of chemically-tested sand and water. The sand sinks to the seabed with the oil adhering to it, and stays there as the oil is dealt with by natural processes. Furthermore, in place of the detergents used in the past, there are now available new dispersants of low toxicity. These promote the easy and permanent division of oil film into small droplets which can be dispersed by natural means.

3.2. The Milford Haven system

Oil companies co-operate in providing and supplying measures to prevent oil pollution and in cleaning-up accidental spillages when they occur. An excellent example of voluntary action in preparing for emergency and taking the necessary measures is the Port of Milford Haven anti-oil pollution plan.[13] The oil dispersal service at Milford Haven—second in importance as an oil port

only to the Thames estuary complex—is controlled by the Milford Haven Standing Conference on Anti-oil Pollution, which consists of the Milford Haven Conservancy Board and the five oil companies using the port. The policy is to clean up first before deciding the source of the spill.

All members hold stocks of approved dispersants; a specially equipped launch is available for immediate operation on minor spillages; and other equipment is available when required. The harbour master has overall supervision of the service. The principle of operation is that when a spillage is positively attributable to a particular tanker or company operation, the company concerned pays the cost and supervizes operations under the overall direction of the harbour master. In the case of non-attributable ('unattached') spillages, all participants in the scheme share the cost.

This is an ideal situation in a case where the trade of the port is almost entirely oil. In ports where general shipping forms a large part of the total compared with tankers, the difficulty of tracing the origin of oil spillage, and the number of ship-owners involved, make such schemes less attractive. Milford Haven is, nevertheless, a good example of the type of scheme for emergency action in which the oil industry readily collaborates with local authorities.

3.3. Meeting clean-up costs

After the Torrey Canyon incident the major oil companies set themselves the task of ensuring that governments and people anywhere in the world who suffer damage from oil pollution caused by tankers are reimbursed or compensated fairly and promptly.

The companies entered into two voluntary agreements:
First, the Tanker Owners Voluntary Agreement Concerning Liability for Oil Pollution (known as TOVALOP);
Second, a Contract Regarding an Interim Supplement to Tanker Liability for Oil Pollution, known as CRISTAL.

TOVALOP was originally sponsored by seven major groups of oil companies.* Every tanker owner who signs Tovalop—and more than 99% are now in the scheme—makes a voluntary public acknowledgment which carries with it the assurance of financial backing at the rate of $100 per gross registered ton of each ship up to the full $10 million per tanker limit of Tovalop liability.

TOVALOP is a tanker owners' scheme. The more recently established CRISTAL is an oil companies' scheme related to the cargo on board. It will increase the amount payable in compensation for any one pollution incident up to a limit of $30 million by paying the difference between combined Tovalop and legal liabilities and the actual loss or expense. It will pay compensation where appropriate to individuals—fishermen, hoteliers, boat owners, and so on—as well as to governments who suffer damage from pollution.

Both TOVALOP and CRISTAL are voluntary measures taken by the tanker and oil industries in the absence of—or to supplement—liability for oil pollution. They ensure that if an accident does happen, dispersal and clean-up

---

*British Petroleum, Gulf, Mobil, Shell International, Standard Oil of California, Standard Oil of New Jersey (now Exxon Corporation), and Texaco

operations can start immediately in the knowledge that adequate compensation will be promptly paid.

4. REFINING
4.1. Effluent control
Effluent control is perhaps the most important problem with which refineries have to deal, especially in densely populated inland areas. Refinery locations are decided by access to crude oil supplies (by tanker or pipeline) and distribution of the products to the market by road, rail, waterway or pipeline.

The plants vary considerably in relation to their type of feedstock, their size and their cooling system and, of course, their age. In separating the various components of the crude oil, heat is applied, all the light oil fractions are vaporized—using steam to assist the evaporation—and the mixture of steam and hydrocarbon vapours separated by stepwise cooling and condensation.

To achieve the cooling and condensation older refineries used large volumes of cooling water, generally on a 'once through' basis. Refineries built after 1960 use air-fin coolers in combination with re-circulated cooling water. The effect of this on total effluent is that a refinery of the older type might use from 10 to 15 or, in exceptional cases, even 20 times the volume of cooling water in relation to the volume of oil processed. In a modern refinery the effluent water may be no more than one-fifth of the volume of oil processed.

Effluent water is made up of three possible components. Refineries use considerable amounts of process steam; this is mixed with the oil vapours and on condensation produces a mixture of oil and water which separates out, the water being drawn off from process vessels and treated before discharge. Rain water is another factor to be taken into account. Drainage waters from the paved areas around the process units, roads and tank farms have to be treated to separate the oil they have picked up en route. Thirdly comes the variable quantity of water employed for cooling purposes. This cooling water, not having been in direct contact with oil, can normally be returned to the river or estuary from which it came after very simple treatment.

In all cases the three water streams—generally treated separately—are subject first to gravity separation; the oil content of the effluent from a well-designed and operated

API type gravity separator is normally in the range of 50 ppm or less.

Where effluent is being discharged into inland waters more sophisticated treatment methods are employed. These may consist of flocculation with chemicals added to coalesce minute oil droplets to produce a drop size capable of gravity separation. Where requirements are stricter still, biological treatment may be needed, activated sludge units being installed to maintain a very dense and active bacterial population in the aeration basin. This reduces oxygen demand in the effluent and gives a further reduction of oil content. Alternatively the ordinary API gravity separator may be improved by the use of parallel/corrugated plate techniques. Air flotation plant may be used to promote separation of oil and water.

The degree of water treatment required should take into account the volume of water, the quality, and the size of the receiving water—whether this is open sea, a closed harbour, estuary, or inland river. The Royal Commission on Environmental Pollution,[4] while emphasizing the need to avoid overloading estuaries, stresses the desirability of using sensibly their absorptive capacity. Any discharge to an inland river or an estuary is subject to consent from the appropriate river authorities, and the new regional water authorities will insist on equally strict criteria.

Oily sludges resulting from the cleaning of process equipment and storage tanks must either be incinerated or disposed of at points approved by the local authorities so that land or water pollution is avoided.

Modern refineries using air cooling in combination with cooling water circulation techniques have achieved a marked reduction in effluent contamination. Tables 2 and 3* cover 82 refineries in 17 countries of Western Europe (12 in the UK).

4.2. Air emissions
In the UK, petroleum refineries are located mainly on the coast, and all processes, including the storage and handling of crude oils, are subject to control under the Alkali Act.[5] The principal pollutant is sulphur dioxide from the burning of sulphur-containing refinery fuel. Other emissions are smoke and dust, nitrogen oxides, and odours. Noise and glare can also create problems.

At present nitrogen oxides are not a serious problem in the UK, since their chief nuisance is as a contributor to smog involving ozone and hydrocarbons in the presence of intense sunlight, a rare occurrence here; but is more of a problem in parts of the USA and Japan. The formation of nitrogen oxides can be reduced by modification of combustion systems. The emission of smoke from refinery stacks is very infrequent because of the close control kept on efficiency of combustion in a highly cost-conscious industry.

4.2.1. SO$_2$ emissions
In the UK, the obligation laid on processes controlled under the Alkali Act[11] is that the 'best practicable means' shall be used for preventing the discharge of noxious or offensive gases to atmosphere, and for rendering such gases where so discharged harmless and inoffensive.

In most cases 'presumptive limits' on concentrations of any emissions are laid down at the discretion of the Chief

TABLE 2   Volume of liquid effluents discharged

Construction date Location	1960 or before Coastal		Inland		Post 1960 Coastal		Inland	
Treatment	A	B	A	B	A	B	A	B
Av. vol. m³/h	10 700	2 480	4 900	4 200	1 700	960	400	820
m³ per ton crude	11·8	3·9	9·7	3·5	2·5	2·9	0·52	0·15

A = gravity separation only
B = gravity separation plus chemical and/or biological treatment

TABLE 3   Mean and weighted average loads kg/1 000 t crude oil

Construction	1960 or before Coastal		Inland		After 1960 Coastal		Inland	
	Mean	Weighted	Mean	Weighted	Mean	Weighted	Mean	Weighted
BOD	293	420	398	375	40	41	54	51
Oil	130	155	366	361	58	62	18	11
Phenols	3·8	2·2	0·4	1·8	0·9	1·3	0·9	0·32

*Sources: Blokker and Marcinowski (I.P. Review Feb. 1971).

Alkali Inspector. They are arrived at only after taking into account the technical possibilities, costs and current information on the effects of pollutants on human and animal health, vegetation and amenities in relation to the specified limit.

In the case of refineries burning sulphurous fuels, the 'best practicable means' of carrying out the obligations of the Alkali Act is accepted to be the installation of a chimney stack or stacks, of such height(s) as will result in a sufficiently low ground level concentration (glc) of SO_2. The Inspectorate does not at present require that fuel oils or flue gases be desulphurized. This can be taken as a recognition of the capacity of the atmosphere to dispose of SO_2 *once it has been sufficiently diluted and dispersed* from high chimney stacks.

In the UK the actual SO_2 glc recorded over long periods (say one year) from a single stack, in the nearby area where maximum concentration is expected, averages one-sixtieth to one-hundredth of the presumptive limit of 17 parts per hundred million (477 µg/m³), i.e. 1/2 000 to 1/3 000 of the maximum recommended by the Factory Inspectorate for occupational exposure.

As an indication of the chimney heights under discussion the following table based on the standard calculation is useful:

TABLE 4   Sulphur dioxide: rate of daily emission to basic chimney height

Rate of emission (tons SO_2/day)	3·6	7·5	13	21	30	40
Basic chimney height (ft)	100	150	200	250	300	400

A 4 million tons p.a. refinery could consume say 5% of its intake as fuel oil or gas; if this contained an average of 3% sulphur the SO_2 emission rate would be 35 tons per day. If this were to be the sum of the SO_2 emitted from several furnaces and stacks, the calculations would be different, and the combination of several stacks close together is a further complication. There is a clear incentive for a refinery to duct all its flue gases from individual furnaces into a common stack, and so take advantage of the extra thermal buoyancy to reduce SO_2 glc (Fig. 8). At one Milford Haven refinery the flue gases from five furnaces are ducted to a single chimney 300 ft high. The Teesport refinery has one 400 ft chimney serving all process and utility combustion chambers.

While dispersion is normally adequate, there may be special weather conditions (such as temperature inversions) which would make an early warning system desirable so that large plants could switch to low sulphur fuel over the critical period. Such a system operates in some continental European industrial areas.

4.3. Noise control
The oil industry has anticipated by several years the present trend towards noise control requirements. The Oil Companies' Materials Association (OCMA) set up a Noise Working Group in 1967. A year earlier CONCAWE established a Noise Control Group to study the effect of refinery plant noises in neighbourhood areas.

Among other tasks, the CONCAWE Noise Control Group has conducted a survey of refineries to relate noise complaints to neighbourhood levels, and it finds that the recommendations of the British Standard 4142 provide a good guide to acceptable noise levels in various types of residential areas. It now has a number of task forces engaged in studying various noise problems in refineries, while individual companies also conduct their own surveys to pinpoint and eliminate objectionable sources of noise.

The primary task of the OCMA Noise Working Group was to prepare standard procedures for specifying noise control requirements.[17] These were published in 1972, but the recommended procedures had previously been used for several years by major oil companies so that there is now considerable experience in their application. The specification is used as the basis of agreements on noise limitation between oil companies and equipment/supply contractors. Its main features are:

(1) A standard form for defining noise limits in working areas (for hearing protection), in offices and control rooms (for speech communication), and in community areas (to avoid neighbourhood nuisance).

(2) A standard form for calculating estimated noise levels (so that purchaser and contractor use the same method). This includes sets of attenuation curves which are at present unique and considerably in advance of any other published curves.

(3) Standard methods for assessing noise from items of equipment (such as furnaces and fin fan coolers) which is unique to OCMA and again in advance of any published methods.

(4) A test procedure for determining whether the finished plant meets the contractual noise limits.

As a result of the increasing emphasis placed by oil companies on noise control many contractors have appointed engineers to specialize in the problems

FIG. 8   A 700-ft stack to combat air pollution from the world's largest refinery. It does the work of 17 smaller chimneys

involved; others employ acoustic consultants as advisers. Many improvements have been made in the design of plant equipment. For example:

(1) Furnaces used to be the noisiest items in refineries. Changes in wall construction, the use of plenum chambers around natural draught burners, and the increased use of forced draught furnaces have made dramatic reductions in noise emission.

(2) Banks of fin fan coolers, because of their size, are significant contributors to community noise. The last few years have seen the development of quieter fans with the same efficiency, although their cost is very high.

(3) Electric motor manufacturers now offer quieter motors, and manufacturers are doing research into improved noise reduction methods.

(4) Control valves, particularly those handling high pressure drops for gases and steam, can be severe sources of local noise. Quieter valves and other techniques are now available and, although there is scope for considerable development, some improvement is evident.

(5) Techniques are available for reducing the noise from high power machinery, such as gas turbines and large compressors, usually in the form of acoustic enclosures, although these often present as many technical difficulties as they solve (for example in cooling and ventilation).

Many refineries have set up a special section for prompt handling of public complaints regarding noise. Managements are increasingly aware of the need to minimize noise nuisance during special operations such as the commissioning of new plants. Relations with local inhabitants are likely to be much improved if operational problems are explained to them in advance.

4.4. Flare glare
In a refinery the flare is essentially a safety device to dispose of large volumes of gas which cannot be utilized commercially or as refinery fuel. Large volumes of gas are difficult to burn under high flaring conditions and steam must be injected to ensure smokeless and less luminous combustion. These conditions can cause neighbourhood nuisance through noise and fluctuating glare. Considerable research is being undertaken to find a solution to these problems, especially in the direction of low-level flares which can be shielded by surrounding walls.

4.5. Dust
Mechanical cleaning of furnace tubes in refineries, which produced dust at infrequent intervals, has been superseded by controlled burn off and chemical cleaning. The only oil manufacturing process where dust can be emitted in any appreciable quantity is catalytic cracking. The objective is to crack down large hydrocarbon molecules into smaller ones, and thus improve the yield and quality of more valuable light fractions from a given quantity of crude oil. This system utilizes a fluidized bed, in which catalyst powder is kept in continuous motion in a vaporized oil stream, under conditions in which the cracking reaction can take place.

In the process the catalyst becomes coated with carbon, and part of it is continuously withdrawn from the reactor into a regenerator for controlled burning of the carbon with air/steam mixture. The regenerated catalyst flows back to the reactor while the waste gases containing a considerable amount of carbon monoxide are burned to raise steam. Before leaving the regenerator, however, any entrained catalyst powder is removed by passing through a series of multi-stage cyclones, which reduce the emission of catalyst fines to negligible proportions.

4.6. Odours
The principal malodorous compounds existing in crude oil or formed during its processing into products are H_2S and mercaptans, a class of hydrocarbons containing sulphur. Should any H_2S or volatile mercaptans escape from a refinery there is a risk of smells in the neighbourhood. Ethyl mercaptan has a perceptible smell when present in a concentration of only 1 part per thousand million, thus even a very small loss can create an unpleasant smell over several miles in certain meteorological conditions.

The control of odours remains one of the most intractable of refinery air pollution problems and is taken very seriously by refinery personnel. Everything is done to minimize the evaporation of hydrocarbons or the release of gases to atmosphere. Volatile products are stored in floating roof tanks which do not require to 'breathe' vapours as is the case with fixed roof tanks. Pump gland leaks are virtually eliminated by the use of mechanical seals, and oil interceptors for effluent cleaning are covered to prevent vapour loss and oily odours.

In all refinery plant operations any gas produced which is not required as a product is normally routed to the fuel gas system, and all odours are destroyed by combustion. Emergency venting of malodorous gases is to the refinery flare system, again with destruction of odorous components.

Some effluent water streams contain odorous components, e.g. 'sour' water which contains dissolved H_2S or volatile mercaptans. These are deodorized by steam stripping before discharge to the refinery sewer and the odour in the evolved gases destroyed by burning.

4.7. Monitoring of effluents and emissions
Control of effluents and emissions from oil manufacturing operations is more meaningful when appropriate monitoring equipment and systems are available. Continuous sampling devices for laboratory analysis are often used to check such properties as oxygen demand, but there is need for devices which will meet the exacting conditions encountered in the inlet to a refinery's waste water separator.

The design of air pollution monitoring schemes for refineries is covered in a recent CONCAWE report.[18] It gives, with simple examples, details of techniques involving both fixed and sequential measuring sites. This study confirms the local nature of the pollution and explains why the most rigorous measures are necessary for refineries in highly industrialized and densely populated areas.

5. PRODUCT STORAGE AND TRANSPORT
Pollution of ground and surface water can stem from two sources: the oil industry and the oil consumer.

Legislation does not differentiate between the manufacturer and user of oil products, although the responsibility of the latter is apt to be overlooked. Oil companies follow rigid codes of good practice in protection against failure of storage tanks and means of transport for oil and oil products. These are not necessarily the results of legislation and often go beyond legal requirements. Leakage or spillage is a safety as well as a pollution risk. Experience indicates that the principal cause of oil pollution in inland waterways is spillage from the installations of industrial oil fuel consumers.

5.1. Storage tanks

Storage tanks may vary in size from some 160 000 m³ to small underground tanks of 10 m³ (at petrol service stations) or 1 m³ (for domestic heating oil). Structural steelwork design of tanks is generally satisfactory and catastrophic failure in service virtually unknown. However, alternative materials such as glass reinforced plastic are being tried out in order to minimize still further the possibilities of corrosion. In the UK there are specific standards and codes of practice which must be met. Spillage by overfilling can be effectively eliminated by use of suitable control equipment.

Smaller tanks at retail outlets, industrial premises, and homes give more reason for concern than the larger oil company storage tanks. Concrete tanks are sometimes used by consumers—an undesirable practice.

Tanks are sometimes over-filled by human failing, and both oil companies and governments are interested in liquid level control equipment which cuts off the fuel supply automatically when a prescribed level in the tank is reached.

5.2. Pipelines

Transport of petroleum products by pipeline rarely causes pollution. Design. constructional, and operational standards are in accordance with nationally, or internationally accepted codes of practice which are being harmonized on an international basis.

The design of a pipeline takes account not only of normal operating pressures, but of static and dynamic (surge) pressures. In areas where the land is in use pipelines are buried wherever possible. Precautionary provisions include controls for valve operation, flow and pressure controls and alarms, and automatic shutdown and relief devices. Every cross-country pipeline system is monitored from a central control station, manned 24 hours a day, and regularly inspected along its entire length to guard against the consequences of accidental damage by third parties. Pump stations and terminals are similarly designed to secure the safety of personnel, equipment, and product.

Leakages from pipelines seldom occur, and the amount of spilled oil is much lower per ton mile than with other oil transportation methods.[19] Pressure testing is regularly carried out; efforts to control methods of detecting leaks are continuous and primarily based on instrumentation at pumping stations, patrolling of the pipeline, and third party reports. One method that has proved effective is the 'ultrasonic leak detector,' which can pick up acoustic signals produced when fluid escapes through a small leak in a pipeline.[20]

5.3. Road, rail and water transport

Most countries have regulations to ensure acceptable safety standards in construction and use of oil carriers by road, rail, and water. The majority of incidents arise from human error and traffic accidents and not from faulty or inadequately maintained equipment. It is of the first importance that the operating staff should be well-trained and fully aware of their responsibility to the community. In practice, there is a combination of 'on the job' training by skilled operators and formal training to make sure that procedures to ensure safety and avoid spillage are thoroughly understood.

Traffic accidents represent a constant hazard and the consequences where large quantities of oil products are concerned can be especially serious. To eliminate possible causes arising particularly from driver fatigue, careful attention is given to vehicle equipment such as seating, layout of controls, ease of access to and from the cabin, range of vision, noise levels and power-assisted steering. For larger capacity water transport, aids to minimize risk of accident include radar equipment, radio-telephones, and echo sounders.

Vehicles are checked for leaks before leaving the depot; spillages that do occur usually happen during loading and discharge. Hence a high standard of housekeeping is essential whatever the product, on safety as well as pollution grounds. To prevent overfilling when loading vehicles, volumes are accurately measured by flow meter (usually of the positive displacement type). These meters incorporate cut-off devices to ensure that only pre-determined quantities are loaded. Volumetric pre-set

FIG. 9 An oil-skimmer with automatic compensating weir for clearing oil spills from inland waters and non-tidal docks. A simulated exercise involving police and fire services, a river conservancy and oil company representatives

mechanisms have a slow-closing device which eliminates shock pressures—particularly desirable when flow velocities of 3 m per second or more are involved.

Several types of overfill prevention devices are being developed to cut off the flow when the level reached begins to exceed the normal ullage levels. Similarly on discharge, liquid level control equipment may be fitted to customers' tanks, either permanently or by the operator making the fuel delivery. The customer must share the responsibility for ensuring that there is enough space in his tank to receive the volume delivered.

To contain any spillage, loading bays at installations and depots are normally drained by oil-water interceptors, which are regularly emptied: here again, good housekeeping practice must become second nature. Spilled oil, at depot or customers' premises, should never be hosed down so that it passes to a surface water drain that is not fitted with an oil interceptor with closed outlet before reaching public drains or water courses.

No matter how carefully oil movement and use are controlled, the possibility of leakages and spillages cannot be wholly eliminated. They will be further reduced by continued improvements in equipment and operating procedures, and by training. Contingency plans have been devized, and are in operation, and will be made still more sophisticated as knowledge is gained about the behaviour and effects of oil in the ground and on water. Co-operation between the oil and water industries has already resulted in useful progress and joint published recommendations[21] for dealing with oil spills. This is a development to be welcomed and extended.

5.4. Dealing with spillage

If a product spillage does occur, there are four phases in dealing with it: limiting it, containing it, removing it, and final clean-up.

Immediate limitation of the amount of product spilled is essential to prevent a minor spill from becoming a major pollution or other hazard. This involves closing emergency quick-action shut down valves on loading points; shutting down pumping; isolating a leaking pipeline section by closing control valves; and transferring the contents of a leaking tank or pipeline to other storage as quickly as possible.

Containing an oil spillage to prevent it from spreading is vital. Prompt action can prevent oil spilled on land from reaching a water course, where the risk of rapid widespread pollution is greatest, and can minimize the spread of oil on water (Fig. 9).

Whether in sea or inland waters the problems of clean-up are broadly similar, and have already been described in the marine transportation section. In rivers and lakes, however, wind and current velocities are generally lower and it is possible to make more use of booms and even air barriers. The quantities involved are also usually smaller and comparatively simple methods and equipment are effective, such as manual removal or vacuum trucks.

6. PRODUCT USAGE

6.1. Car exhaust emissions

Of all the problems connected with the consumption of products, the one which attracts most attention is gasoline in the motor car. Motor gasoline is a key

FIG. 10 Analysing car exhaust emissions in an oil company laboratory. In the right foreground, an analyst is operating a series of infra-red analysers which give an instantaneous record of carbon monoxide, unburned fuel, and nitrogen oxides in the exhaust. The remainder of the exhaust is piped to large plastic bags, from which composite samples can be taken for further analysis

petroleum product, without which modern communities could not function. During 1972 the car population of the United Kingdom increased from about 12 to about 13 million; in addition there were about 2 million gasoline-powered commercial vehicles on the road. Between them they consumed more than $15\frac{1}{2}$ million tons of gasoline and travelled more than 130 000 million miles.

Some years ago diesel trucks and buses were considered objectionable in conservation terms, principally because of their exhaust smoke. More recently this problem has been substantially mastered by improved design (and especially by better maintenance) to such an extent that the diesel's inherent economy may secure its place in the future compared with other prime movers, none of which is perfect.

In the pollution context, car exhaust emissions are properly a subject of public concern.[22] Gasoline is basically a combination of carbon and hydrogen which, when burned, produces carbon dioxide and water. Complete and ideal combustion is never achieved by the motor car engine. In consequence, some unburned hydrocarbons and carbon monoxide are discharged from the exhaust (Fig. 10).

In some concentrations these may be toxic, irritants to the eyes or nose, or just unpleasant contaminants of the air. Additionally, oxides of nitrogen are produced from oxygen and nitrogen contained in the air supplied to the engine; this is unaffected by the composition of the fuel. In sunlight the nitrogen oxides and hydrocarbons in combination may create a photochemical reaction resulting in the production of ozone and certain oxygenated substances, leading to aerosol formation, eye irritation, and plant damage. These are symptoms of smog, notorious in Los Angeles. The intermediate or end products of the postulated chemistry may be used to measure the intensity of smog or indicate its imminence.

The nature of the emissions from the car depends upon the design of the engine rather than upon the fuel it burns. Nevertheless, considerable research has been undertaken by the oil industry to determine how undesirable effects can be minimized. The industry's first concern was to establish whether any function of fuel composition could be relevant to the problem. Work done over a number of years has shown no positive indication that the hydrocarbon composition of the gasoline significantly affects exhaust emissions. Whether aromatics, paraffins or olefins are used in the fuel makes little difference to exhaust emissions.

In addition to this work, the oil industry has co-operated with motor car manufacturers in research to determine ways in which exhaust pollutants can be reduced. Some of the programmes, such as IIEC* and FEEMAS† still continue; they have involved the expenditure of millions of pounds. There is also a continuing exchange of technical information in such committee activities as BTC,‡ CEC§ and CONCAWE.

The fact that there are emissions from a car, other than from the engine, that can contribute to pollution should not be overlooked. There is always some evaporation of fuel from the carburettor and fuel tanks that escapes into the atmosphere. In this case gasoline composition may be significant, since it may be that unsaturated hydrocarbons make a greater contribution to smog than paraffins and aromatics. The motor industry, with some help from the oil industry, is overcoming the problem of hydrocarbon emissions from the fuel system by equipment which absorbs gasoline vapour evolved from the system, and re-directs it back to the engine. Similarly crankcase blow-by, which was formerly directly discharged to the atmosphere, is now recycled to the engine. Thanks to the growing adoption of the various emission control devices it can reasonably be re-affirmed that the hydrocarbon composition of a gasoline will exert an insignificant effect on the pollutants generated by the motor car.

There is a further substance used in gasoline which is relevant in the pollution context. Lead additives have been used in gasoline for a number of years to reduce the tendency of engines to 'knock.' They have been instrumental in enabling car engines to develop high power in small unit space. As the compression ratio of an engine is raised to produce higher engine power and efficiency, peak cylinder temperatures and pressures increase. Without adequate anti-knock quality in the fuel these conditions give rise to uncontrolled spontaneous combustion which causes 'pinking'; at high speed it can lead to catastrophic engine failure.

The oil industry has progressively raised the anti-knock quality of gasolines to match the advance in engine compression ratios. Since the end of the war the quality of top-grade gasolines has been raised from 79 RON (research octane number) to better than 101. Correspondingly car engine compression ratios have been raised from about 6 to 9·5. At a given speed this means an increase of about 15 per cent in power from the same engine.

Lead compounds have been effective in helping to raise anti-knock quality. Throughout the period during which such compounds have been used, going back some 50 years, great care has been taken regarding possible health hazards. It was well understood that tetraethyl lead (TEL) is toxic and precautions were taken in manufacturing processes. Before tetramethyl lead (TML) was introduced some 13 years ago a re-check of the possible hazards of different lead alkyls was made.

Strong reassurance was given that no health hazards were involved providing precautions were taken in the handling of concentrates in manufacturing the anti-knocks themselves and leaded petrols.

Throughout these 50 years of lead use there has been no clear medical evidence of any hazard in the normal handling of gasoline, or from the air lead levels resulting from car exhaust emissions.[22] With more motor cars, however, the levels of lead in the air in city centres are becoming noticeably higher than in rural areas, and governments are seeking to limit further increases in air lead levels. In Britain the oil companies have voluntarily agreed to reduce the lead content of petrol from 0·84 g per litre to 0·64 g per litre, with further reductions to 0·55 g at the end of this year and 0·45 g by the end of 1975.

Further decrease will be implemented by agreement with other countries in the European Economic Community and establishment of a maximum lead level in premium gasolines of 0·4 g/litre is aimed at within the next few years.

Additional reduction of lead could be called for to meet legislative measures restricting the emission of carbon monoxide, hydrocarbons, and oxides of nitrogen. Regulations proposed in the United States for implementation in 1976 could not be met in practice without the option of fitting various devices to cars. These devices—which will bring about a significant increase in petrol consumption—would be installed to reduce exhaust hydrocarbons, carbon monoxide, and oxides of nitrogen to extremely low levels. Some of them consist of reactors containing catalysts which promote burning of combustible materials and destruction of oxides of nitrogen. To date, many catalyst materials comprise ceramics coated with platinum or other metals. If the gasoline contains lead the life of the catalysts will be shortened. Hence, the intention in the United States to eliminate lead from one grade of gasoline to facilitate the operation of the catalytic reactors.

There is serious argument as to whether the emission standards proposed in the United States are necessary, or desirable. The oil and motor industries believe that they are unnecessarily severe, unjustifiably costly, and contrary to the principles of energy economy, since cars meeting these requirements will use around 15% more fuel than at present. While legislation is expected in Europe to reduce the current levels of carbon monoxide and hydrocarbon emissions, neither the standards nor the catalysts proposed for the United States are widely contemplated.

Given time for engineering developments, reactors—which can only be regarded as a stop-gap solution—may not be required, and adequate standards of emission control secured without them. If the measures proposed for the United States are ultimately recognized as unnecessarily rigorous, then it will be possible to look forward to a gasoline engine car of the future with a continuing improved performance over present-day cars; running with cleaner exhaust emissions on a gasoline with the minimum lead content to provide adequate octane quality, and flexibility in manufacture. Should it be necessary further to reduce the emissions of lead from such cars, there are good possibilities that well-advanced development work now in progress will provide a lead filter in the form of a simple silencer which could stop nearly all the lead getting into the air.

---

*IIEC (Inter Industry Emission Control).
†FEEMAS (Fiat, ENI, Mobil, Alfa Romeo, Shell emission control research programme).
‡BTC (British Technical Council of the Motor and Petroleum Industries).
§CEC (Co-ordinating European Council of the Motor and Petroleum Industries).

6.2. Industrial pollution

The principal pollutant affecting industrial usage of fuels is sulphur oxide. Sulphur is present to some extent in most primary fuels (Table 5).

Nitrogen oxides (NO_x) from combustion sources are another pollutant associated with all forms of combustion, whatever the fuel. Above about 1 500°C atmospheric nitrogen and oxygen react together to form nitrogen oxides. Although this reaction is reversible, in practice because of rapid cooling there is little opportunity for the reverse reaction to take place. The amount of nitrogen oxides produced depends upon the combustion temperature and availability of oxygen. In addition, the combustion system type and design and the fuel type can have a considerable effect on emissions (Table 6).

NO_x is an important contributory factor to the formation of smog, but the atmospheric conditions which contribute to smog in Los Angeles rarely occur in the UK. Research is being undertaken to modify combustion systems to minimize NO_x. Recirculation of flue gas, and the use of staged combustion techniques, can bring large reductions of NO_x emission. In some cases, the continued application of such techniques can give reductions as large as 80%.

6.3. Desulphurization of products

The control of sulphur contents of distillate products is systematically undertaken. In many cases the flue gases from kerosine and gas oil combustion for space heating are emitted near ground level and satisfactory glc's of SO_2 are only achieved with low sulphur products. The same situation applies for heavier fuel oils burned in many industrial applications, but normally not in large power stations where—as in refineries—high stacks disperse the emissions thoroughly.

Product desulphurization is accomplished by catalytic hydrogenation, when the sulphur is removed from the hydrocarbon molecule as H_2S. This in turn is converted to elemental sulphur by controlled combustion in Claus kilns, in which the sulphur recovered is a very high proportion of the total content.

There is some indication that pollutants may produce regional effects—that is to say, air movements could carry a pollutant emitted in one region to ground level in another. Two projects, one British and the other in co-operation with other members of OECD, are being undertaken to study long range transport of sulphur pollutants.

The UK project is concerned with the fate of sulphur pollutants emitted in the UK to determine their contribution, if any, to ambient levels in neighbouring continental countries. The project consists of a number of parts involving (a) sampling of sulphur pollutants using aircraft; (b) sampling at ground level sites; (c) determination of the uptake of sulphur dioxide by vegetation; and (d) estimation of total sulphur dioxide emitted in the UK. Airborne sampling will be made near the east coast, and several hundred miles further east in westerly winds, to obtain more information on the lifetime of sulphur pollutants. Measurements of sulphur dioxide, acids associated with solid or liquid particles, sulphate, and chloride, are proposed over flight paths of 200 to 300 km at a number of heights. There are also plans to take ambient SO_2 measurements on North Sea platforms,

TABLE 5 The 1972 UK average sulphur content for refined automotive and industrial oil fuels

	Sulphur content %
Motor spirit (BS 4040)	0·05
Vaporizing oil	0·16
Premium kerosine (BS 2869 Class C1)	0·03
Regular kerosine (BS 2869 Class C2)	0·06
Automotive diesel fuel (BS 2869 Class A1)	0·3
Gas oil (BS 2869 Class A2 and D)	0·7
Light fuel oil (BS 2869 Class E)	2·4
Medium fuel oil (BS 2869 Class F)	2·4
Heavy fuel oil (BS 2869 Class G)	2·8

TABLE 6 Typical values of NO_x emission from combustion of different fuels are as follows:

Source	Fuel	NO_x emission kilograms per 10^9 kilocalories	ppmv under typical operating conditions
Domestic and commercial equipment	Natural gas	200	70
	Oil	140–700	50–200
	Coal	600	150
Industrial combustion appliances	Natural gas	370	120
	Oil	860	280
	Coal	1 500	750
Electric power generation boilers	Natural gas	670	300
	Oil	1 200	600
	Coal	1 500	750

since this area presently constitutes a gap in the sampling network.

6.4. Disposal of waste oils

Virtually all industrialized countries are worried about the problem of disposal of waste oils. Conventional treatment processes do not remove oil discharged into sewerage systems. Oil can significantly reduce the oxygen dissolved in water, and so lead to biological damage.

In several European countries, though not in the UK where the high proportion of do-it-yourself motorists makes waste engine oil especially important, there are incentives to encourage its collection and satisfactory disposal. Where no form of subsidy is available from government or municipal sources, companies setting themselves up to collect and dispose of waste oils—by re-refining, for example—often have not prospered. Over the past seven years more than half the re-refiners in the USA have gone out of business, and experience in Sweden and Canada has been similar.

Re-refining has not been considered a worth-while means of disposal because:

(a) it creates other pollution problems in disposing of acid sludge and spent clay often used in the process;

(b) its overall economics have been unattractive;

(c) there are possible hazards to health from heavy metal compounds (which may be present in used crank-case oils through contamination from combustion chambers, and which may not be adequately removed in the re-refining process), and

(d) there are misgivings about the product quality.

However, refining is a matter that has to be kept under review in the light of new processes becoming available. On the industrial side, some users are believed to regenerate their own used oil (engine, metal working, industrial lubricants, etc.) while others burn it either in a high-temperature incinerator designed to minimize pollution, or as a fuel in their heating or power-raising systems. Other firms arrange for their used oils to be collected by contractors, but some is undoubtedly dumped, perhaps illegally, though the greater facilities and expertise available to industry should give more scope for unobjectionable disposal.

Some firms and automotive service stations may well need help in the collection and disposal of their waste oils. One view is that this should be the responsibility of the municipal authority that normally undertakes refuse disposal. This is another matter requiring regular review. However, all automotive service stations should aim to provide for quick and free oil drainage from customers' cars, and for adequate used oil storage pending collection. Local or national legislation prohibiting casual disposal of used oils by the retail customer—and its enforcement —should be supported.

In the other principal market segments, except perhaps commercial road transport, there is usually less knowledge of what customers actually do with their used oil, and virtually no means of control available to the supplier. This applies particularly to disposal of plant maintenance products (e.g. hydraulic and gear oils) and production engineering products (e.g. cutting oils) by static industry (which, in Europe, accounts for at least 25% of lubricants consumption); to current agricultural practice (some 9% in Europe); and to disposal of marine lubricants by vessels on the high seas or in coastal and inland shipping areas. Present practice should be closely studied with a view to offering appropriate guidelines to users and municipal authorities.

Meanwhile, the oil industry believes that:

(1) Legislation on control of waste oil disposal is desirable.

(2) Collection of waste lubricating oil might logically be the responsibility of the local authority in the first instance, as for other refuse. If private contractors are employed to collect waste oil they should be eligible for a local or national government subsidy, provided the ultimate fate of the waste oil in the contractors' hands is above reproach.

(3) To facilitate collection of waste oil, service stations and industrial outlets should provide sufficient storage for discarded oil until it is collected in convenient lots. Ideally this should be underground in order to avoid spillage which may occur in transferring the oil from sump to above-ground container or from container to the disposal vehicle. It is appreciated that cost and local regulations may in some cases make underground storage impracticable.

(4) After collection, if re-refining remains undesirable, used oil should be burned to produce heat or power, or at least without causing air pollution. This step, too, might be undertaken by municipal refuse disposal organizations.

(5) Oil companies should be willing and able to give guidance to all users on methods of collecting waste oil in convenient lots for pickup, also advice on disposal methods to municipal authorities, or at service stations or industrial sites where, for example, burning on the spot for heating may be preferred to loss as refuse.

6.5. Health aspects: additives

In the manufacture and marketing of petroleum products the industry accepts, in addition to any legal constraints, an obligation to provide guidance for customers and staff on their safe use. There is probably no such thing as a totally 'safe' material; everything can be misused in some way. The user himself has therefore a considerable responsibility. The industry's objective is to eliminate any risk associated with the use of the products by providing practical advice on handling, use, and, increasingly, disposal.

Additives pose special problems by their proprietary nature and by their diverse properties and applications. Apart from some well-known examples where the problems are being tackled on an industry-wide basis, the approach being used is that of obtaining toxicity and chemical data on each individual material from the various manufacturers.

After studying such information in relation to end-use, medical authorities give expert opinion both on the overall product and on the additive itself. If complete clearance is not given (with or without precautionary advice), recommendations may be made either to withdraw or reformulate the product, or to ensure that it is used only for certain applications.

In line with the policy of giving customers adequate advice on the safe handling of products, the Institute of Petroleum has issued two memoranda for guidance of users of cutting oils and aromatic process oils.[23] In both cases the suppliers sent these statements to all their customers in the UK with the recommendation that their contents should be brought to the notice of employees required to handle these materials. In addition, the Institute has formulated a code of safe practice for the precautionary labelling of petroleum products.[24] Finally, to ensure that the oil industry maintains its reputation as a responsible manufacturer and seller of products, the IP is encouraging research on techniques to improve knowledge on the toxicity characteristics of all petroleum products, including the additives contained in them.

6.6. Helping the consumer to minimize pollution

A marketing company with a good technical assistance and development programme comes to be regarded as an expert on the design and installation of equipment to whom the potential customer can turn for practical guidance. The skills involved are principally the study and choice of the equipment to be installed, and the supervision of fitting, start-up and maintenance.

Stringent British Standards are enforced.[25] In addition, the major companies conduct extensive research to develop better techniques of fuel utilization. In the process a high degree of technology has been developed to promote cleaner air and anticipate possible legislation. Some examples are cited below:

New developments in oil-firing techniques, many of which have been initiated by the oil industry, have done much to reduce emissions of smoke and particulates. One important aspect is the measurement of such pollutants arising from incomplete, or poor fuel combustion. A smoke meter developed many years ago is now used internationally for measuring smoke from boilers and furnaces.

A small trailer-mounted monitoring station has been designed to measure and log environmental variables on magnetic tape in a form suitable for computer processing. The measured variables are wind speed and direction, temperature, total hydrocarbons, sulphur dioxide and noise. These variables are logged at pre-selected intervals (one, two, five, or 20 minutes) and any two can be logged continuously. The equipment should be particularly useful to monitor air in the neighbourhood of plants and complexes, and also as a means of measuring city pollution.

Unburnt hydrocarbons and partial oxidation products can be objectionable pollutants. Flame ionization detectors[26] have been introduced to estimate quantitatively the concentration of hydrocarbons for use in the evaluation of equipment. This instrument is used by the Domestic Oil Burner Evaluation and Testing Association (DOBETA)* and specified in their standards.[27]

It is foreseen that plume opacity regulations will become more restrictive in the future and that visual methods will be inadequate, especially for faint plumes. In-stack monitoring is coming into use, but again it is limited in that it does not predict the opacity of detached plumes. Efforts are being made to develop an instrument to measure opacity of a plume by means of back scattered laser light. Considerable success has already been achieved, and with further development it is believed that the technique could be established as a remote plume monitor/alarm device.

6.6. Car emissions
A complete working knowledge now exists of the development and operation of exhaust gas analysis equipment and test techniques for all current pollutants, and a close watch is kept on developments in service station test equipment. Considerable work has been carried out using conventional analytical equipment and techniques for assessing the content of exhaust emissions and the merits of the different techniques for measuring air/fuel ratios from exhaust gas analysis fully evaluated.

For analysing car exhausts a hydrogen flame ionization detector has been evolved with a specially designed sampling system. The equipment operates continuously, is accurate, and has a delay time of less than half a second.

A unique relationship has been found to exist between the air/fuel ratio supplied to the engine, and the dielectric constant of the fully oxidized exhaust gases. This relationship has been used to develop a fast response air-fuel meter which measures the difference between the dielectric constants of air and the exhaust gases (after they have been catalytically oxidized in a small furnace). The output is continuously recorded on a calibrated chart, and the instrument responds to changes in mixture strength in $0 \cdot 1$ second with extreme accuracy. Considerable interest in this device has been shown by the car industry.

6.6.2. Gas and oil leaks
The necessity to find means of identifying natural gas leaks from buried mains has led to the invention of a portable detector which picks up abnormal hydrocarbon concentrations at ground level above the leaks. It is

based on a flame ionization detector and cannot therefore be used in inflammable atmospheres. It has, however, great sensitivity and can detect hydrocarbon concentrations as low as a few ppm.

An oil sheen floating on water can now be detected by means of a simple and sensitive instrument which sounds an alarm if oil slicks are detected. In this device an incandescent lamp and collimating lens generate a light beam incident on the water surface; a photodetector measures the intensity of the reflected beam. Thin films of oil give rise to more reflected light than very thick films. Adoption of this measurement principle for use in waters susceptible to tidal fluctuations, wave action, high winds, large changes in flow direction and rate, and miscellaneous floating solids has met with success. A model suitable for refinery effluent streams is available, and the design may be modified to make an instrument suitable for use in the more harsh environment prevalent in large open bodies of water.

7. CONCLUSION
It is scarcely possible that any paper could deal with all conceivable aspects of pollution that concern even a single industry. Some, such as land usage, apply to all industry and in a sense to all human activity; others are in the marginal area between suppliers of energy and the designers of the equipment using it. In the present state of knowledge others again may be problems not yet fully comprehended. This points to two particular desiderata for considering the whole question of the quality of our environment: first, alertness to potential problems in advance of damage; secondly, readiness to stimulate and to share in joint consultation with authorities, other industries, and other disciplines.

The record shows that pollution confronting us can be rendered innocuous by technology, often on a co-operative basis. Draconian solutions, such as ceasing a productive activity altogether, are rarely, if ever, justified. In judging what it is best to do, we should perhaps think more often than we do of the balance between natural and man-made sources of pollution; for, with all the reservations to be made, it is a wasteful usage of limited resources to concentrate on reducing man's contribution so rigorously that it becomes utterly insignificant in relation to the total. It is a matter of perspective, bringing into harmony as many of the relevant factors as one possibly can; and this again underlines the need for consultation to achieve agreement on a framework of sensible legislation.

8. ACKNOWLEDGMENTS
This paper could only have been written with very extensive assistance from many individuals in many companies who have given most generously of their time and knowledge. They are too many to name individually, but the author takes this opportunity of expressing his warm thanks to them.

9. REFERENCES
1. National survey of air pollution, 1961-71, Volume III, HMSO 1973.
2. What is CONCAWE? (CONCAWE Report No. 1/72).
3. U.S. Senate committee for interior and insular affairs (1972).
4. Instructions for using the WSL dispersant spraying equipment (Warren Spring Laboratory, 1970).
5. BP Statistical Review of the World Oil Industry 1972.

---

*DOBETA is sponsored by the UK oil industry to improve standards of domestic oil firing.

6. International conference on pollution of the sea by oil, held in London, 26 April-12 May 1954. Final act of conference and text of the International Convention for the Prevention of Pollution of the Sea by Oil. London, HMSO 1954 (Cmnd. 9197).

Final Act of the International conference on prevention of pollution of the sea by oil, and of the conference of Contracting Governments to the Convention, signed in London on 12 May 1954. London, HMSO 1962. Miscellaneous No. 23 (1962).

7. New steps against oil pollution of the sea. *Petroleum Times*, *68*, 26 January 1964, 313. Load on Top system will halt sea pollution. G. Sterry. *Oil and Gas International*, *4*, 10, Oct. 1964, 67-8.

8. Clean seas guide for oil tankers (The operation of 'load on top'). International Chamber of Shipping/Oil Companies International Marine Forum—1973.

9. Manual on Oil Pollution—IMCO 1972 12 (E).

10. Oil Spill Working Group (Marine). *IP Journal*, January 1971.

11. European Experience in the Identification of Waterborne Oil (E.R. Adlard, May 1973).

12. Recommended treatment of oiled seabirds (1972); Second and Third Annual Reports of the Research Unit on the Rehabilitation of Oiled Sea Birds 1971-2 (UK Advisory Committee on Oil Pollution of the Sea).

13. Anti-oil pollution plan (Milford Haven Standing Conference on Anti-Oil Pollution, March 1972).

14. Pollution in some British estuaries and coastal waters, September 1972 (para 14 *et al*)

15. Alkali &c Works Regulation Act 1906.

16. Annual Reports of Chief Alkali Inspectors 1966 (Ap. 5); 1969 (Determination of chimney heights in Britain).

17. OCMA, NWGI procedural specifications I-II-III

18. No. 14/72, Dec. 1972.

19. Spillages from oil industry cross-country pipelines in W. Europe (CONCAWE, Report No. 2/73).

20. Listen for leaks in liquid lines. H. Bosselaaf, *Pipeline and Gas Journal*, **198**, 7, January 1971, 96-7.

21. Inland oil spills—Emergency procedures and action (prepared by the Oil and Water Industries Working Group and published by the Institute of Petroleum, 1972).

22. The problem of engine exhaust control (and supplement) (CONCAWE Report No. 12/72).

23. IP lead in the environment (1972). IP letter to members (available on request).

24. IP code of safe practice in the precautionary labelling of petroleum products (with a foreword by the Minister for Industry, Department of Trade and Industry)—available early 1974.

25. *British Standard 799*—Oil burning equipment.
Part 1: 1962 Atomizing burners and associated equipment.
Part 2: 1964 Vaporizing burners and associated equipment.
Part 3: 1970 Automatic and semi-automatic atomizing burners up to 36 litres per hour and associated equipment.
British Standard CP 3002—Oil firing
Part 1: 1961 Installations burning class D fuel oil and CTF 50.
Part 2: 1964 Installations burning class C and class D fuel oils for vaporizing burners.
Part 3: 1965 Installations burning pre-heated fuels; Class E, F and G fuel oils and CTF 100 to 250.

26. MCWILLIAM, I. C. and DEWER, R. A., Proceedings of Conference on gas chromatography, Amsterdam (1958), page 142.

27. DOBETA Approvals Manual, DOBETA, London.

10. BIBLIOGRAPHY

1. IP Code of safe practice for drilling and production in marine areas.

2. IP International oil tanker and terminal safety guide.

3. An experimental investigation of the dispersion of hydrocarbon gas during the loading of tankers.

4. IP Refining safety code.

5. IP Liquefied petroleum gas safety code.

6. IP Marketing safety code.

7. IP Storage and piped distribution of heating oil safety code.

8. IP Petroleum pipeline safety code (with supplements).

A F HETHERINGTON, DFC*

4 The gas industry

SUMMARY

This paper reviews a range of the less obvious attributes, and potential problems of natural gas as a domestic and industrial fuel, with particular reference to the environment. The paper seeks to describe what the natural gas industry has done, and to what practical further action it is committed, to honour its responsibilities to the community in respect of environmental performance. The criteria employed in the fields of judgement, and in those actions within the Corporation's control are explained *vis-à-vis* the main environmental feature of the product: its contribution to clean air.

1. BACKGROUND OF THE INDUSTRY

From its foundation over 150 years ago, the gas industry was traditionally urban. It had many works, producing low pressure gas close to the towns where its markets lay. The towns soon expanded to engulf the works, which became embedded in centres of population or industrial employment. Today town gas is being phased out, and nation-wide conversion to a high calorific value gas system is more than half-way through. Last year 92% of the gas used in this country was derived from the North Sea. 73% was natural gas supplied directly to customers; the other 19% was town gas produced in reformers using natural gas as a feedstock.

Natural gas is delivered at high pressure to coastal terminals for transmission and distribution throughout the country. Major gas industry installations will in future usually be built in rural locations, thus avoiding the main pollution problems of the past. There will, however, be new environmental and conservation problems in the new gas industry and it is with these that this paper is mainly concerned.

2. ATTITUDES TO THE ENVIRONMENT

A country whose sole source of energy was natural gas would have no significant pollution problem arising from fuel usage. This reflects no particular credit on the natural gas industry—it is simply a feature of its product. But

---

* Chairman, British Gas.

only a part of the United Kingdom's energy is derived from natural gas. What then can, and should the gas industry do; and what should others do, towards improvement of the environment through the use of natural gas? It seems that these are two distinct areas of interest and opportunity, and should be considered separately.

First, the gas industry regards it a duty to look critically at its environmental performance in the widest sense, and to show rather more than commercially reasonable concern in its improvement. This term is chosen with some thought. It would be easier to say 'the greatest possible concern,' but this would disguise the nature of environmental expenditure—the translation of concern into business terms. Like any other expenditure, it has to be looked at in terms of need, return and priority within the industry and within the community generally. A useful criterion is 'how much would an intelligent customer want us to spend if he knew all the facts?' That is the judgement we try to make. This paper describes what the gas industry has done, and is doing.

The second area of interest is in the use to which the natural gas is put. It would be nice to say that natural gas is used only where it best serves the interest of the environment. But even if it were not too far from the truth, it would be misleading to make such a statement. It would imply that this is the main, or the only criterion for usage of the always limited amount of natural gas available to this country. There are, of course, many other social and economic considerations. They are often put very strongly to us in British Gas by potential customers: keeping down prices, improving exports, serving development areas—all of them worthy objectives. After a lot of thought, the view has been taken that it is not for a fuel industry to set up as expert in the evaluation of such factors, and certainly not to make qualitative judgements. We have to assume that the relevant pressures are, through legislation and the attitudes of the day, reasonably reflected in the fuel market; in other words, that the usage most valued by the community is that for which the product commands the highest price. The fuel market is, of course, not wholly within the control of the gas industry. The fuel users and the community generally, through its central

FIG. 1   Natural gas: transmission system

and local government, are also involved through value judgements and actions taken upon them.

One might have expected the price difference between pollutant and non-pollutant fuels to have widened in the last decade, as social concern, translated into legislation, increased the demand for the one at the expense of the other. This is not so; it has generally narrowed. Social attitudes have certainly moved on, but legislation has not always followed close behind.

Sometimes it seems we hold on rather dearly to our freedom to pollute. As to natural gas, it may be questioned whether, as a country, we are getting quite the best environmental advantage from a product of such quality. It would be nice to see it as used mainly in places where the most people live and work, and in applications where the pollution control of other fuels would be impracticable or costly; but we in the gas industry do our best (other things being equal) to achieve this; although we in the industry cannot ourselves create the legal and market situation in which it will automatically occur.

Sulphur control would probably have had the most significant effect on the pattern of use of natural gas through legislation. A paper to the Clean Air Conference at Southport in 1968, when natural gas was first becoming available, gave tables of comparative sulphur figures. It called for a limitation on the sulphur content of fuel burned in towns and cities at 2% immediately, and 1% in five years' time. Five years have now passed, but without effective sulphur legislation. After providing for continuing domestic and non-domestic load growth, the residual annual supplies of natural gas have now been placed in the industrial contract market, as stated in the paper that they could be—'in any quantity likely to become available.'

Is it too late now for sulphur legislation? It would inhibit the use of pollutant fuels in towns and cities, and, through the price mechanism, lead to a gradual redistribution of commercial and industrial gas usage. It would certainly affect the disposal pattern of further supplies of natural gas; it could even, through consequent changes in market prices, i.e. in community evaluation, bear upon the probability of purchase of further supplies.

It is trite, but probably true, to say that an industrial community gets the pollution it deserves. The quality of air it breathes will depend mainly on the value it puts upon it. In respect of freedom from sulphur, this does not seem to have been very high. The slow rate of reduction achievable without direct effort seems acceptable to most authorities, with enlightened exceptions such as London and Sheffield.

2.1. Market research on social attitudes
It is important to know what the public really thinks about pollution. The gas industry has carried out, or taken part in several surveys to establish the public rating of current problems. Researches* indicated that, in 1972, pollution of rivers came third and air pollution fifth. Unemployment and drug addiction came first and second; traffic was fourth. Concern for pollution was, however, on the increase and pollution was expected to get worse. In 1969 the gas industry had been thought of as second in air pollution, equal to the chemical industry, and behind the car industry. In 1972 it came fifth: the coal and steel industries having moved up. This confirmed

*Market and Opinion Research International.

the view that our bad image was mainly associated with town gas works. The prime responsibility to control pollution was seen by most people to lie with national government; by far fewer to lie with local government, and by fewer still to lie with private companies. Most people are sympathetic to the idea of tax reductions or subsidies to help companies to install pollution control equipment.

In late 1970, a survey told us what people meant by the word 'pollution.' To most people it meant chimney smoke, exhaust fumes, sewage and detergent effluents in sea and rivers, and oil deposits on beaches. The great majority thought that publicity on pollution was desirable, and that the Government and the National Society for Clean Air did most to make the public aware of pollution. The gas and coal industries were thought to be doing most among the fuel industries; there was also a high recall of TV commercials. A survey in mid-1971 confirmed these findings, but with rather less emphasis on smoke and rather more on sea and rivers. A special survey of a structured sample of 128 industrial fuel users was made in 1971. About half had taken recent action on pollution control. This included all those in metal manufacturing, pottery and cement. Most firms did not, however, consider taking action until it was forced on them. According to the subsample giving cost figures, 10% of their costs on pollution control arose from their fuel, and 90% from other aspects of their operation. The survey supported the view that expenditure on the control of pollution is a small fraction of the cost of pollution to the community.

Satisfaction in terms of pollution was high for gas and LPG compared with oil and solid fuel, but no company expected financial savings through environmental improvement. Grit, dust and effluent gases (other than smoke and sulphur) were regarded as most important in pollution terms; dark smoke was seen as less important and sulphur dioxide least of all. The significant conclusion from British Gas point of view was that, although the pollution-free properties of natural gas were generally realized, the choice of fuel still depended mainly on price. This confirmed our own experience that most firms are not against pollution control, but will not spend money on it until they and their competitors are forced to do so. Our research and our experience suggest that industry and the public might be more willing to accept anti-pollution measures than is generally believed; but only legislation, or the profitable anticipation of it, will lead them to spend money on 'other people's environment.'

3. ATTRIBUTES OF NATURAL GAS
It is central to the purpose of this conference to take note of the environmental aspects of each fuel. These are summarized below for natural gas:

3.1. Smoke
None is produced.

3.2. Sulphur
Sulphur emission is negligible, i.e., about one five-hundredth of that of the other main fossil fuels used in industry. The total national emission is about 5 million tons per annum. The limited amount of natural gas in use in 1975 should reduce the emission by $\frac{1}{2}$ million tons. If

natural gas were the only fuel the emission would be only about 10 000 tons.

3.3. Nitrogen oxides
High temperature combustion can produce nitrogen oxides, even with natural gas; but this does not imply any serious utilization problem, or limitation.

3.4. Noise
In a few special industrial applications, natural gas is noisy in use. This can be corrected; or alternatively, the particular uses avoided.

3.5. Toxicity
Natural gas is non-toxic. Accidents and suicides from unburned gas fall to negligible numbers in converted areas.

3.6. Smell
There is little or no smell. A distinctive odour is introduced only for detection in the interest of safety.

3.7. Danger
If, in its widest sense pollution is taken to include hazards, then, like any other primary fuel, gas has a potential problem in use. Operational control, backed by extensive research and development, keeps this to the minimum.

3·8. Visual
Modern gas supply installations are much less disagreeable on the eye than in the days of town gas. They are less likely to be close to where people live; and they are carefully landscaped. Transmission mains are, as is well known, usually under grazing pastures or fields of waving corn!

3.9. Traffic
Natural gas puts no burden on the road system.

3.10. Water
There is no pollution of sea or rivers.

4. GAS SUPPLY OPERATIONS
More detailed comments are now made on the various environmental aspects of the operations of the gas industry. They can be divided into four major sectors:

(a) production and reception,
(b) transmission and distribution,
(c) storage,
(d) communications.

4.1. Production and reception
The gas as it comes ashore is not suitable for onward transmission. A terminal is required at the coastline to remove the liquids, to dry the gas, to remove impurities

FIG. 2   The Gas Council's section of the installations for the reception of natural gas from the North Sea at Bacton, Norfolk

such as hydrogen sulphide and to control the flow. These terminals are very large in terms of energy throughput, but very small in terms of land usage. For example, the total area of land in use at Bacton by both the producers and British Gas is only 185 acres, although when fully developed the throughput is equivalent to that of a 1 000 ton coal train every 10 minutes. The terminals have to be at or very near the coast; at present there are three in operation in England and one at the planning stage in Scotland. The policy in this regard is:

(*a*) no proliferation of terminals,

(*b*) each site to allow for more than one producer,

(*c*) each site to be developed as a whole,

(*d*) the design of buildings to fit in with the local environment; therefore no standard design,

(*e*) an overall development plan from the outset; no buildings to be left half-completed for any length of time,

(*f*) landscaping and visual appearance to be considered at the site selection stage by 'environmental impact analysis,'

(*g*) installations cannot be hidden and so must be made acceptable, the architect to be a good plant designer, not just an 'industrial cosmetician,'

(*h*) noise suppression to be dealt with at the design stage.

As a practical example, at Theddlethorpe, where the preferred site was close to a nature reserve, the method of bringing the pipeline to the site was agreed, together with barrier zones, with the Nature Conservancy and with the Lincolnshire Trust for Nature Conservation. Acceptance by both parties that the other had a legitimate, although different interest, led naturally to an amicable working relationship.

4.2. Transmission and distribution

There are three categories of pipe-conveying gas. The natural gas transmission system operates in the 1 000 to 550 psi pressure range and is about 2 000 miles in length. The regional distribution system, of over 5 000 miles, is in the 450 to 20 psi range. The low pressure distribution system serving customers directly operates at less than 20 psi and is approximately 100 000 miles in length. To minimize the number of transmission pipelines, the largest possible diameter of 36 in was adopted for the main feeder lines and these are buried with the top of the pipe at least 3 ft deep. Although there is disturbance while the pipelines are being laid (less than six months at any one point), careful reinstatement ensures that, except to the practised eye, visual evidence of the line rapidly disappears. In open country all that remains are marker posts to help the helicopter pilot fly the route for fortnightly inspection. Care is taken to avoid woods and forests wherever possible. Where this is impossible then consultations with forestry owners can result in the pipeline swaths of cleared trees being used as fire breaks or access roads, to the benefit of the forest as well as British Gas.

4.3. Compression

The transmission system can require compressor stations at 40 to 50-mile intervals along its route. Because of the large power requirement (10 000 to 30 000 hp), these are usually driven by gas turbines. The installations are almost all in remote areas. Problems of both appearance and noise are solved by the acoustic cab principle.

FIG. 3 'Lowering in' by side-boom pipelayers of a section of a 36-in diameter feeder main from the Bacton terminal to the backbone of the national transmission system

Instead of mounting the compressors in large conventional buildings, a small cab is built around each unit. This incorporates the silencing equipment and is much more amenable to a satisfactory architectural treatment. A night-time ambient noise level test is carried out near the site, and before the station is designed. A design specification is then drawn up to ensure that the night-time ambient level is not exceeded at the nearest residence by more than 5dBA (which is about the limit detectable by the human ear). This is a very severe specification, and there were doubts expressed by local planning officers that it could be achieved. The results have proved that it is possible to silence a gas turbine-driven compressor of 15 000 hp so that, at the site boundary, the ear cannot detect whether it is running or not.

4.4. Gas storage

The storage of gas is essential to gas industry operations. The traditional gasholder has been a feature of most industrial landscapes and, on a smaller scale, of rural scenes as well. No such gasholders are now being constructed, but many will remain in service for many years. The main impact is visual, but despite some inherent problems of external painting, this can be ameliorated by careful use of colour. (It is, however, hard for a fuel industry to please everyone. Despite the slighting references often made to gasholders in the past there are conservationists now protesting about proposals to demolish certain holders—even in the heart of London).

Present practice is to use high pressure gas storage, either in buried pipes of large diameter, such as that at Biggin Hill, or in horizontal storage vessels of 10 to 12 ft in diameter. A much more difficult problem is the storage of gas on a large scale, to provide security of supply and peak load cover over long periods. This requires storage volumes measured in thousands of millions of cubic feet, rather than the millions used in conventional diurnal storage.

An ideal solution would probably be to store the gas in a depleted gasfield underground. This can only be done when the geological conditions are right and a suitable storage structure has yet to be located. A project to store large quantities of gas in underground cavities produced by leaching-out salt is being planned at Hornsea in Yorkshire; also, the small gas field at Lockton may be used at some stage for gas storage.

A more promising general solution is the storage of gas in liquid form, with its consequent 600-fold reduction in volume. This enables large quantities of gas to be held in reserve at strategic points in the system, usually extremities, or centres of major demand. The quantities involved at each point of use, 20 000 to 100 000 tons, mean that the tanks are bulky and can create a visual amenity problem. Other environmental problems are minor, as there are virtually no noxious emissions of either liquid, solid or gaseous materials; noise is dealt with in our usual manner.

A tank to contain 20 000 tons of liquefied natural gas at the atmospheric boiling point of $-160°C$ is about 150 ft high and 150 ft in diameter. It should be remembered, however, that one such tank contains the equivalent of the whole storage capacity of the former town gas industry. The tank has to be surrounded by bunds to contain any spilt liquid, and vapour barriers to disperse any accumulation of cold gas in the remote event of a major leakage. This makes site selection difficult: considerable map and field work is required to locate the optimum site. To improve the visual impact of the tanks, the use of colour is of vital importance. Each installation is treated suitably to its particular site. At Glenmavis, the tanks are banded horizontally in shades varying from dark blue at the base to pale blue and white at the top. At Partington, vertical asymmetric bands of white, grey, orange and purple are used to alter the visual impact against different foregrounds and backgrounds.

For fully effective operation, the transmission system calls for complex monitoring and control. This entails a comprehensive communications network, much of it by VHF radio with the inevitable use of high masts. The policy is to keep their number to a minimum by sharing with other users wherever possible; in fact, it is hoped to achieve full grid coverage with less than ten new masts.

5. ENVIRONMENTAL ASPECTS OF NATURAL GAS IN USE

The environmental factors of a fuel in use can be divided into those of fuel input, fuel combustion and products output. Natural gas has no particular problems of input, either in delivery or storage, to or within the home or factory. Its combustion is not usually space-consuming and is generally neither dirty or noisy. The products are non-toxic and, in most applications, conveniently removable.

The significant effect which the use of natural gas can have in reducing air pollution is well illustrated by the pioneer research work of Professor Alice Garnett, Director of the Air Pollution Research Unit at the University of Sheffield. In a paper to the 1972 Annual General Meeting of the Institution of Gas Engineers it is demonstrated that air pollution levels in some industrial areas of Sheffield for the winter of 1970/71 show close relationships between decreases in SO_2 and increases in the use of natural gas, consequent upon a changeover from fuels that are not sulphur free. This study took into consideration the climatological conditions of the 1970/71 winter, and the geography of the Don Valley. At points, adjacent to industrial areas of substantial changeover,

there was a reduction of SO_2 at some sampling sites of as much as 46%, in a winter when higher (rather than lower SO_2) levels would have been expected. We look forward to learning much more from the results of this continuing work.

In the matter of domestic usage, the gas industry has the great advantage of control to standards, both of performance and durability and of safety, of the design of gas appliances in use. This is exercised through the Seal of Approval given by the Watson House testing station. Close, formal liaison with the manufacturers ensures that the standards are reasonable and acceptable. This control is backed by continuing experience of installing and servicing the appliances of 13 million customers and an extensive Research and Development programme on utilization.

A large proportion of the work of the 50 000 employees on the Marketing side of the industry, and many of those in other Divisions, bears directly or indirectly on the environmental performance or the safety of gas and gas appliances. Convenience, cleanliness, control and other attributes must, on commercial as well as human grounds, be firmly based on safety of the product.

5.1. Safety services

Natural gas improves the individual's control over his environment but, as with any fuel technology, at some risk of misuse or mishap. The gas industry takes this risk very seriously and supports its operations by an advanced emergency organization.

Customer contact facilities ensure that any incident, such as a suspected gas escape, which could endanger or alarm the public, can be reported at any time of the day. Telephone equipment is planned on the basis that even during the busiest hour of the day, no more than one caller in 25 receives the engaged tone. Teams of specialist fitters equipped with vans and R/T mobiles are maintained around the clock to respond immediately to the customer's call.

The industry also takes steps to ensure the safety of the environment even before a gas appliance is in the hands of a user. The Seal of Approval scheme subjects appliances to rigorous testing for safety: Codes of Practice, often more stringent than legal or British Standards requirements, are maintained and constantly updated. When the appliance installation is outside the direct control of the industry, CORGI*, the gas installers' watch-dog on safety standards, ensures that, as far as possible, the gas industry traditions of safety are maintained in the private sector. The industry has taken conversion as an opportunity to promote a general updating of standards of gas installation. Substandard installations, often with inadequate flueing and ventilation, some old, some more recent, are coming to the attention of the industry for the first time. Recommendations are made for modifications to meet present technical standards. Promotion of regular servicing for gas burning appliances should go far in ensuring that many unsatisfactory situations discovered during conversion to natural gas do not recur.

The approach to safety in the industrial use of gas involves the extension of basic principles to a wide range of specialized plant. The principles are set out in the industry's Industrial Gas Controls Handbook and the

---

*Confederation of Registered Gas Installers.

FIG. 4 Britain's first North Sea gas liquefaction plant at Glenmavis in Scotland at the northern end of the national transmission system. The plant liquefies natural gas so that it can be more readily stored in bulk. The liquid is stored in a tank with a capacity equal to 1 000 million ft^3 of gas

specialized areas are covered by Codes of Practice, particularly where no British Standards are available. The considerable communication problem is dealt with by publication of documents, by courses and public meetings, and by meetings with customers and manufacturers. Apart from direct expenditure on safety at the Midland Research Station, almost every investigation into performance has an element involving assessment of its safety features.

These comprehensive safety services came out well from the independent Morton enquiry of 1971. Since then there has been continued improvement in the industry's safety record. Deaths from carbon monoxide poisoning (as many as 1 300 in 1963) fell to 120 in 1972. Only one of these was due to leakage from mains or services, and this was from manufactured gas. Apart from the tragic Clarkston Toll incident, there have been only two deaths from mains or service explosions in the last four years. The fire record reveals that gas was involved in 1970 in as few as 23 of the 158 000 fires that were outside buildings, and about 7 000 of the 90 000 fires that were inside buildings.

6. RESEARCH AND DEVELOPMENT

So far this paper has been concerned with the impact on the environment of the natural gas industry, and with what the industry has already done to reduce or overcome such environmental problems as have been incidental to gas supply and use. The rest of the paper is concerned with practical proposals for future action, mainly by British Gas to dispose of any remaining problems of supply and use, but also by Central and Local Govern-

ment to obtain better environmental advantages from its product.

Although natural gas supply is not beset by environmental problems, a considerable proportion of the gas industry's Research and Development budget is allocated to areas of possible concern, or possible improvement in environmental performance.

Over £500 000 per annum is spent on projects bearing directly on environmental improvement. The approximate allocation is £100 000, on the outdoor, and £150 000 on the indoor environment; £175 000 on noise reduction, £75 000 on visual and £25 000 on land and vegetation.

About £1½ million per annum is spent on projects whose primary object is the improvement of supply security and £1¾ million is spent on projects concerned with safety mainly in use. There is other expenditure on projects which have security and safety as a secondary object. Thus about £5 million per annum of the Corporation's R & D programme has a bearing on the environment itself, on safety, or on the security of gas supplies.

6.1. Safety in transmission

During the 10 years over which the UK transmission system has been constructed, research has been directed first towards design philosophy, specifications, standards and codes of practice and, as construction proceeded, to continuing improvements and better standards. Now the direction of research is moving over to aspects of maintenance for long-term safe operation. The research budget for transmission this year is £1 million. One major project is the on-line inspection of pipelines in operation at full load, for the purpose of finding, and removing any small defects before they can give cause for concern.

6.2. Safety in distribution

There is a potential hazard when the circumstances of gas/air mixture, environment and ignition sources combine unfavourably. The distribution chain to the consumer, involving pipe systems of varying ages, laid in widely different types of soil, which have carried gases of different compositions over the years, produces problems that have to be solved to maintain the system. In particular, the materials used at the joints of pipe lengths undergo dimensional changes that require remedial action. Methods and materials are being used, and further developed, for internal and external application to prevent or repair joint leakages. The effectiveness of these methods is evaluated by field trials.

Where gas mains lie close to other services, e.g. GPO ducts and sewers, gas can travel considerable distances from a source of leakage. Research effort is directed towards developing methods and equipment that will locate the position of a gas leak rapidly, resulting in greater safety, fewer holes made in searching for escapes, and less noise and interference with traffic. Studies continue on the correlation between the safe life of a gas pipe in the ground and soil environmental factors, so that renewal may be to the most economical plan. The properties of the newer polymer materials are being developed for pipeline system use, both in fresh ground and as lining in old mains.

6.3. Tree damage

Though natural gas is not itself toxic to plant life it can, by leaking into soil, deprive tree and plant roots of oxygen needed for respiration by displacing the gases normally present in the soil. If methane-oxidizing bacteria are present in soil penetrated by natural gas, they use oxygen to oxidize methane, and thus they can cause an even greater shortage of oxygen. The products of the oxidation are water and carbon dioxide, and the latter, if it accummulates, can also harm the roots of trees and plants. Leakage detection and control has ensured that cases of damage to urban trees and other vegetation in Great Britain have been few. Deaths of trees partly or directly attributable to natural gas leakage in 1972 were less than 200. Records are kept and information and data obtained from abroad; methods for investigating suspected damage have been developed; special methods are used for sampling and analysing soil gases. Ways of aerating soil affected by leaking gas are being considered, especially where oxygen concentrations are or may be likely to remain very low. A further aspect is that over 250 000 trees have been planted as part of the landscaping on the terminal compressor stations and other sites on the national transmission system for natural gas.

6.4. Substitute natural gas

The gas industry may need to manufacture gas again, to supplement natural gas supplies for peak shaving or other uses. This in no way implies the reconversion of appliances for town gas operation. There is full commitment, in both distribution and utilization, to the new higher calorific value of natural gas. If gas be made, or bought in the form of liquefied natural gas, it will be indistinguishable in use from natural gas.

There is a programme of research into the pollution aspects of manufacture of 'substitute natural gas.' The industry itself could be a significant fuel user in providing steam or other sources of heat for such plant. The fuel would tend to be the cheapest liquid fuel available for purchase, or derived from the feedstock. But this would create no problems not shared by other commercial plants firing with oil. Sulphur would be rigorously removed from the major part of the feedstock to be converted to substitute natural gas. In terms of overall fuel treatment, it should be expected that the pollution position in any new works would be far better than in the gas works of former days; and, of course, there would be far fewer of them.

6.5. Oxides of nitrogen

The information available so far on emission levels of NO_x (nitric oxide and nitrogen dioxide) from industrial and large commercial equipment is encouraging for gaseous fuels. In some applications the NO_x emitted by a flame is formed above the cone by entrainment of air and participation of the atmospheric nitrogen in the flame reactions. In general the higher the flame temperature, the more NO_x is produced. Simple methods of NO_x reduction therefore involve temperature reduction and control of secondary aeration. The industry is making a significant effort to produce methods of accurate measurement of these oxides suitable for its own fundamental studies, and for use in atmospheric measurements.

6.6. Noise in use

The combustion of solid, liquid or gaseous fuels necessarily results in some small fraction of their energy, about 10^{-8} to 10^{-7}, being liberated as noise. This does not usually create a problem, but there are occasions when combustion-generated noise needs to be brought down to an acceptable level. There are two categories, turbulent combustion noise (more or less random in nature) and periodic combustion oscillations.

Combustion noise, with a broad frequency spectrum, will always exist to some extent, the level depending on the combustion intensity of the burner, its heat input, whether or not it fires into a partially closed appliance (providing acoustic absorption) and, most important, the acoustic characteristics of the room in which the heating plant is located. Additional to the noise produced by the combustion process there is often a contribution from the air supply fan in forced draught systems. In industrial gas fired heating plant, other more intense noises are present and the combustion noise is not significant. A more critical case is where the plant is installed in a quiet environment, such as a school, hospital or hotel and is the only noise source in the vicinity. Levels are usually low, and substantial reductions are difficult to achieve. So far there have not been too many problems in this field and the noise reduction achieved by acoustic insulation is often sufficient. To improve the position further, information is awaited from more basic studies on the mechanism of combustion noise, on the matching of burners and air supply fans, and on the location and design of chimney terminals.

Resonant noise is produced when there is a coupling between heat release from the flame and the pressure wave due to an acoustic resonant from the burner and/or plant. Noise of this type is characterized by presence of discrete frequencies in harmonic relationship. Although noise levels can be very high, actual occurrence is fortunately rare, and effective empirical cures have almost invariably been found. It is anticipated that studies aimed at a better insight into the noise generating mechanism will enable trouble to be avoided at the design stage with new plant.

6.7. Visual aspects

Natural gas requires chimney stacks of only modest size. If oxides of nitrogen rather than sulphur dioxide are used as the critical pollutant in assessing chimney height, this results in heights for gas fired plant of one-fifth to one-third of those for liquid fuels. Even when chimneys are placed on top of buildings considerable savings in height are possible. Discussions on this topic have taken place between the industry and the Department of the Environment. General agreement has been reached, subject to clarification of the provision which must be made for the effects of adjacent buildings on the low chimney heights to be recommended.

6.8. Indoor environment

The strong trend in gas usage is towards flued appliances. Central heating units have their own flues; new bath heaters use 'balanced flues,' and gas fires are flued into chimneys in special ducts. Former portable fires are no longer used on natural gas. Despite these facets, even though natural gas is non-toxic, there is still a sizeable research programme devoted to indoor environment.

In studying the formation and concentrations of carbon dioxide, carbon monoxide and oxides of nitrogen produced by cookers in ventilated kitchens, and by water heaters in bathrooms, information is being obtained about possible health hazards from the flueless operation of appliances, and on the ventilation required to limit the concentrations that may occur. An investigation of the formation of carbon monoxide and nitrogen oxides under laboratory conditions is also being undertaken in order to recommend suitable designs for burners and heat exchangers to limit the production of these pollutants. The correct and safe operation of flued gas appliances calls for an adequate level of ventilation to the room. Weather-stripping, room sealing, and new methods of construction can have an effect which may demand special

FIG. 5 *Natural gas being flared off during testing aboard the drilling rig 'Orion,' while operating in the North Sea*

consideration of installation Codes of Practice. Current investigations are designed to provide the necessary data.

Subjective response to their thermal environment is also being studied by determining the response of people to variations in temperature, radiation, air movement and the type of heat emitter. The effects of temperature gradients and air movements are not well defined, and the present work is designed to produce these in a controlled environment and to discover the subjective reactions arising from them.

The manufacturers of gas appliances and other equipment also contribute to their own development effort. An important project was recently completed by a manufacturer,* working in close association with the Department of Mechanical and Electrical Engineering of Greater London Council. This improved the environment within an occupied two-bedroom flat, in eliminating the severe condensation and mould growth from which the flat had suffered for some six years as a result of the intermittent heating routine followed by the tenant in the interest of economy. The improvement was achieved cheaply and with minimal increase in fuel consumption, without change in the heating routine of the occupants and to their entire satisfaction. Good progress has now been made towards a method of automatic prevention of potentially damaging condensation at minimum cost in any dwelling, under varying conditions of occupancy, activity, tightness and type of structure, heating, natural ventilation solar gain, and wind chill. Some larger dwellings are now being converted to 'heater-ventilator controlled environment,' a system readily applicable to both existing and new buildings, without critical design features.

The basis of 'heater-ventilator controlled environment' is to induce a proportion of fresh air through the thermostatically controlled warm air heating system, together with automatic control of the rejection of moisture laden air from the dwelling. This provides ventilation no more than sufficient to maintain the dewpoint of the air below the minimum inside wall surface temperature.

As to noise in domestic use, sponsored research is being directed towards the development of small and quiet devices suitable for incorporation in fan-assisted combustion systems. The combustion process produces a proportion of the noise from domestic gas appliances, and if appliances become more compact, noise problems are likely to increase. Research is being directed towards obtaining information on the noise produced, from which control measures may be evolved.

6.9. Safety in use
Incidents involving gas explosions have focussed attention on devices designed to give warning of the build-up of

*Thorn Heating.

concentrations of unburned gas in buildings and elsewhere. A large number of models are available, but many have inherent disadvantages. Performance criteria against which to assess them are being developed and the various methods evaluated.

In the approval testing of gas appliances a major part of the tests applied is directed towards the safety of the installed appliance, under all conditions of normal use or misuse, both towards the user and his property. The same interest is reflected in the appliance Standards and Codes of Practice, both British and Continental with which Watson House is actively concerned. The quality assurance of approved appliances is also aimed at safety in use, as is reliability engineering for maintaining satisfactory performance of appliances in use.

Appliances operated under vitiated conditions have lower levels of performance and safety. This is being studied to give a better appreciation of the factors controlling performance of burners under adverse conditions. Safety is also an underlying consideration in the research of flame stability, ignition, propagation, flue design and almost all the development projects of the industry.

7. CONCLUSION
The gas industry, from a pollution standpoint, has had a great stroke of good fortune: the discovery of natural gas in the North Sea. Its exploitation, transmission, distribution and utilization have inherently fewer environmental problems than most other fuels.

Nevertheless, in its operations and its supporting Research and Development, the industry is putting substantial effort in terms of money and manpower into improving what cannot be other than a good environmental record.

But there is a limit to what British Gas can do alone. Natural gas is, in the main, reaching the premium markets. It is dominant in the domestic heating market and prominent in commercial and industrial usage. It would not, however, be quite true to say it is always used to best environmental advantage, particularly in respect of its freedom from sulphur. The absence of sulphur legislation implies a random element in the pattern of use where there could be a pressure towards applications in places where people live and work, where other fuels would pollute. The drive against pollution cannot, of course, be left to fuel suppliers and users alone. Central and Local Government must weigh all the factors and decide whether, and how, to limit the permissible content of sulphur in fuel, especially where there are large concentrations of people. This is perhaps the one area where, at the moment, there is some lagging behind other countries in those efforts to make the most of natural gas.

A E HAWKINS, BSc(Eng), CEng, FIMechE, FIEE, FBIM, MInstF*

5   The electricity supply industry

SUMMARY

An undertaking the size of the electricity supply industry cannot be run without some impact on the environment but, with care, that impact is kept small compared with the benefits electricity also brings. Electricity generation consists of a series of energy conversion stages at each of which some losses occur which interact with the environment. When chemical and nuclear energy is converted to heat the effect on air quality is the main concern. When heat is converted to mechanical energy the principal problem is the effect on water quality. Visual amenity is carefully considered in both the generation and distribution of electricity. A number of lesser impacts on the environment are also itemized and discussed. At the point of use electricity is wholly beneficial and non-polluting.

The electricity industry operates under a statutory duty to minimize its effects on the environment. In various ways it goes beyond the mere avoidance of nuisance and makes a positive contribution to environmental betterment. For the future, care will continue to be exercised, by anticipating potential environmental problems, and by seeking timely solutions.

1. INTRODUCTION

By almost every standard electricity business is big business. In all the technically advanced nations of the world electricity supply dominates the industrial scene in the scale of its capital investment requirements; the huge tonnage of primary fuels it consumes; and in its ever growing demands for large complex and technically sophisticated plant and equipment.

In Britain electricity supply has grown in 90 years into an industry investing capital at a rate exceeding £1 million per day. It is an industry in which the value of a single major power station rivals the total assets of such industrial giants as Esso Petroleum or British Leyland. Power station boilers consume one-half of the coal, and one-third of the residual fuel oil burned in the country. Electricity supply reaches out to practically every home, every farm and every office and factory in the nation

*Chairman, Central Electricity Generating Board.

and without it the present social fabric could not be maintained.

It would be nonsense to suppose that an industrial undertaking on this huge scale could be run without some impact on the environment but, by good management, that impact is small indeed compared to electricity's economic and social benefit. The industry is not hungry for land. The area required for new power station and substation sites is only a trifling percentage of the land developed for all other purposes. Sites acquired in the earlier days of the industry are used, and re-used, for different operational purposes as the original plant upon them is superseded. The improvement of amenity and elimination of pollution made possible by the ready availability of electricity for all domestic, commercial and industrial purposes far outweigh the adverse effects caused by its production and transmission. Thus the industry may claim a very considerable credit entry in the nation's environmental balance sheet.

Electricity generation and supply is an energy conversion industry. It is not concerned with the exploitation of natural resources, but it operates upon the fuel supplied by other industries represented at the Conference. Generation consists in essence of a series of energy conversions where the chemical and potential nuclear energy represented by the fuels is first converted to heat energy (in steam boilers or nuclear reactors) then to mechanical energy (in steam turbines) and finally to electrical energy (in alternators). The fundamental laws of physics decree that energy cannot be created or destroyed, only changed in form, and furthermore that no such change can be 100% efficient. Each energy conversion stage, therefore, implies some losses of greater or lesser magnitude, some of which may affect the environment. In succeeding sections of this paper each energy conversion step is examined in detail, the corresponding impacts on the environment are tabulated, and an account given of the manner in which the Generating Board has tackled the problems that emerge so as to eliminate or (where this is not possible), to minimize, their adverse effects on the locality.

The environmental problems of the industry do not end with the generation of electricity. The power has still to be transmitted at high voltage to the major load centres and distributed at successively lower voltages to the multitude of individual consumers throughout the country. While this distribution system contributes almost nothing to pollution in its conventionally accepted form, the impact on visual amenity is such that without the most careful planning, and execution, it could be wholly unacceptable to a discerning public. Even with this care the overhead transmission lines of the Generating Board and the Area Electricity Boards are a main focus of criticism of the industry's interaction with the environment.

At the point of consumption, however, electricity is totally clean, silent, invisible and intrinsically non-polluting. It is therefore possible to claim quite un-equivocally that the use of electricity is wholly beneficial to the environment and offensive to none of the senses.

Apart from the inherent cleanliness of its product the electricity supply industry has found ways of making a positive contribution to environmental betterment. Dealing with the large tonnages of ash and dust that result from coal burning is a case in point. By determination and ingenuity, a waste disposal problem of embarrassing magnitude has been transformed into a valuable source of by-products of proven usefulness in a surprising range of applications. On a number of sites, tree-belts and similar features maintained for screening purposes and to improve visual amenity, have been further developed to provide unique cultural and educational habitats highly praised by the local communities to whom they have been dedicated. The Board has sought ways of turning its effluents and by-products to profitable, or socially beneficial use. It has also exploited ways in which its operations may provide socially important fringe benefits in the form of conservation, recreation, amenity and educational facilities.

The concept of the balance sheet therefore provides a basis on which to examine the overall impact of the electricity supply industry on the environment. A number of positive credits can be claimed to offset the debits that arise where, even after all due concern has been exercised, some residual effects cannot be avoided. Many of these effects derive from the sheer scale of the industry's works and the problems will not lessen as the supply system continues to expand to meet the public's ever-growing demands. The paper concludes with some comments on likely development in the next decade or so, and on the way in which these environmental considerations influence the industry in its day to day working. These influences arise even before fuel is delivered and the process of converting it to electricity begins; and the first task, therefore, is to examine the environmental issues involved in the planning of new power stations and transmission lines, and in selecting the sites where these activities are to be undertaken.

2. SITING

Until the recent economic recession electricity consumption had grown fairly steadily at an annual rate of increase of 7%, so that the total demand doubled every ten years. This scale of growth is quite common in all the developed countries; indeed other Common Market countries have achieved rather higher rates, especially during the past five years.

These days a good deal of concern is expressed about the problems posed by rising population and economic growth. It is worth examining the causes of the increase in electricity demand. Over the past 20 years 85% of this growth has been due to an increase in consumption per head, and only 15% of the increase has gone to supply new consumers. This means that we are all using more electricity for comfort, convenience, recreation and production. Opinion is divided on whether our energy resources and fuel reserves can bear this kind of growth created by a rising standard of living. But even a reasonable and controlled improvement in the living standards implies some additional requirement for electricity, and even a small annual increase in so large a resource represents a substantial investment in new generating plant, transmission lines, and distribution equipment.

The CEGB has to be ready to satisfy public demand for electricity; and to meet that demand instantaneously and in whatever quantities required. The CEGB would like to be able to do this by building the necessary power stations and routing the transmission lines where technical considerations dictate; where they do not conflict with other land uses and where they are wholly acceptable to the public at large. That, however, is an impossibly tall order. The best is done to meet these conflicting criteria under the firm direction of the two statutory duties imposed upon the CEGB when it was formed in 1958:

(1) To develop and maintain an efficient co-ordinated and economical system of electricity supply.
(2) To take into account any effect which their proposals would have on the natural beauty of the countryside and on flora. fauna. features. buildings and objects of special interest.

The second of these duties which, in today's language may be summed up as 'showing a concern for the environment,' was unique in its time as a statutory goal for a major industry. The implications of having this dual responsibility are far reaching. Both must be regarded as of equal rank. The Board is neither required to develop the cheapest possible electricity supply system nor to protect the environment at all costs—if indeed either goal were achievable. A balanced judgement for each development must be reached: one that seems the most appropriate in all circumstances.

Planning a power station involves making an assessment at a given point in time, of many technical, economic and social factors. It is necessary to draw up a kind of balance sheet of costs and benefits that will range over the life span of the station. In a coldly impartial assessment, these factors would be quantified and reduced to present day values for comparison, with price tags placed on all the natural resources employed, including all aspects of amenity. But there is no way of costing qualities such as natural beauty and tranquility, so subjective judgements have to be made in drawing up our plans; but the statutory planning and consent procedures to which these proposals are subjected ensures that they are adequately tested against the opinions both of experts and of the general public.

2.1. Site selection and development

The technical requirements for the site of a modern power station are exacting. Among the necessary features are:

(a) Proximity to an economic source of fuel in large

quantities. This includes good rail access in the case of coal-fired stations, and access to a deep water anchorage or positioning within pipeline range of a refinery in the case of an oil-fired station. (The source and distance of fuel supplies for a nuclear station are not a critical factor.)

(b) Proximity to an adequate and unfailing supply of cold water for condenser cooling, requiring over 60 m³ per second for a direct cooled 2 000 MW station. Cooling water tunnels for quantities of this order are expensive and deep water close in shore is a distinct advantage. Where insufficient flows are available for direct cooling, cooling towers must be used—requiring about 2 m³ per second at a 2 000 MW station for making good evaporation losses and providing purge.

(c) A reasonably level area of land of 50 to 150 hectares

(depending on the type of station) sited above flood level, but not so elevated that heavy cooling water pumping costs are involved. The foundation conditions must be suitable for very heavy loads.

(d) Access from existing roads must be good, or capable of improvement to deal with the heavy flow of construction traffic.

(e) The site must be so located that the height of the power station chimneys and cooling towers or nuclear reactor buildings do not conflict with aircraft safety requirements in the vicinity of an airport.

(f) For coal-fired stations, nearby facilities for large scale utilization and disposal of pulverized fuel ash are desirable.

(g) For nuclear stations the site must be sufficiently remote from large population centres to comply

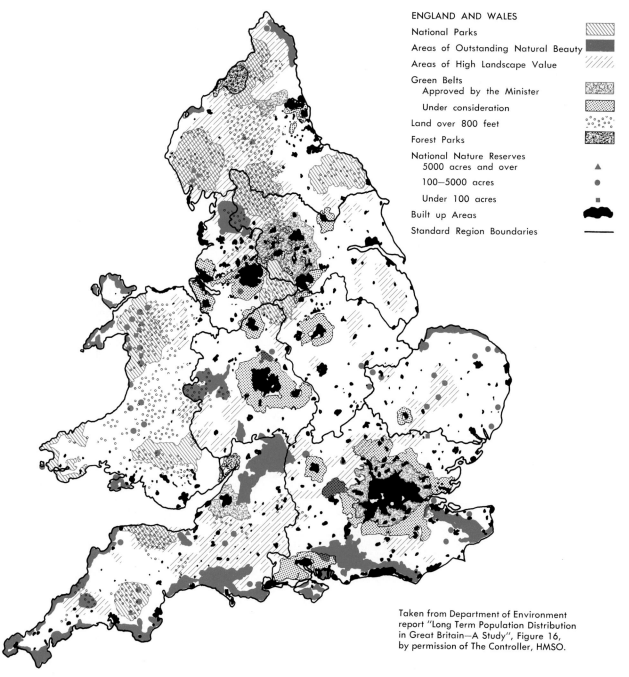

ENGLAND AND WALES

National Parks

Areas of Outstanding Natural Beauty

Areas of High Landscape Value

Green Belts
    Approved by the Minister

    Under consideration

Land over 800 feet

Forest Parks

National Nature Reserves
    5000 acres and over

    100—5000 acres

    Under 100 acres

Built up Areas

Standard Region Boundaries

Taken from Department of Environment report "Long Term Population Distribution in Great Britain—A Study", Figure 16, by permission of The Controller, HMSO.

FIG. 1 National Parks; Areas of Outstanding Natural Beauty; Green Belts; and other high amenity areas that present major restraints to development

E

FIG. 2   West Burton Power Station: '. . . an immense engineering work of great style . . .' (Civic Trust Award citation)

with the requirements laid down by the nuclear licensing authorities.

Examining the technical features of a proposed power station site is usually a long and detailed process. It involves negotiations with many other bodies, the commissioning of specialized reports from expert consultants, and the study of hydraulic and wind tunnel models. At the same time consultations will be held with planning and amenity interests so that the second part of the Board's statutory responsibility is discharged, and the project planned from the start to create the least possible disturbance to the environment.

2.2. Environmental restraints

In a heavily populated country like Britain conflicting demands for land use inevitably arise, and the preservation of suitable tracts of countryside for amenity and recreation is essential. This preservation imposes considerable restraints on industrial development. Over 40% of the land area of England and Wales is protected to form National Parks, Green Belts, Nature Reserves, etc. Similarly, some 60% of the coastline is protected, and a further 10% of the land area is already built up urban development. (Fig. 1) The remaining white areas on the map include some upland regions together with much of the agricultural land in the country where industrial development is, to say the least, unlikely to be welcomed.

The degree to which the amenity areas are preserved

from development varies. Often there is no absolute prohibition but a more searching planning procedure has to be followed before decisions are reached on individual proposals. Nevertheless there is keen public pressure to preserve such areas and only in exceptional cases would proposals to site power stations within amenity areas be put forward by the Board.

At an early stage in the planning of all power station sites available information on the natural quality of the locality is sought and studied. Leading architects and landscape consultants are engaged to advise on site layout, orientation and shapes of buildings, and the use of texture and colour in the construction of major station components. Landscape design and maintenance proposals are formulated, including land contouring and tree planting both on the site and, where suitable arrangements can be made, in strategic positions off site. Sensitive siting and design is a policy that the CEGB is required to follow. It is one that the Generating Board would in any case strive to pursue, for a good-looking power station is an excellent advertisement for its owners' sense of public responsibility.

The design of the 2 000 MW coal fired station at West Burton (Fig. 2) involved extensive studies by the architects of models of the main elements of the station in a heliodrome. The design was viewed from every quarter of the compass under lighting conditions which were varied to simulate the whole range of diurnal and seasonal lighting conditions that would be experienced in practice. Particular attention was given to the grouping of the eight cooling towers and the use of coloured cement in the construction of some of the towers to obtain the best visual composition. This station has been granted a Civic Trust Award for its outstanding contribution to the surrounding scene—in fact the Civic Trust describe the station as 'an immense engineering work of great style, which far from detracting from the visual scene

acts as a magnet to the eye from many parts of the Trent valley and from several miles away.'

Making power stations more acceptable to the public does not necessarily involve spending large sums of money. Simple and inexpensive acts of good neighbourliness can often create extremely valuable bonds between the power stations and those who live, or have interest in the locality. Some examples of this kind are given later in the paper.

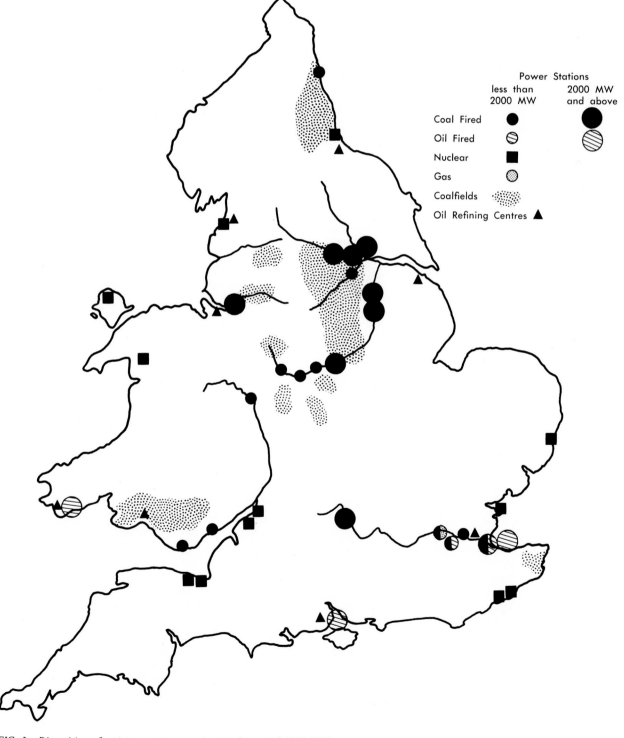

FIG. 3   Disposition of major new power stations in the period 1961–1975

In the case of nuclear power stations the Government decided at the outset that, as a matter of policy the first such stations should be sited in areas remote from substantial populations. The numerical rules that govern this remote siting policy have been modified from time to time but generally they have restricted the Generating Board to the consideration of sites in sparsely developed areas where conflict with amenity interests is almost inevitable.

Quite apart from this remote siting policy the cooling water requirements of the early Magnox stations were much greater than at contemporary fossil-fuelled power stations so that it was logical to reserve the limited inland water supplies for coal-fired plant with cooling towers, and to site the nuclear plant on the estuaries and coasts where unrestricted quantities of water were available. The cooling water requirements of the next generation of advanced gas-cooled reactors are much more closely comparable with present conventional plant, and these stations can be considered for both inland and coastal siting.

Nuclear power stations remotely sited from large populations are inevitably remote from the loads they serve. Long transmission lines are therefore required to connect these sites to the National Grid network which, again, introduces environmental conflicts.

2.3. Siting strategy

The developing pattern of electricity generation in this country demonstrates the way in which the Board has set about its delicate balancing act between technical requirements and environmental pressures.

The disposition of major new power stations in the 15-year period from 1961 to 1975 is shown on Fig. 3. About two-thirds of the coal to be used for electricity generation comes from the highly mechanized East Midlands and adjoining Yorkshire coalfields. (The largest of the inland coalfields indicated on Fig. 3.) Most of the new coal-fired plant has therefore been located as close to these coalfields as the supplies of cooling water would permit. The rivers concerned are too small for direct cooling, so that all of these inland coal-fired stations are provided with full cooling tower capacity.

The preponderance of coal-fired generation in the North has been partly balanced by the development of large oil-fired stations in the South, close to the major oil refining centres. Most of the Board's nuclear stations have also been sited in the South and West where they are best suited to supply demands that would be costly to meet with coal or oil. The two latest nuclear stations have however been sited in the North to improve the diversity of plant in that area and to help reduce the heavy off-peak power flows from South to North that results from having the majority of the low-operating-cost, nuclear and oil-fired plant in the southern half of the country.

It is a tribute to CEGB skill that only two of the 36 power stations developed in this 15 year period have had to be sited in areas of high amenity value. The coal-fired station at Ratcliffe-on-Soar is within the proposed Green Belt around Nottingham; a site that is uniquely placed to draw on the coal resources of the South Notts area, and in so doing provide employment for some 10 000 miners. The Trawsfynydd nuclear power station is within the Snowdonia National Park on the shore of a lake that forms the reservoir for a hydro electric scheme built many

years ago. In both cases, a lengthy Public Inquiry was held at which the Board's proposals, framed only after considering all possible alternatives, were searchingly cross-examined before consent to these developments was granted. In the case of Trawsfynydd gloomy forecasts were made of the adverse effect that this development might have on the popularity of the area for tourists. In fact this nuclear power station now entertains so many visitors from the general public (over 60 000 since the station opened), that special facilities to receive them have been provided.

At many other sites a lot of time, trouble and money has been spent to reduce the impact of a power station on its neighbourhood. The oil-fired power station at Fawley, for example, was sited not in the best technical position, but in an area that lay completely outside the proposed Hampshire Green Belt. Special architectural treatment was afforded to the buildings because of their dominant position at the entrance to Southampton water.

The Sizewell nuclear power station also was sited well back from the most economic location so as to preserve the beach and sandhills for public use, while existing woodlands were reinforced and extended by new planting to provide a visual screen.

At Hinkley Point and Aberthaw the proposed layouts of new power stations were turned through a right angle for architectural and landscaping reasons in order to accord with the expressed views of local planning experts: a modification which sounds simple on paper, but which can give rise to many technical complications and increased costs.

Despite the size of the electricity supply industry and its extension to almost every corner of the country, the total area of land owned and developed by the Board for power stations and substations amounts to only 0·1 % of the area of England and Wales, with a further area of similar size in the Board's ownership but remaining underdeveloped, mostly as the catchment basins of hydro electric schemes. The annual requirement for additional land is minute compared with that needed for new urban development, and the provision of other industries and services. Nevertheless, the Board has never underrated the need for careful stewardship of the land in their control and remain confident that the record will show how the dual responsibility placed on them by Parliament has been exercised with care and consideration for other interests.

3. FUEL TO HEAT

Once sites have been acquired, and power stations built, the first stage of producing electricity in conventional plant involves the combustion of fuel, and the conversion of its energy content into heat in the form of steam at high temperature and pressures or, in the case of gas turbines, as very hot combustion gases. Although this energy conversion step is of high efficiency (approaching 90%) the effects of burning fuel are one of the main sources of interaction between the industry and the environment. These effects arise from the handling and storage of fuel before combustion, and the disposal of the flue gases and particulates after combustion. The actual process of burning the fuel has little impact other than minimal energy losses in the form of radiated heat and noise local to the boilers. In addition, there are some secondary effects on the environment associated with

FIG. 4 Scraper on coal storage area, fitted with noise-reducing panels and improved silencer

boiler ancillary plant, or arising consequentially in the operation of large modern boilers.

With nuclear power stations the process is broadly similar, but instead of fuel combustion, which is a molecular process, uranium is fissioned at the atomic level to produce the required heat energy. The two processes differ fundamentally in the nature of the effluents they produce and in the effects of these on the environment.

3.1. Conventional fuels: handling and storage

Some of the environmental problems at this stage are:

(a) Dust blowing from coal handling plant and stock piles.

(b) Noise from coal handling plant.

(c) Fuel oil spillage particularly into estuaries and waterways.

(d) The visual impact of large coal stores and fuel oil storage tanks.

The prevention of dust nuisance from coal-handling and storage is mainly a matter of enclosing as much as possible of the plant to exclude the wind. Conveyor belts and hoppers are fully enclosed, and the unloading of coal wagons at modern stations using the 'merry-go-round' system of non-stop fuel delivery takes place within a roofed enclosure. The coal stock itself may cover several hectares and is therefore too large for a roof, but windblown dust is minimized by careful layering and compaction of the coal as it is put out to stock. This process is aided by the use of mobile equipment, such as bulldozers and scrapers for coal-handling since the wheels and tracks compact the surface of the coal stock to prevent erosion by wind and rain.

The use of mobile coal-handling plant brings with it the problem of noise. Electricity generation is not a particularly noisy process by normal industrial standards and public complaints are few. Some that do arise relate to this use of heavy mobile equipment on coal stores. The machines used may have engines of as much as 600 bhp; as purchased, their noise levels are inherently high due, principally, to 'diesel knock.' In conjunction

FIG. 5 Ironbridge Power Station showing the landscaped banks that screen the coal storage area from view

with the equipment manufacturers the Generating Board
has investigated the problem of quietening these machines
mainly by enclosing the engine compartments with stiff
body panels to attenuate the noise levels, and by closer
attention to the design of the exhaust silencers themselves
(Fig. 4). Significant improvements have been achieved
in this way, and all future equipment of this kind will
need to comply with acceptable noise level standards
specified by the Board.

Where fuel oil is delivered, particularly by sea, close
attention is paid to the design and operation of the fuel-
handling plant so as to minimize the risk of spillage
which (quite apart from its threat to the environment)
is a fire hazard, and an economic loss to be avoided.
Great strides have been made in these techniques in
recent years and the Generating Board is closely in
touch with all the developments and equipment for
containing and mopping-up oil spills. As improved
methods are developed they will be adopted.

Oil spillage is rare, but not unknown; and if the residue
should contaminate the shoreline, there is as yet no
clear cut answer on the best procedure to adopt. When
public access and bathing beaches are involved there will
obviously be justified pressure to remove the oil as
completely and as quickly as possible. In other situations
there is a significant body of scientific opinion that believes
that action to remove or destroy the oil may be more
harmful to the ecology of the area than if natural
weathering is given time to take effect.

At a coal-fired station, the fuel storage area may cover
several hectares and represent an unwelcome intrusion
on the landscape. Compared with the tall structures of
the power stations themselves, and the cooling towers
often associated with them, coal stores are of low profile
and some attempt can be made to screen them from
public view. This has been done at several power stations
using landscaped banks of spoil excavated from other
parts of the site during construction of the station. At
Ironbridge power station (Fig. 5) this screening has been
particularly successful.

3.2. Conventional fuels: the products of combustion

Combustion products can broadly be classified into
gases and particulates and arise wherever fuel is burned;
whether in power station boilers, in industrial furnaces
or in domestic grates. The environmental effects associated with combustion products include:

(a) Smoke, which is the readily visible carbonaceous
    emission arising from inefficient combustion.
(b) Grit and dust, which result from the mineral
    impurities in the fuel as purchased.
(c) Sulphur oxides, also resulting from impurities in
    the fuel.
(d) Nitrogen oxides, which arise partly from the fuel
    and partly from the air used for combustion.
(e) The disposal of ash and dust collected by arrestor
    plant.

The dispersion through chimneys into the atmosphere
of combustion gases and particulates and their subsequent effects at ground level, form one of the major
areas of concern in any examination of the effect of
electricity generation on the environment. The exhaustive
studies undertaken by the Generating Board over more
than two decades have made a significant contribution
to international scientific knowledge in this field and an

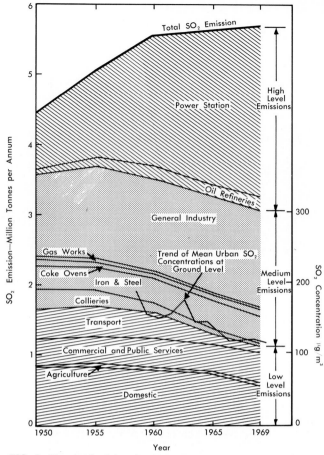

FIG. 6 Trends of sulphur dioxide emissions and urban ground-level
concentrations since 1950. (Courtesy: National Society for Clean Air)

extensive literature on this subject has been published
by Generating Board authors.

The one outstanding lesson that has been learned
over the years in Britain is that the scale of an air pollution problem cannot be judged solely in terms of the
quantities of pollutants emitted into the atmosphere.
Other factors that need to be taken into account are the
time, the place and the manner of the emission. These
are vital factors that can determine whether the pollutant
is dispersed harmlessly, or whether it is released with the
maximum risk of creating a hazard. Elsewhere in the
world there is evidence that this simple lesson has still
to be learned and, all too often, a confusion results over
the objectives and methods of air pollution control. To
some extent this confusion is compounded by the am-
biguous use of the word 'pollution' to describe both a
substance emitted into the environment, and the consequential effects of such an emission. This observation
is not a mere pedantic quibble over terminology; the
lack of understanding involved is already causing huge
technical resources throughout the world to be channelled
into projects that at best fall short of the optimum course
of action, and, at worst, could lead to vast expenditures
without significant improvement to the environment.

To illustrate this point, Fig. 6 shows the trend of
sulphur dioxide (SO_2) emission in Britain since 1950
broken down into different sources. Total SO_2 emission
has risen steadily over this period although the rate of
increase since 1960 has diminished, due principally to a
welcome fall in the sulphur content of heavy fuel oil.
The diagram also shows, as a more direct measure of

environmental quality, the average SO_2 content of the air in British towns and cities. The latter indicates a distinct downward trend amounting to a reduction of about 40% during the last ten years. This has been deduced from the results of over 1 300 measuring gauges maintained under the National Survey of Air Pollution (some 15% of which are CEGB gauges' sites). The corresponding trend of rural SO_2 pollution levels is less well defined since there are far fewer gauges in these areas. The available information suggests that there has been no consistent change over the same period of time. Judged on these objective measurements of air quality there is no possible doubt therefore that SO_2 pollution in British urban areas is markedly decreasing, and that in the country areas it has remained substantially un-

changed. Had a judgement been based solely on the annual quantities of SO_2 emitted into the atmosphere it might well have been concluded that 'pollution' was getting steadily worse.

The reasons for the lack of direct correlation between the total SO_2 emission and the ground level SO_2 concentrations are by now well known and require little emphasis here. It is sufficient to point out the obvious relationship between the falling urban concentrations (Fig. 6) and the fall in the fraction of the total SO_2 emission that is made from low level chimneys. What is perhaps less widely appreciated is the remarkable extent to which large scale emissions, such as those from a major power station, can be dispersed from a tall chimney without creating any significant addition to the SO_2 concentrations measured in the vicinity of the station. Fig. 7 shows the location of an actual group of SO_2 measuring gauges chosen from the National Survey records, and indicates their relationship to neighbouring urban areas. Table 1 summarizes the published readings from these gauges over a period of three years. During this period a 2 000 MW power station was brought into operation within the area shown and its consumption of heavy fuel oil increased from zero to a maximum of some 10 000 tonnes/day. This rate of fuel consumption implies an SO_2 emission of 600 tonnes/day, which is substantially in excess of the combined emission from all other sources in the area.

It is an interesting but fruitless exercise to try and deduce from the tabulated SO_2 concentrations alone approximately where the power station is situated in relation to the gauges, and when it commenced to operate. Had it been practicable to tabulate in this paper all the individual daily readings of these gauges instead of monthly averages, the same situation would result; a statistical analysis of daily readings, in conjunction with the concurrent wind records, still fails to reveal the presence of the power station. There is on the other hand no difficulty at all in identifying from the records the gauges which lie in urban areas and those which lie outside.

This example is in no way exceptional. Similar results have been recorded at power stations in all parts of the country. The lack of a discernible effect on local SO_2 concentrations is no mystery to the Board's engineers,

TABLE 1 Monthly-mean concentrations of sulphur dioxide at measuring sites shown in Fig. 7 (micrograms per cubic metre)

Year	Month	Site A	Site B	Site C	Site D	Site E	Site F
	Jan.	57	72	94	73	46	61
	Feb.	41	71	69	70	46	43
	Mar.	40	59	56	52	34	45
	Apr.	N	67	70	72	63	60
	May	77	71	66	59	45	N
	June	N	73	62	45	39	61
1	July	N	86	74	50	35	55
	Aug.	53	79	95	64	50	N
	Sep.	70	131	137	95	50	96
	Oct.	55	134	108	63	48	55
	Nov.	63	91	132	84	45	90
	Dec.	63	101	111	74	53	75
	Annual means	57	88	91	67	46	66
	Jan.	45	83	85	72	48	64
	Feb.	44	73	80	63	52	58
	Mar.	29	83	59	47	41	46
	Apr.	13	39	40	40	29	32
	May	14	43	35	31	26	35
2	June	26	36	63	55	40	48
	July	15	49	68	41	17	70
	Aug.	29	69	100	57	35	72
	Sep.	N	N	N	N	N	N
	Oct.	30	122	115	101	34	N
	Nov.	29	105	118	72	30	84
	Dec.	37	135	N	67	38	72
	Annual means	28	N	81	N	34	59
	Jan.	45	N	102	N	54	75
	Feb.	31	92	71	64	55	62
	Mar.	29	64	62	56	48	49
	Apr.	28	92	64	60	60	53
	May	20	50	41	49	36	23
	June	29	77	79	54	37	N
3	July	35	65	92	68	30	62
	Aug.	30	79	114	63	26	N
	Sep.	41	93	118	85	29	71
	Oct.	39	97	114	73	35	58
	Nov.	52	88	120	83	N	67
	Dec.	29	137	92	75	24	61
	Annual means	34	N	89	67	N	59

N = Insufficient readings for valid averages to be calculated.

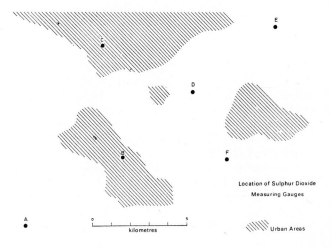

FIG. 7 *Location of sulphur dioxide measuring sites in the vicinity of a new 2 000 MW power station*

Location of Sulphur Dioxide Measuring Gauges

Urban Areas

kilometres

who planned and designed the power station with this express aim.

The total SO_2 emission from all British power stations does not on the basis of Fig. 6 appear to influence the average SO_2 concentrations in town and country to any measurable extent. It is, however, the corollary of this conclusion that compels the most sober reflection. Had the total amount of SO_2 emitted from tall power station chimneys been reduced by some means to a small fraction of the actual figure, or even if it had been eliminated altogether, then the available evidence strongly suggests that there would have been virtually no improvement in the quality of the environment to set against the undoubtedly huge expenditure that would have been involved. Yet in some countries where 'pollution' is measured in terms of the tonnage emitted this is exactly the way in which the regulatory authorities are seeking to move. There are surely more deserving ways to spend these large sums of money and produce a greater benefit to the environment of those who must bear the cost.

It may well be suggested that if the conclusions reached above are accepted within the context of Britain alone, then there is still the problem of long distance drift of SO_2 which may adversely affect other countries, possibly hundreds of miles away. Despite the complete lack of scientific data to support claims that British SO_2 emissions are already having such effects, and despite the fact that on our present knowledge of atmospheric dispersion such claims lack conviction, the Generating Board is nevertheless concerned that the true situation should be adequately investigated. They support the United Kingdom contribution to the present OECD studies of long distance drift, and are themselves undertaking a number of related studies on the transport and ultimate fate of SO_2 once it has left the near vicinity of the power stations. All this work is in its early stages, and while nothing yet has emerged to suggest that long distance effects are other than negligible, the Board recognizes that a final judgement on the efficacy of tall stack dispersion must await the completion of these studies.

The other main lesson learned from experience with chimney emissions is that the problems that loom large in the eyes of experts and regulatory authorities are not necessarily those to which the general public raise most objection. The whole philosophy of stack gas diffusion and chimney design has hinged on the study of sulphur dioxide, and most regulations throughout the world pay particular attention to this pollutant. There is, on the other hand, no possible doubt that the public in general are chiefly concerned with particulate emissions from chimneys; visible dust that soils and discolours their homes and belongings, and visible chimney plumes that offend their eyes. Over 90% of the objections received about power station chimney emissions relate to these two factors; and these can justifiably be regarded as questions of amenity rather than pollution since, other than in the case of 'acid smuts,' there is little or no evidence that particulate emissions lead to physical damage. These objections are nonetheless important, and no responsible body which is concerned about relationships with its neighbours will want to treat them lightly.

While the removal of SO_2 from flue gases remains an intractable problem, the removal of dust is a well established procedure for which all the necessary equipment has long been available. The CEGB has helped in no small measure to develop the large electrostatic precipi-

tators that are universally provided today on coal fired power station boilers (and in many other industries as well). The CEGB has also developed monitoring instruments to warn boiler operators when the dust burden in the flue gases is increasing, and has designed new and more sensitive gauges for measuring dust fallout in the vicinity of power stations.

Against this background it might well be queried why problems of dust emission still occur. One circumstance that can never be wholly eradicated is the unexpected breakdown of some item of equipment that permits the boiler to continue functioning but with an increased emission of dust to the atmosphere. The most common failure of this type is with the dust collecting plant itself. The only countermeasure is for the plant operators to reduce load or shut down altogether—as quickly as possible consistent with the safety of the plant—and to repair the fault at the earliest opportunity.

More difficult decisions arise in the case of very old power stations where the dust arresting equipment is deteriorating with age and was, in any case, designed to a lower standard than would be considered acceptable today. Unfortunately many such power stations are the Board's legacy from former local electricity supply authorities, whose plants were often sited within urban areas and, hence, where potential objectors are most numerous. The preferred remedy is to scrap these old plants but this is not always immediately possible. There is on the other hand some reluctance to spend large sums of money in rehabilitating plant that may only have a very limited remaining life, but, nevertheless, this remedy has been applied to many of the CEGB's older power stations. In a number of cases, old coal-fired stations have been converted to burn distillate fuel oil of low sulphur content as a means of preventing dust emission.

A third class of particulate emission problem arises with 'acid smuts' from some oil fired stations. This a very local, but objectionable type of fallout that can cause discolouration and damage to paintwork and textiles. Much has been done to minimize these emissions, mainly by closer control of combustion conditions within the boiler in order to prevent the initial formation of these acidic particles. Arrestor plant is not a solution since many of these particles form within the chimney itself. Whilst the occurrence of acid smuts has been greatly diminished it cannot be claimed that the problem is wholly solved, and the CEGB is continuing to examine all possible counter measures as a matter of priority.

Nitrogen oxides are a class of pollutants that have received less attention in the past than the corresponding oxides of sulphur, but which are coming under increasing scrutiny today, mainly because of their part in the formation of the 'photochemical smog' traditionally associated with Los Angeles. It is clear that the major source of nitrogen oxides remains the motor vehicle whose emissions, close to the ground and in crowded thoroughfares, must be the matter of first concern. Measurements of nitrogen oxide concentration in a wide range of power station boilers have confirmed that they are present in much less quantity than sulphur oxides; seldom exceeding 25% of the SO_2 concentration. It follows that chimneys designed for the dispersion of SO_2 to negligible proportions in the atmosphere will be more than adequate to disperse the nitrogen oxides as well. The ability of tall chimneys to disperse all the products of

FIG. 8   Member of environmental monitoring team taking radiation readings at Sizewell Nuclear Power Station

combustion to the same degree is one of their more valuable features.

3.3. Nuclear fuels

The process of nuclear fission gives rise to emissions that are wholly different in character from the emissions of conventional fuel burning. While they still consist of gases, liquids or particulates, their physical form and chemical composition are of secondary importance to their radioactivity. The potential effects on the environment that have to be considered when uranium undergoes fission to produce heat energy, include:

(a) Activated gases and particulates discharged to the atmosphere.
(b) Activated liquids discharged into the sea, or into water-courses.
(c) Radioactive waste from fuel reprocessing.
(d) The remote risk of serious radioactive discharges following a major accident or breakdown within the plant.

The risk of a nuclear accident is as remote as human ingenuity can devise, and these weighty matters form the backbone of the Government's nuclear siting policy. The problems of nuclear waste disposal arising from fuel reprocessing are the primary responsibility of the UKAEA.

Since the first CEGB nuclear reactor started operation in 1962 an impressive total of 120 reactor-years of operation has accumulated in Britain. The safety record is unexcelled—there has been no fatality from ionizing radiation, nor has there been any accidental release of radioactivity producing significant effects beyond a site boundary. The operators of nuclear reactors have to comply with stringent and legally binding conditions laid down by Government Departments, and the main safeguard for the public throughout has been the very high standards of design, construction and operation in these stations. The conditions imposed on radioactive releases to air and water take into account not only possible direct effects on human beings but the whole complex chain of physical, chemical and biological events that govern their dispersion in the environment. Maximum allowable limits for such releases are determined by the Ministry of Agriculture, Fisheries and Food and the specific authorizations for each discharge are set as far below these limits as the design and operation of the station will allow.

Nuclear emissions to the air include activated gaseous impurities extracted from the coolant gas (carbon dioxide) together with irradiated particulates arising from corrosion and erosion of the nuclear circuit components. Particulate emission is reduced to the minimum by filtering the gases before discharge into the atmosphere. The efficiency of the filter is governed by the nature of the particles involved and their degree of radioactivity. In the earlier Magnox stations with steel pressure vessels cooling air was required between the pressure vessels and the concrete biological shields surrounding the reactors, and there was still sufficient neutron activity in this area to activate the minute percentage of argon in the cooling air. The final levels of radioactivity in discharges to the atmosphere are contiouously monitored and are so low that they represent no material hazard even in the near vicinity of the discharge ducts.

As a back-up to the measurement of airborne radioactivity in the plant itself, a strict programme of monitoring the environment in the vicinity of each power station is followed involving periodic samples of herbage, milk· from selected farms, measurements of particulate fallout and gamma radiation dosages at various distances from the station. (Fig. 8). These surveys have proved that nuclear power station emissions are virtually undetectable against the background radiation arising from natural and other causes.

Liquid radioactive emissions arise partly from the moisture extracted from the coolant gas, and partly from the process of purifying the water in the fuel element cooling ponds. These liquids are small in volume, and are mixed with the main condenser cooling water discharge where they immediately undergo dilution by a factor of up to 10 000, even before they are discharged from the station. No such discharge is authorized before careful examination has been made of the subsequent movement of the radioactivity within the aquatic environment involving studies of tidal regimes, and currents, and the possible uptake of the material in biological food chains. Fish, molluscs and other key species are periodically monitored to ensure that no accumulation of radioactivity is occurring.

The results from all these measurements to control discharges to air and water and to monitor the local environment are reported to the authorizing Departments and are also discussed with local liaison committees voluntarily set up by the CEGB at every nuclear site. Committees of this kind, in which representatives of the authorizing Departments, the local authorities, River Authorities, farmers and other interested organizations are invited to participate, have proved an invaluable forum for the exchange of information with local people. They have undoubtedly served to demonstrate the extreme care taken by all concerned to control radioactive discharges, and in doing so have allayed any residual fears in local communities. The enviable safety record of nuclear power stations in Britain has earned public acceptance for them and this country has been able to avoid the unreasoned opposition that has unfortunately become endemic elsewhere.

3.4. Secondary environmental effects
Some secondary outputs and losses associated with the conversion of fuel to heat energy that may affect the environment are:

(a) Wastes arising from boiler feedwater treatment.
(b) Wastes arising from the internal cleaning of boilers and from periodic blow-down.
(c) Steam releases from boiler safety valves and other vents.

The feedwater required for modern high pressure

boilers, both conventional and nuclear, has to be of extreme purity necessitating the use of ionic exchange water treatment plants. The periodic regeneration of these plants releases a quantity of hydroxides and other harmless chemicals which are normally discharged under high dilution in the condenser circulating water.

The water passages of newly constructed boilers have to be cleaned internally with a strongly acidic solution to remove scale and debris. Similar treatment may be necessary at intervals in the life of the boiler plant. The quantity of solution required is fairly large and arrangements are made to hold the used liquid in temporary storage tanks on site, where it is neutralized with alkali prior to discharge. All such discharges are organized in consultation with the local River Authority or sewage undertaking, as appropriate, and are made in compliance with any conditions laid down by these authorities. Similar conditions are also attached to the regular blow-down of accumulated precipitates within the boiler.

The discharge of steam through drains and safety valves can create high noise levels in the vicinity of a power station and, by their nature, these noises are frequently of a sudden and unpredictable character. Silencers are now fitted on all such vents to attenuate the noise levels as far as practicable. A problem of design in this connection is that the silencer must not impede the performance of safety valves in releasing the over-pressure in the boilers that causes them to operate, so that full noise attenuation cannot be achieved.

During the construction phase it is also necessary to blow steam through newly completed boilers before they are commissioned in order to remove miscellaneous debris. This is undoubtedly a most noisy process but one which can be planned and operated to an advance schedule. This enables the CEGB to issue warnings to local residents explaining the need for the steam release and giving a timetable so that they need not be alarmed when the noise occurs. As a further measure the CEGB has had constructed a transportable silencer (Fig. 9) which can temporarily be attached to the appropriate vents and reduce the noise level by 30 dB or better.

4. HEAT TO MECHANICAL POWER
After the energy in conventional and nuclear fuels has been converted to heat it must undergo another transformation—into mechanical energy—before electricity can be generated. Steam turbines extract the heat from high pressure steam and convert it into mechanical energy of rotation. For maximum efficiency the steam is expanded down to a low vacuum and is then condensed for re-use in the boiler. The condensation process requires a large flow of cooling water at ambient temperatures and almost half of the total heat energy available in the steam is lost to the environment in this way. Such losses are an inevitable result of the universal laws of physics that govern these energy transformations. Research under way in many parts of the world on magnetohydrodynamics, on fuel cells, and on other alternative processes may lead to a more direct utilization of heat energy for electricity generation some time in the future; but for the period ahead covered by this review the condensing steam cycle will remain the major source of energy for this purpose.

In gas turbines the condensing losses are avoided since the heated gases from fuel combustion are converted

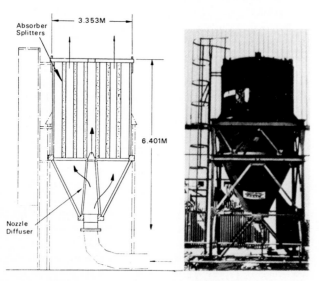

FIG. 9 Transportable silencer used to reduce noise from safety valves during commissioning of new power station boilers

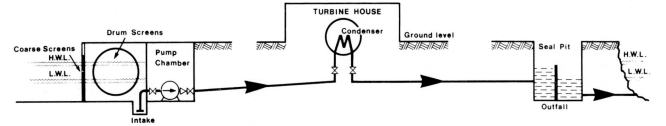

FIG. 10 Main elements of a direct-cooled power station circulating water system

directly into mechanical power in a different form of turbine. But again, basic physical laws limit the thermal efficiency to a figure well below that of the condensing steam cycle. Peak load gas turbines have an increasingly useful part to play within the overall pattern of electricity generation, but they are not a substitute for the large steam-powered generators which will continue to form the backbone of the supply system.

The environmental effects to be considered at this stage of the generation process are almost wholly concerned with the heat rejected in the cooling water which, apart from the latent heat from the condensing steam, includes some very minor additions from ancillary plant such as oil coolers. The other identifiable loss of energy is noise from the rotating machinery. This poses some internal problems in relation to personnel protection, but the buildings which house the turbines are designed to attenuate noise levels sufficiently to prevent any external environmental effects.

Rather greater care to avoid noise nuisance is necessary with gas turbine generators where the attenuating effect of the building structure is augmented by soundproof enclosures around each generator. In addition, complex silencers have been developed to avoid noise from the air intakes and exhaust ducts that handle the extremely large flow of fast-moving gases characteristic of this type of plant.

The heat rejected from steam turbine condensers is discharged to the environment in one of two ways; where there is plentiful cooling water available it is used only once in a 'straight through' system and the warmed water is then returned to its source (Fig. 10). Alternatively, where the cooling water supply is limited the warmed water is recirculated through cooling towers and the heat is rejected to the atmosphere in the form of warm air and evaporated water. The recooled circulating water is returned to the steam condensers thus limiting the demand for fresh water to a small fraction of that required for direct cooling (Fig. 11).

4.1 Thermal pollution

With direct cooled systems the major concern for the environment centres on the possible effects of a large and usually continuous discharge of water heated some 9° to 12°C above the natural temperature of the surrounding water body. In recent years much has been written, and much conjectured about the effects of such discharges on the chemical and biological processes that take place in natural waters. The influence of temperature on the dissolved oxygen content of water has led to forecasts of oxygen shortage caused by heated discharges which would reduce the water's ability to support life. Raised water temperatures were said, also, to have detrimental effects on fish and other life forms altering their metabolic and reproductive functions, or in the extreme, causing death. An alternative suggestion (somewhat at variance with the foregoing fears) has been that raised temperatures would encourage the proliferation of marine pests such as the common ship-worm (Teredo Navalis). Lastly it has been postulated that raised temperatures, by upsetting even one minor link in the complex life chain that exists in the aquatic environment, could lead to wholesale and detrimental changes in the natural ecology of the area where the heated discharge is made.

None of these problems is a novel disclosure and the

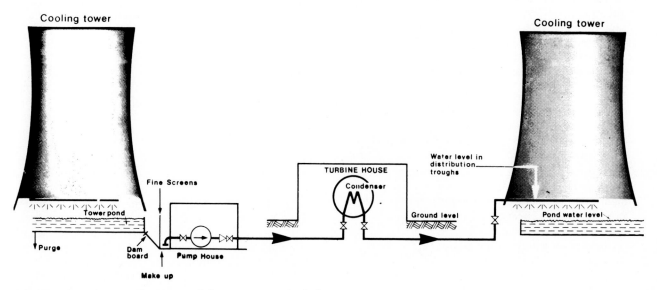

FIG. 11 Main elements of a tower-cooled power station circulating water system

necessity to protect aquatic life from the effects of raised temperatures has long been in the forefront of the minds of those concerned with the design of power station cooling water systems. There have in fact been legislative measures to control heated discharges from power stations from as long ago as 1919.

To assess the true extent of these problems and to gather the scientific data on which to base any remedial action necessary, lengthy and detailed ecological surveys have been made at a number of power station sites in different parts of the country covering both fresh water and coastal cooling systems. In some cases the relative abundance of life forms and their seasonal changes have been tabulated both upstream and downstream of the point where the heated discharge is made. In other cases similar surveys have been made over a period of many years starting well before the power station concerned came into operation, and continuing long enough to detect any long term changes in the ecological patterns of the locality.

As with the Generating Board's air pollution surveys, the data collected have been widely disseminated and discussed by the scientific community. The overriding conclusion has been that the postulated effects of 'thermal pollution' do not in fact appear to be occurring in Britain; it has proved virtually impossible to establish any case where a significant and adverse effect on local life forms has resulted from a heated power station discharge. An exception can possibly be made of the area extending a few tens of metres around the actual point of discharge, but even here it is by no means certain that temperature alone has been the cause of the observed disappearance of certain species. Changes in water flow and turbidity may be equally important.

Outside these limited areas effects are either too small to be measurable, or are of a kind which it would be difficult to label as harmful. It has, for instance, long been known to keen anglers that fish are often plentiful in the vicinity of a power station discharge but whether they are attracted by the warmth or by the steady supply of fresh food in the flow of cooling water is not known for sure. What is certain is that the fish are not adversely affected, and there have been serious proposals to create commercial 'fish farms' using the warmed water from power stations to stimulate the growth of the fish to a marketable size at a much faster rate than in their natural habitat.

These observed effects about the extent of 'thermal pollution' are so much at variance with the fears expressed that it is instructive to list some of the reasons why the influence of a heated discharge is so minimal in practice.

(a) Direct cooling is nowadays restricted to coastal and estuarine sites where even the large flows of cooling water required by modern power stations are dwarfed by the huge volumes of seawater passing back and forth with every tide.

(b) The warmed discharge is reduced in density so that it tends to float in a thin layer on the surface of the main body of water. This prevents the warm water from having any effect on organisms living on the sea bed or at mid-depth. A surface layer also dissipates the heat more rapidly to the atmosphere.

(c) In temperate climates the seasonal and diurnal ranges of temperature to which local species are adapted often exceed the temperature rise in cooling

water by a considerable margin. This is particularly the case in shallow waters where solar radiation and tidal movements have a marked influence on littoral temperatures. The natural ecology of these areas is tougher and more adaptable than is generally realized.

(d) Much of the available data on the temperature tolerance of different marine species relies on long-term laboratory tests where the experimental conditions do not reflect the normal habitat. The mobility of many species is ignored in these studies—a fish can swiftly move away if the conditions in a restricted area cause it distress.

(e) The relationships between oxygen content and the temperature of water are sometimes misunderstood. Moderate heating of water will not 'drive out' the oxygen unless the water is already nearly saturated with it, in which case the small reduction is not important to biological life. Only where the oxygen content of a waterway is seriously depleted by organic pollution (e.g. sewage effluent) can the additional effect of heating cause problems with deoxygenation; when the organic pollution is removed, as in the River Thames in recent years, the effect of heat becomes negligible. Finally, the turbulence created by pumps and screens in a cooling water system acts to aerate the water and compensate for oxygen losses due to the temperature increase.

All these alleviating factors help to ensure that power station cooling water discharges do not adversely affect the marine environment, but it is nevertheless essential that the cooling water system should be designed with this end in mind if harm is to be avoided. Fortunately, the economic requirements for electricity generation tend in some respects to harmonize with environmental protection; there is, for instance, a technical incentive to minimize the temperature rise in cooling water in order to achieve more efficient generation. For the same reason it is also necessary to avoid recirculation of the warmed water back to the power station intake which could lead to successive increments of temperature rise in the same body of water.

In practice, the design of cooling water systems is a long and painstaking process involving hydrographic surveys, measurements of tides and currents; hydraulic model tests and sometimes aerial infra-red photography. Where knowledge of the marine life in the vicinity is lacking a faunal survey is often undertaken as well.

4.2. Cooling towers

Cooling tower systems are more costly to build and operate than direct cooled systems, and are used only when there is an insufficient quantity of water available for direct cooling. With the present size of power stations, however, this limitation applies to almost all the inland waters of Britain.

Within a cooling tower about one-third of the heat contained in the cooling water is transferred to the air passing up the tower, and the remaining two-thirds is absorbed as latent heat in evaporating about 1% of the water flow. This water vapour also passes up the tower and is discharged to the atmosphere. The net water requirement for a cooling tower is therefore only one-hundreth of that required for direct cooling, but in practice an amount rather greater than this is drawn from the river and the difference in quantities is returned as purge water. This is done in order to remove from the system the salts and suspended matter which were present in the initial intake of water, and which would otherwise gradually accumulate as a result of the evaporation. Purge water is normally drawn off from the cooling tower ponds at the lowest temperature in the system so that it is returned to the river only a few degrees above ambient temperatures.

Two main effects on the environment are associated with cooling towers; the physical effect of the warm air and water vapour discharged from the top of the towers, and the impact on visual amenity of the large tower structures and their associated plumes. The amenity problem is a material one which continues to exercise the minds of power station designers to find satisfactory solutions; whereas the emission problem has been found to be of negligibly small proportions.

The quantity of water vapour discharged from the group of cooling towers serving a large power station might be expected to have some effect on the local climate—for example, an increase in humidity downwind of the towers, an increased tendency for fog to form in the locality, or possibly an increase in rainfall. Careful measurements have failed to indentify any such changes near to Generating Board power stations. This is reminiscent of the lack of observed ecological effects from direct cooled stations. Again it is worthwhile examining the reasons.

(a) The quantities of water vapour emitted, although large by comparison with other industries, are still puny in relation to natural phenomena. For instance, the cooling towers serving a 2 000 MW power station evaporate some 50 000 m³ of water a day. During an average day, however, the ambient air passing over the power station site would contain the equivalent of at least 5 million m³ of water or 100 times the cooling tower emission.

(b) The humidity of air is a function of both the water vapour content and the temperature. The effect on humidity of the water vapour released from cooling towers is compensated by the large quantity of heat also emitted.

(c) The natural draught cooling towers commonly employed in Britain are 100 m or more in height so that the plume is emitted well above the ground. Furthermore, the heat content of the plume gives it buoyancy so that it rises well above the level of the tower top as it disperses downwind. Any effect of the plume at ground level is thus likely to be minimal. (The large batteries of low level mechanical draught cooling towers often used in other countries do not have this advantage and sometimes create undesirable conditions of fog or high humidity near to a power station).

A problem that occurred in the past with cooling towers was the emission of water droplets which were carried up the tower and deposited in the vicinity of the station. In freezing conditions this emission could lead to icing of nearby roads and thus create a hazard to traffic. This problem was solved some twenty years ago by the development of efficient droplet eliminators and more recent studies have confirmed that these still provide an adequate protection against droplet emission, even for the much larger cooling towers now being employed.

4.3. Cooling towers and visual amenity

While their physical effects on the environment are negligible cooling towers undoubtedly represent a major visual feature in the surrounding area. Natural draught towers, of the type used in Britain up to the present, have been restricted in size on structural grounds so that as many as 8 or 12 are needed for the largest power stations now being built. These groups of towers tend to dominate

FIG. 12   Eggborough Power Station: cooling towers dominate the visual appearance

the visual appearance of inland power stations to the extent that the main power station buildings—in themselves very large structures—are almost insignificant in comparison. (Fig. 12).

The Generating Board seeks to develop pleasing layouts and architectural treatments for these large structural developments, using all the devices of form, texture and colour available to the modern industrial designer. Public recognition, such as the Civic Trust Award for West Burton power station is a measure of the success of these efforts. Visual amenity is a matter of subjective judgment, however, and even a sophisticated architectural achievement may be less pleasing to some eyes than to others. There is little doubt that the main visual problem with tower-cooled power stations is the sheer physical bulk of the tower structures so that diminishing this bulk will ease the problem.

With this in mind the Generating Board has recently developed an alternative type of cooling tower expressly intended to reduce the numbers required on a specific site, and hence reduce the visual impact. The main tower shell is retained but instead of relying solely on natural draught a ring of electrically driven fans at the base augments the airflow up the tower and greatly increases its cooling capacity. As now developed, the individual rating of these assisted draught-cooling towers is some four times that of similar-sized natural draught towers so that a power station which required eight of the latter, can now be adequately served by only two of the new towers (Fig. 13(a) and (b)). The first of this new generation of cooling towers is now under construction for Ince B power station in Cheshire.

Further developments of this principle are already in hand which may lead to another doubling of cooling capacity by increasing the shell size and so permit a single cooling tower to be used where previously eight were needed. (Fig. 13(c)).

The resulting improvements in visual appearance is not achieved without cost. The capital cost of a single assisted draught-cooling tower of the type now being introduced is not very different from the cost of the four towers it replaces, but the running costs over the life of the plant will be increased, mainly in providing power for the fans.

FIG. 14 Prototype 120 MW dry cooling tower at Rugeley power station, compared with normal evaporative cooling towers of the same individual cooling capacity

There is, of course, no formal way to equate these extra costs with the value of the improvements to visual amenity, but the Generating Board is convinced that the benefit to the environment in this instance is worth the additional expenditure, and that the reduction in the number of cooling towers will be welcomed by the public.

The plume of water vapour that emerges from the cooling tower forms another problem of visual amenity, but is one which is more difficult to solve. The condensation of water vapour which creates this plume takes place in the atmosphere and is, in fact, a visible indication that the tower has performed its proper function. The length of the visible plume is determined by the prevailing meteorological conditions; short and wispy on warm sunny days; long and dense on days of high humidity. It is fortunate that the latter condition usually coincides with dull overcast weather, so that the cooling tower plume tends to be lost to view against a cloudy background.

There are no technical measures available to the Generating Board that would significantly reduce the length of plume from evaporative cooling towers, whether natural draught or assisted draught. A complete answer to this problem would be the adoption of 'dry' cooling towers in which the heat from the cooling water is all dissipated in warming the air flowing up the tower and no evaporation of water takes place. The Board has investigated 'dry' cooling and has built and operated a 120 MW prototype tower at Rugeley power station. (Fig. 14). There are, however, two disadvantages to this type of tower. As can be seen (Fig. 14) they have to be considerably larger than evaporative towers for the same cooling duty, and they cost considerably more both to build and to operate. The extra size, of course, is a

FIG. 13 Visual appearance of cooling towers: present, and possible future, assisted-draught cooling towers at a 2 000 MW power station, compared with the previous complement of natural draught towers

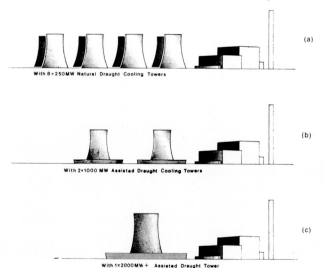

With 8 × 250MW Natural Draught Cooling Towers   (a)

With 2 × 1000 MW Assisted Draught Cooling Towers   (b)

With 1 × 2000MW + Assisted Draught Tower   (c)

move in the wrong direction as far as visual appearance of the towers is concerned; and the cost differential is such that no reasonable case can be made to adopt 'dry' cooling solely to avoid the occurrence of long visible plumes on humid days.

5. MECHANICAL POWER TO ELECTRICITY

The final energy conversion in the sequence leading to electricity production takes place in the power station alternators. All the stages so far described culminate in the mechanical rotation of the alternator shaft so that electrical current can be generated in the stator windings. It is something of an anti-climax that after all the weighty problems of pollution control, and amenity protection associated with the preceding stages, the actual production of electricity takes place with virtually no effect at all on the environment. A small heat loss in the alternators is dissipated in hydrogen coolers and forms an insignificant part of the heat discharged in the power station cooling water. A further small energy loss appears as part of the overall noise emission from the power station machinery and is dealt with by the general acoustical treatment. With this stage, completed the electricity is available for distribution to the public and eventual utilization.

It is, however, appropriate at this point to introduce an alternative method of electricity generation that plays a small, but important part in the country's supply system.

Hydroelectric power represents a case where mechanical energy can be made directly available and converted into electricity without any of the preliminary stages necessary at thermal power stations, both nuclear and fossil-fuelled. The production of hydroelectric power is however strictly limited by the natural resources available—in particular topography and rainfall. In England and Wales these resources are few and there are no further opportunities for the economical production of water power by the classical method—that is by impounding water temporarily behind a dam, and allowing it to flow through hydraulic turbines on its way down to the lower reaches of the original water course.

Where hydroelectric stations have been developed in the past, as at Rheidol in mid Wales, they have of necessity been sited in those mountainous regions most likely to be prized for amenity reasons. With good planning however the impact on the environment of such schemes can be much less than some individuals and amenity bodies may fear at the outset. It is a matter of record that large numbers of the public take pleasure in visiting the dams and lakes associated with these works and the Rheidol plant has earned for itself a citation from the Welsh Tourist Board. Before this development, the water of the river was badly polluted by metals leaching from old mine workings. The Generating Board built into the scheme settling ponds and filtration trenches which intercepted these pollutants. As a result, both salmon and trout have now returned to the river. The Rheidol

FIG. 15   Rheidol Hydro-Electric Scheme. Dam and weir constructed of local materials and designed to blend with the surroundings

scheme is also notable for the way in which local materials have been used in the works to help them blend into their surroundings and for the avoidance of harsh geometrical outlines in the design of some of the structures (Fig. 15).

Following the exhaustion of the limited resources for traditional hydroelectric generation an alternative use of water power has come into prominence with the development of pumped storage schemes. In these, the availability of low-cost electricity during the night enables water to be economically pumped 'up hill' to replenish the reservoir of a hydroelectric generating station, and enable it to produce power during the time of peak demand on the following day. This reduces the amount of thermal peak load capacity that it would otherwise be necessary to maintain on the supply system, and which is relatively costly to operate. Pumped storage schemes can also be designed to produce electricity at very short notice and so act as a safeguard against the unexpected loss of generation elsewhere on the system due to plant breakdown.

A 360 MW pumped storage scheme has been in operation at Ffestiniog in North Wales for several years. Like the Rheidol scheme before it, the Ffestiniog plant attracts a large number of interested visitors and has received a similar citation from the Welsh Tourist Board. At the present time the Generating Board is promoting a Parliamentary Bill to enable it to construct a much larger pumped storage scheme at Dinorwic in North Wales, for which the Board is proposing exceptional measures to protect the amenities of the area.

A pumped storage scheme calls for two reservoirs; an upper and a lower one that are linked to each other through a power station. At Dinorwic, the Generating Board proposes to adapt an existing lake and reservoir to serve the scheme, and to construct a 1 400 MW power station completely underground.

In order to preserve the natural beauty of the scheme's surroundings the Board has engaged leading landscape consultants to advise on aspects of this major civil engineering project such as the design and screening of dams, and the disposal of large quantities of spoil. The storage capacities of the lake and reservoir will have to be increased by the construction of dams. However they would not be concrete or masonry dams but natural rockfill embankments made of local materials.

The lower reservoir, an existing lake, Llyn Peris is already dominated by a disused quarry with ranks of slate terraces and spoil heaps rising to about 600 m above water level, evidence of earlier exploitation by man. This lake will require two embankments, one at each end, and both about 300 m long and about 10 m high. These embankments, although an intrusion in the valley scene, would resemble natural banks rather than conventional dams, especially after being carefully grassed, and planted with trees.

Adapting the existing reservoir, Marchlyn Mawr 500 m above the lower lake, would require the construction of another rockfill dam about 40 m high and 600 m long. This would be shaped and faced so as to merge with the surrounding large scale scenery.

Water levels in the reservoirs would fluctuate. The visual effects of this fluctuation would be relatively insignificant in the remote upper reservoir but more noticeable in the lower reservoir, lying in a tourist area at the foot of Snowdon. Here changes in water level would be in the order of 14 m, with an early morning water level of about 6 m below that at present.

To reduce the visual impact of this fluctuation existing terraces, on one of which there are some derelict quarry buildings, would be removed and after careful treatment of the lake's margins, to make them appear irregular and varied, the rise and fall in water level would pass almost unnoticed beneath the massive quarry terraces.

Following some initial success with trial plantings at Ffestiniog, the Board is promoting research into the selection and development of species of trees and shrubs that would withstand the rigours of daily submergence under water. This intriguing work, if successful, would offer another possibility for screening the banks of similar pumped storage reservoirs between the higher and lower water levels, and improve the natural appearance.

The two reservoirs would be linked by underground tunnels through the power station itself, housed in a vast cavern excavated beneath the quarry. Power would be transmitted from the station by underground cable and existing overhead line to the Pentir substation about 9 kilometres away. The additional cost of undergrounding would be about £5 million, but would avoid the construction of overhead power lines in the area.

Snowdonia is a considerable tourist centre and on the basis of previous experience the Generating Board confidently expects that Dinorwic—the only underground power station in England and Wales and the largest power scheme of its kind in Europe—would become a tourist attraction.

Facilities for visitors would be provided and minibuses would take them through a tunnel previously used for construction purposes to the turbine hall 1 km deep inside the old quarry. In such ways the Board believes that the scheme would add to the existing attractions of the area rather than diminish them.

6. ELECTRICITY DISTRIBUTION

The constraints on the siting of power stations were discussed in section 2 and it was shown why they can seldom be ideally located near the loads they serve. Furthermore both the growing demand for electricity, and the economy of scale in construction have encouraged the use of larger individual generating units concentrated into fewer sites as time goes on. The number of individual power stations operated by the Generating Board has decreased from 300 in 1948, when the industry was nationalized, to 183 in 1973, while the total output capacity has increased more than five times in the same period. This reduction in numbers has in itself an obvious benefit to the environment, but has also necessitated the development of a major arterial network of high voltage transmission lines to carry electricity away from where it is made, and to deliver it to the main centres of demand. From these centres a further network of successively lower voltage lines distributes the electricity to each town and village; to each street; and ultimately to each home, office and factory served.

6.1. High voltage transmission

There are no energy conversions involved in this part of the electricity supply system but small energy losses occur giving rise to:

(a) Radio and television interference from high voltage lines.

(*b*) Noise from transformers and switchgear.

(*c*) Heat losses, which are only of environmental consequence in the case of large underground cables.

Apart from these, the remaining important impact on the environment created by high voltage electricity distribution is the effect on visual amenity.

6.2. The effect of energy losses

Radio and television interference from high voltage transmission lines is a difficult problem but fortunately a rare one. Interference to the reception of radio signals in the long and medium wavebands is usually due to small electrical discharges at the insulators and conductors. The level of this interference tends to increase in rain and fog, particularly in the presence of pollution. Television interference on the other hand is often due to acts of vandalism—a piece of wire or chain deliberately thrown on to a line conductor—insulators make attractive targets for air guns. The prevention of interference presents technical and economic difficulties and the problem could be made much more difficult if a higher transmission voltage is adopted in the future. Research is in hand both here, and overseas, to find satisfactory solutions.

Noise problems with transmission networks arise mainly from transformer 'hum'; audible vibration related to the 50 Hz supply frequency is generated within the core of the transformer itself. Advances in core design have reduced this hum but it has not been possible to eliminate it altogether. Where the noise might disturb local residents transformers are built within acoustical enclosures designed to absorb and attenuate the noise to the necessary degree. A different noise problem occurs with high voltage switchgear of the air-blast type which can create a loud and abrupt noise on the infrequent occasions when these circuit-breakers are operated. Silencers have been developed and are used wherever a nuisance to residents might otherwise be created. More recent developments in switchgear for extra high voltages will further reduce the risk of these noise problems.

A high voltage underground cable conveys a considerable flow of energy so that only a very small percentage lost as heat can nevertheless create a serious cooling problem. Where cables have been laid in tunnels (as in the former railway tunnel at Woodhead) a system of troughs surrounds the cable and water is sluiced through these to carry off the heat. Similar cables buried underground in the countryside require parallel water conduits to absorb the heat that might otherwise sterilize the ground for other uses for some metres distance. At intervals of about 2 km it is necessary then to build surface installations to pump and cool the water—adding another factor to the problems of putting cables underground.

6.3. Overhead or underground?

The high voltage transmission network of the Generating Board includes 11 000 km of overhead routes and 1 100 km of underground cable operating at potentials of 132 000 to 400 000 V. But Britain is a relatively small and crowded island, so it is almost inevitable that the appearance of overhead transmission lines and the selection of routes for new construction should often be contentious and that demands will be made for the electricity transmission system, or larger portions of it,

to be neatly hidden underground. The current preponderance of overhead lines is however understandable when the cost of underground cables for these high voltages is at least 16 times that of overhead construction. Nevertheless, there is already a greater length of underground high voltage cable in Britain than, for instance, in the whole of the United States.

The relative economy of overhead construction lies at the centre of this visual amenity problem. Whilst there are special cases where the Generating Board agrees that transmission lines should be buried to avoid spoiling a particularly beautiful stretch of countryside, the adoption of underground cables on a wider scale is far more difficult to justify when the cost differential is so great. To place this matter in perspective, a 400 kV underground cable costs over £600 000 per km to lay. A relatively modest proposal to cable, say, a 10 km length of route, instead of building it overhead, would involve an extra cost equivalent to:

(*a*) The total annual expenditure by all the National Park Authorities in England and Wales, plus

(*b*) The total annual expenditure on environmental research at all the Government laboratories in Britain, plus

(*c*) The total contribution over seven years to 'Operation Neptune,' for protecting prized stretches of coastline.

FIG. 16 400 kV oil-filled cable of 1936 mm² copper cross section showing precision lapped construction. This cable is used for underground transmission lines and is 126 mm in overall diameter. (Courtesy: BICC–Callender Cables)

FIG. 17   Excavations in progress for a 400 kV underground cable route at Goring Gap, Berkshire

As in so many other matters associated with environmental protection the material problem is not simply cost but 'cost effectiveness'; the difficulty of deciding whether a particular expenditure is worthwhile when there are so many alternative ways in which the money could be spent and perhaps reap greater benefits to more people in doing so.

One way of solving this dilemma would be a marked reduction in the costs of cables compared to overhead lines. To understand why this is so difficult to achieve,

it is worthwhile considering what is involved in making and laying a large high voltage cable. The central conductor—up to 25 cm² of stranded copper—has to be precision-lapped with over 200 layers of very special paper to a final thickness of 2½ cm then dried in a vacuum and impregnated with oil (Fig. 16). The cable is then enclosed in a seamless aluminium tube, corrugated to make it flexible enough to wind on a 4 m diameter drum for transport, and then finally protected by a tough plastic sheath. By comparison the bare overhead line carrying the same power requires no insulation or protection at all: 3 m of free air does the same job as the 2½ cm of expensive paper insulation in the cable.

These cables must be laid, six or 12 at a time, in a 3 m wide trench cut across the countryside which, even if temporary, is no small intrusion on the landscape (Fig. 17). A large jointing bay has to be constructed about every 300 m since cables of this size cannot be manufactured and transported in greater lengths. Then, at intervals of approximately 2 km a surface structure (about the size of a bungalow) has to be built to house the water pumping and cooling plant. Tunnels or bridges have to be provided wherever the cable route crosses roads and streams.

Research goes on to improve the construction of cables and to reduce the cost. Plastic tape insulation is being studied and also gas-filled cables using sulphur

FIG. 18   Bishops Wood 400 kV substation is visible from the air; but the site, surrounded by natural woodland, is almost entirely screened from view at ground level

hexafloride. Much further off lies the possibility of super conducting cables, cooled with liquid helium, which would exploit the singular property of metals at temperatures near to absolute zero in conducting electricity without any losses. Whether the smaller cables that would result from the successful outcome of this research will prove as cheap to construct and lay as an alternative overhead line, is very much a matter for the future. A comparable objective would be to design and produce a good quality saloon car for no more than £100.

6.4. Overhead lines and substations

In the meantime much is being done to reduce the impact of overhead lines on the environment. Careful and considerate choice of routes is the starting point. Consultations are comprehensive and involve not only the statutory planning authorities but often the Countryside Commission, the Nature Conservancy and other bodies concerned with amenity. Every possible measure is taken to fit the lines unobtrusively into the landscape; by avoiding skylines, by detours to avoid spoiling fine views; by taking advantage of broken country as background; and by using the natural screening of woodlands and hedges.

The siting of substations is also given similar consideration. Advantage is taken of the natural contours of the ground and existing woodlands to site substations where they are least visible (Fig. 18) and, if the topography alone is not sufficient for this, landscaping schemes and tree planting are used as well.

The current development of high voltage metal clad switchgear now offers the prospect of a quite remarkable reduction in site area and in the height of structures for future substations (Fig. 19).

6.5. Distribution at lower voltages

Distribution in England and Wales is carried out by the 12 Area Electricity Boards who take supplies from the CEGB at 132 kV. In Scotland two Electricity Boards are responsible for generation, transmission and distribution in their respective territories. Technically their arrangements are similar. Networks at 33 kV or higher voltage feed substations at which transformation to 11 kV takes place. An 11 kV local distribution network supplies some industrial customers directly: the majority of customers are, however, supplied at 415/240 volts, so that a further stage of transformation is required. The lengths of mains at different voltages are shown in Table 2.

It will be seen that more than half the distribution mains at 11 kV are overhead lines. The cost differential between high voltage cables and overhead lines is unlikely to be eliminated. In addition to the cost of trench work the cost of the insulation and impervious sheath that

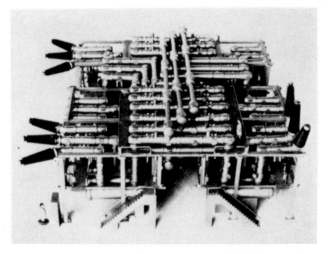

FIG. 19 Model of 300 kV metal-clad switchgear being constructed for Neepsend Substation. (Courtesy: Reyrolle-Parsons)

surrounds the cable conductors greatly outweighs overhead line costs, even when allowance is made for additional maintenance costs of overhead lines and their supporting structures. Although the higher cost of underground cables, as compared with overhead lines, is less than for the higher voltage transmission lines the difference is important because of the difficult economics of the electrification of rural areas. A major effort, which was possible only because of the use of overhead distribution circuits, has resulted in the availability of electricity in every village and hamlet. Since 1948 the proportion of farms connected has risen from 33% of the total to 97%.

However, Electricity Boards are very conscious of their responsibilities in acting to preserve the amenity of the countryside and are continually faced with reconciling these with the need to provide supplies by the most economical means. The 11 kV overhead lines are normally carried on wood poles and the three conductors are arranged horizontally giving minimum height. They are less obtrusive than higher voltage lines and great care is taken in routing the lines to avoid high ground, long runs near roads and locations generally detrimental to amenity. Not only are routes carefully chosen but selected parts of these routes are sometimes placed underground at additional expense. Situations where a number of overhead lines converge on a major substation can be improved by these methods.

As shown in Table 2 the major part of the low voltage distribution system which is predominantly close to dwellings and new roadways is underground.

The distribution Boards maintain close liaison with County planning authorities and with local authorities on the effect of their lines. The greatest benefit to amenity can usually be achieved by removing old medium voltage overhead lines in village communities and all Boards have such programmes of amenity improvement. About £1 million per annum is spent by Boards, split between undergrounding existing lines that do not meet current amenity standards, and using cable where new overhead lines would normally be used but would be deleterious to visual amenity. Typically the selection of schemes that can be carried out within the available budget is undertaken in collaboration with County planning authorities who, in turn, consult with local authorities in deciding

TABLE 2 Circuit kilometres of distribution mains in service

Voltage	Overhead lines	Underground cables
Over 11 kV	46 000	18 000
11 kV	127 000	97 000
415/240 kV	67 000	205 000

FIG. 20   *East Street, Littlehampton, before and after removal of overhead distribution lines and replacement of street lights*

priorities from a list of costed schemes suggested by the Electricity Board, for the area concerned.

No subsidies are available for this work which is a charge on electricity customers generally, and progress in removing old lines is gradual; but taken with the attention to new works there is a substantial effort. Of course, electricity lines in villages are only one aspect of the amenity problem, and for many of the schemes undertaken there is co-ordination with the Post Office and the local authority who may themselves be involved in replacing street lighting attached to the poles. An example of the progress and improvement being achieved by removal of overhead lines is shown in Fig. 20.

Amenity is not confined to the countryside. In urban and city areas a large number of distribution and some major substations have to be provided to house the switchgear and transformers essential to distribution and these can often best be accommodated in existing buildings; otherwise they are constructed to be unobtrusive in the particular surroundings they occupy by screening and landscaping. Modern high density development requires substations on the premises. For this, and other amenity reasons, efforts have been directed to minimizing the noise emission by transformers and providing very compact designs of switchgear. Major substations in urban areas require careful architectural treatment and are provided with noise enclosures of special material for the transformers with the cooling system outside for better air circulation (Fig. 21).

As a further contribution to improvement of the home

environment, encouragement is being given in the design of new houses for meters to be sited so that they may be read from outside, so avoiding the need for the meter reader to gain access to the house. This has a dual benefit in that regular meter reading can be obtained even though the householder and his wife may be out all day, whilst the need is obviated to set apart house space for siting meters and providing access to them. For the future, investigations are being made into the possibilities of taking meter readings by remote control.

7. ELECTRICITY AND THE USER

At various stages in the production and distribution of electricity, the risk of polltuion and damage to amenity arises. Serious and sustained effort is required to limit such effects to the minimum, and research is a continuing process to seek solutions to the problems that remain. This said, we should be clear on one fundamental matter. Any remaining harmful effects on the environment resulting from the production and distribution of electricity are far outweighed by the environmental benefits that result from its application and utilization.

At its point of use electricity is much more than a raw fuel, it is a premium source of energy, versatile, controllable and utterly clean; and not only is it clean, but it creates no need for flues, combustion air and fuel storage so that architects have far greater freedom in which to design. They are not inhibited by site considerations that must have regard to exhaust fumes, residue disposal or storage.

In England and Wales there are about 19 million electricity customers. Of these, 17 million are householders whose specific energy needs are for environmental change in the home and for housepower to take the drudgery out of housework. In the typical house, these needs can be met to present day levels of comfort by the use of some 15 000 kW hours per year, more than half of which will be used for heating and the remainder for water heating, cooking, lighting and power. Apart from its cleanness at point of use, electricity has many other attributes in the home. For example, some forms of electric central heating such as floorwarming and ceiling heating exploit electricity's unique feature of invisibility, and high thermal insulation and use of fortuitous heat gains mean that electricity is increasingly becoming competitive in the market place on a price basis.

It is not proposed to survey the entire field in which

FIG. 21   *Ladywood 132 kV Substation in Birmingham. A large urban substation cannot be hidden, but can be made visually attractive*

FIG. 22 *London County Hall floodlit*

FIG. 23 *Multi-level roadway interchange at Derby, illuminated by mercury discharge lamps*

electricity meets customers' needs, but to highlight three areas. These are areas where electricity plays a unique part, and where its increased use will not only satisfy customers' needs but will make an important contribution to good environmental conditions.

7.1. Electric lighting

It is almost exactly 100 years ago that light was first produced in a hermetically sealed bottle. Today we take electric lighting so much for granted that people are

FIG. 24 *Avonbank office of the South-Western Electricity Board: the first building constructed to Integrated Environmental Design principles*

rarely aware of the extent to which we all depend upon it. About 80% of our information reaches us visually. Cheap and abundant electric light enables this to continue unimpaired for as many hours as we wish, irrespective of time or season. This century has brought a new art and science—the world of the illuminating engineer and lighting designer—concerned not only with ways of helping us to see, but with how lighting affects us as people, how we feel, how we react, what we prefer (Fig. 22).

The life blood of a nation's economy is its industry; lighting is vital to its factories. Not only is general lighting provided to enable people to work, but it is basic as a production tool. Transparent objects can be inspected when in front of a luminous screen, minute threads can be studied by projection, reflection can be used to show imperfections in specular surfaces, special lamps are employed for colour matching, and light is made to change direction within a confined space by means of fibre optics. The laser is a cutting tool as well as a highly accurate means for measuring distances.

Progress continues both in light sources themselves, and in the equipment and techniques designed to exploit their characteristics and advantages. For example, a comparative newcomer to our environment is the large multi-level road interchange. To light each road by the same techniques as used at ground level would result in a forest of columns, dangerous glare and misleading visual patterns. So the Derby interchange, opened last summer, is lit from only 16 slender 30 m high masts, spaced approximately 80 m apart (Fig. 23). Each headframe supports a cluster of four lanterns, each housing a 1 kW colour corrected mercury discharge lamp and the necessary control gear. A huge 4 kW packet of light high in the air, with no glare to drivers, photo-cell controlled, and aesthetically satisfying.

The invention of the tungsten halogen lamp has made possible a large family of small inconspicuous floodlights. And so we go on, each phase contributing to better utilization of our resources. Our present mode of life would collapse without abundant electric light and, as a passing thought it should be remembered that it is the atmosphere that pollutes electric lighting. Dirt deposited on lighting equipment means an average loss of some 20% of useful light.

7.2. Integrated environmental design

Turning now to a customer need in the commercial market: the need for a completely controlled internal environment in new offices, public buildings and schools. Electricity has come up with an answer to this need which reduces to a minimum the use of energy specifically for heating purposes. This concept is called Integrated Environmental Design (IED). In any new building, a good thermal and visual environment is regarded as necessary to enable the occupier to carry out his tasks efficiently and with the minimum of fatigue. A lighting level of 1 000 lux is recommended. Inevitably the lighting fittings will generate heat and this can be used to warm the building.

In addition, with a good standard of thermal insulation heat is gained from people, office machinery, fans, pumps and computers. The philosophy of IED is to arrange for all people concerned in the design of the building, the client, the architect, the heating engineer, the electrical engineer, the quantity surveyor and others to meet at the outset of the proposal to determine the needs of the building occupants, how these can be translated into terms of the building and how each ingredient can be integrated to best advantage without conflict in the process. The end product will be a building of high architectural merit with a properly considered internal environment at a very competitive cost-in-use. Avonbank, the first IED office building in this country was built in 1971 at Bristol for the South Western Electricity Board (Fig. 24.) It is a spacious, three-storeyed building of 5 200 m² and, although it has a much higher standard of internal environment than its contemporaries, costs were competitive.

Double-glazed vertical vision windows occupy 15% of the outside wall area, about the same as in Georgian architecture. This rationalized fenestration greatly reduces the effects of solar radiation on the internal environment, and the high insulation of the building envelope reduces the effect of external temperature variations. There is no need for an independent heating system. The internal environment is controlled, winter and summer, by a heat recovery air conditioning system. The contribution to the external environment is obvious—with no fuel burned there are no chimney emissions, and the use of otherwise wasted heat helps to conserve energy resources.

IED principles can be applied to almost any type of building including offices, schools, department stores, factories and computer centres. Some 40 buildings are now in use with a further 40 in the design or construction stage.

7.3. Industrial pollution control

In many industrial processes, pollution can be controlled but at a cost. Because of the cost, control equipment is usually applied only to the larger plants, and many fuel-consuming units that are too small to justify the cost of control equipment are left to contribute their pollution products which, though individually of low quantity, can collectively be a major pollution problem.

The application of electrical processes makes a substantial contribution to the reduction of pollution. It is clean at the point of use in that there are no products of combustion. Moreover it offers a variable technology that can, for example, reduce process temperatures by obviating the need for external heating, the heat being generated within the product itself. A particular example is in induction heating for drop-forging. By replacing a multiplicity of fuel heating units with a single automated high speed induction heating furnace in which the heat is induced within the billet, a greatly improved working environment obtains.

In the metal melting field the coreless induction furnace, producing cast iron from scrap, replaces the coke-consuming cupola and obviates the need for grit arrestors, and other gas cleaning plant which are particularly expensive when the use of hot blast is practised. Further, the channel induction furnace applied to holding and super-heating of iron replaces the less efficient fuel-fired rotary furnaces. In the non-ferrous industries producing aluminium, copper, zinc, etc., and their alloys, induction and resistance heated furnaces and crucibles are substitutes for fuel-fired rotary and reverberatory furnaces with the elimination of smoke, fume and grit.

A case in point is the melting of scrap aluminium in fuel-fired rotary furnaces. A particular characteristic of the furnace in this application is that the charge requires a salt cover to minimize oxidation. Salt fume badly affects the atmosphere on the shop floor and corrosion of the roof steelwork is a problem. Forthcoming regulations under the Alkali Act will make it necessary to restrict these emissions to very low levels. An induction furnace that does not require a salt cover eliminates the problem of emission and easily meets the requirements of the Act. The electric melting of glass greatly reduces contamination of the product and grit emissions to atmosphere. The combustion of fuels is further reduced by electric pottery kilns, paper drying by infra-red or dielectric heating, and microwave ovens in the food industry.

This list is not exhaustive. In fact there are no heating processes wherein the energy from electricity has not contributed to the improved environment; whilst outside the heating field, the battery electric industrial truck provides a much healthier environment by eliminating noise and exhaust fumes from confined premises such as warehouses. In the whole field of energy utilization, electricity is unique in being non-polluting at the point of use.

8. WORKING FOR THE ENVIRONMENT

The preceding sections have described the various stages in the generation of electricity and its distribution to the general public. The environmental problems that arise at each stage have also been described, and an account given of the way the electricity supply industry has tackled these problems. However the concern of the industry in these matters is not solely that of avoiding pollution and minimizing impacts on visual amenity; there are other more positive contributions that are made to the benefit of the environment.

8.1. Environmental research and its application

The dual statutory responsibility uniquely placed on the electricity industry has led to organizational policies concerning the environment which may differ to some extent from those in other industries. The need to achieve a high technical and economic performance from power station and transmission facilities is, of course, a matter of importance to all staff at all levels in the industry. The requirement to design and operate plant with a proper concern for environmental effects is equally the general concern of all staff and is not a matter which can be segregated into a special department, or made the responsibility of a single senior officer. Both technical and environmental factors are essential elements in the day-to-day performance of the Electricity Board's duties.

These twofold duties required the electricity industry to develop an internal capability for securing a flow of authoritative information on environmental matters of the same standards as its technical knowledge of electricity production. Such knowledge begins with research. Teams of scientists have been built up over the years to investigate every aspect of power station emissions and to develop a means for controlling them to the required standards. In addition to three central Research Laboratories, each of which undertakes research on one aspect or another of environmental interest, the research effort extends to five regional Scientific Services Laboratories where many studies of national importance have taken place; for example, field surveys of air pollutants around large power stations, ecological surveys of estuaries and

FIG. 25   *Former gravel pit near Hams Hall Power Station, filled with PFA, then top-soiled, grassed, and returned to agriculture*

rivers, the restoration of ash dumps to agriculture. The Generating Board's research in the environmental fields is now of high international repute and, in many aspects, leads the world in scientific knowledge of the behaviour of potential pollutants and their control. This standing is reflected in the interest shown in the Board's environmental work which results in a steady flow of national and international visitors each year who wish to see what we are doing at first hand, and to seek our advice on their own similar problems. Research staff and other environmental specialists employed by the Generating Board are frequently sought to undertake formal consultancies for organizations in all parts of the world—by Governments and international agencies as well as by foreign electricity undertakings.

To assist in the management of environmental research projects, an internal Steering Committee has been in existence since 1961 ensuring full discussion of the needs for new research projects between the Research Department and the 'user' Departments of the Board. It also helps to assess priorities in allocating the available research effort to the best effect. The Board is also assisted by Advisory Panels on air and water pollution, including eminent external specialists who advise on the scientific content of the Board's research programme, and on the interpretation of the results.

In addition to the research effort, a specialist group has been maintained for more than 15 years solely to consider the environmental impact of new power stations from the early planning stages, and to recommend the control measures to be adopted. It is also part of their task to look ahead as far as possible to identify environmental problems, and to take steps to ensure that adequate research and development programmes are in hand to provide any new control techniques required.

This scale of research and engineering development has enabled the industry to make original and important contributions to scientific knowledge in many fields of pollution control that are of benefit not only to the electricity industry itself, but to many other organizations as well.

8.2. The utilization of pulverized fuel ash

Another positive contribution to environmental improvement (perhaps a little unexpected) has been achieved in

the disposal of ash from coal-fired power station boilers. The scale of the Generating Board's coal consumption, and the necessity to operate high efficiency dust arrestor plant has led to an increasingly large tonnage of ash collected each year for which means have to be found for disposal.

Ash production has risen to over 11 million tonnes per annum, of which 85% is in the form of fine powder as a result of firing the coal in a pulverized form. The remainder is coarse clinker or furnace bottom ash, for which there has always been a steady commercial market, and disposal of this fraction of the ash has rarely presented a problem. The fine pulverized fuel ash (PFA) had however no obvious uses when it first started to be produced in large quantities some 20 years ago, and its disposal was a liability on the Generating Board.

When the quantities produced were still fairly small there was usually no difficulty in finding disused quarries or gravel pits in the near vicinity of a power station that could be filled with PFA. Frequently this resulted in a net gain to the environment, since the area, when filling was completed, would be soiled and grassed and returned to agricultural or other uses (Fig. 25). In other cases low lying marginal land was raised and drained by a fill of PFA and its agricultural quality improved.

It was foreseen, however, that the rapid increase in PFA production would soon outstrip local disposal facilities and different outlets for this material were sought. In the early 1950s the Board sponsored two research programmes at the Building Research Station on the manufacture of bricks and lightweight aggregate from PFA. In the same period, production of aerated concrete blocks using PFA as an aggregate was begun by commercial firms. Within a few years it was established that there was a growing market for these new materials and an ash marketing organization was set up within the Generating Board to sponsor these developments. By the end of the decade a commercial plant for the manufacture of lightweight concrete aggregate by sintering PFA had been established at Northfleet. At the same time the use of PFA as a load-bearing fill was being developed.

By 1961 the potentialities of PFA as a versatile civil engineering and construction material were becoming well established. As a fill for heavy construction works such as motorway embankments, PFA offers consistent and predictable mechanical properties that have led to a spectacular increase in demand rising to over $4\frac{1}{2}$ million tonnes in 1970/71. In the same year almost a further

FIG. 26   *PFA being used in the construction of the M5 Motorway in Somerset*

2 million tonnes were absorbed in other uses: the building blocks and lightweight aggregates already mentioned, together with a number of more specialized uses in concrete, in grouting, in soil stabilization, in fired clay products, and as an industrial filler. Together these practical uses for PFA now account for 60% of the total production; demand already exceeds supply in some parts of the country, and only the logistics of its transportation from other areas hamper a further increase in the proportion used. Even so, the constructors of the M5 motorway in Somerset found it worthwhile to transport PFA over 80 miles by rail from South Wales to provide 'fill' on a particularly difficult section of the route (Fig. 26).

On a tonnage basis PFA utilization represents a totally new industry, already one third the size of the cement industry. A complete range of technical literature on PFA, a textbook on its utilization and a series of films showing its diverse uses have been produced. A former waste material has, by determined effort, been transformed into a valued by-product of electricity generation with unique properties and a wide range of uses.

The remaining 40% of PFA production has to be disposed of in other ways. Most of it goes to two special disposal areas that also represent a contribution to environmental betterment. At Peterborough, where clay has been dug for brickmaking since the beginning of the century, over 400 hectares of disused clay pits are being filled with PFA which has been transported from power stations in the East Midlands in specially designed rail wagons. As the pits are filled they will be grassed and trees will be planted under the guidance of an eminent landscape architect. One of the older flooded pits with its adventitious wildlife population is to be retained as a lake and incorporated into the final scheme. By these means a large area of land made derelict by industry is to be restored to beneficial use and pleasing appearance.

The other major disposal scheme is at Gale Common in Yorkshire where PFA from several large power stations is being used to construct a 'hill' on land previously destined to become waterlogged and useless as a result of mining subsidence. Again, expert guidance from landscape architects is being used to create a pleasing artifact which will be a far cry from the usual industrial spoilheap. The hill will be contoured and graded to a natural shape and will reach an ultimate height of 50 m above the surrounding land. The surface will be topsoiled and grassed and research has been underway for many years to develop species of trees and shrubs that can be grown successfully on the PFA substrate. The objective once more is to go well beyond the mere disposal of a waste material and in doing so to create a positive contribution to the local landscape, (Fig. 27).

Nature trails

Although the Generating Board's land holding is modest in relation to its productivity inevitably it includes areas that are not used operationally. Such areas include woodland screens at power stations and major sub-stations planted and maintained as part of landscaping schemes to preserve visual amenity. They also include riverside land that cannot be developed because it forms part of flood-plains designated by the river authorities.

In the past six years, in association with many outside interests including education authorities, local naturalist societies and ornithologists, the Board has sponsored a variety of conservation schemes to exploit the environmental potential of such land. These include nature trails, field study centres and nature reserves.

The pioneer venture of this kind was launched at Drakelow power station near Burton on Trent in 1967. It

FIG. 27 An impression of the Gale Common Ash Disposal Scheme: a 'hill', constructed of PFA, top soiled, grassed and landscaped

FIG. 28 Students working near Buildwas Abbey ruins, on the 'Nature Trail' at Ironbridge Power Station

was, we were assured by the Nature Conservancy, whose support and guidance was of great encouragement to us, the first nature trail to be launched on an operational industrial site anywhere in Europe. Since then it has been widely used by schools from all over the Midlands. Altogether to date 12 000 children have taken advantage of this open-air classroom for the kind of practical field work which is so valued by educationists today, and for which there is a dearth of suitable sites, particularly near large cities.

That trail has been visited by conservationists and educationists from many countries, and has been followed by many more at other power stations and transmission substations up and down the country (Fig. 28).

At Drakelow, an area of derelict gravel workings has been adapted to form a wildfowl reserve that ornithologists say is one of the finest in that part of the country for many varieties of waterfowl. Similar reserves have been launched at other sites.

The cost of these activities to the Board is very small indeed and infinitesimal in comparison with their positive benefit in creating an understanding of the environment. They demonstrate the important truth that with care and concern the natural world and industry on a giant scale can peacefully co-exist side-by-side. They also associate the Board in constructive and productive partnership with amenity groups who, in other contexts, may view some of the Board's development plans with apprehension. The better understanding of each other's point of view that results from such joint enterprises is a help to the Board in formulating its proposals with environmental interests in mind, and to the conservationists in appreciating the demands of the technology of our industry.

9. THE TASK AHEAD

So far we have been concerned with the past and with the present. It is now time to look ahead to identify the tasks that will need to be undertaken if we are to improve still further the management of environmental quality in this country.

At the present time it is particularly difficult to make prognostications since so many factors in the energy supply sector are undergoing rapid change. Estimates of the future availability and above all, the cost of different fuels, are subject to frequent adjustment, not least because of the newly-found aspirations and concerted actions of the oil-producing nations in the Middle East

(to say nothing of the aspirations of other fuel producers much nearer home).

Coupled with this, the effects of joining the European Communities and of integration within the Community energy and environmental policies, still have to emerge. At the technological level there are uncertainties over the next stage of nuclear development, and various options are still open.

Some developments become more clear in the longer term view. There can be no doubt that the future lies with nuclear power, if only because of uncertainty that fossil fuel supplies can always be expanded to meet the ever-increasing public demand for energy. It also seems likely that electricity's share of the total energy market will continue to increase, although possibly at a somewhat lower rate than has been the case in the past. Both these long-term developments would be to the overall benefit of the environment. For the present, and for the fore-seeable future, nuclear power can be used in quantity only for the generation of electricity, where it will replace the use of fossil fuel, conserving the limited supplies for other uses. Whilst this paper has shown that the environmental problems arising from fossil fuel combustion can successfully be solved at a cost, these particular problems are eliminated altogether with nuclear power. Furthermore, expansion of electricity's share of the market means less raw fuel burned in small individual installations where the pollution potential is so much greater than when the fuel is burnt in large well-equipped boilers.

However, within the more limited timescale which is the concern of this Conference, the immediate uncertainties still loom large. The task ahead for those concerned with protection of the environment, and in particular with the impact of the fuel industries, may be summarized as follows:

(a) To keep abreast of the social and political changes taking place both here and throughout the world, and to assess how far these will affect future demands for energy and in what form it is likely to be required.

(b) To anticipate how far energy supplies will be influenced by current social demands for conservation and an improved environment.

(c) To continue to keep before the public up-to-date information on the probable cost, in money and resources, of seeking to establish and maintain a better environment for all to enjoy.

(d) To encourage solution of environmental problems on the basis of knowledge rather than fears.

(e) To disseminate generously our own accumulation of data and experience to other countries so that all may move forward together towards controlled and progressive solutions of environmental problems; avoiding thereby the panic measures and ill-conceived legislation that tend to emerge under the influence of emotion rather than reason.

(f) To seek always to do a little better tomorrow than we do today.

The electricity industry's concern for the environment will remain sufficiently broadly based to cater for changing circumstances and needs. Its past record shows that successes have been achieved that stand comparison with those of other organizations, and in many cases exceed them. Assisted by far-sighted legislation, the electricity industry—at all staff levels—had integrated environmental considerations into its planning and operational policies long before they became a popular concern. This anticipation of needs, and the timely allocation of resources to tackle them, has proved the key to the successes achieved.

10. ACKNOWLEDGEMENT

The progress of the electricity supply industry towards preserving and enhancing the environment, described in this paper, is a product of the labours of all the industry's staff, past and present. Among these there are a lesser number, still too numerous to mention by name, who have made some part of this great work their speciality and who have devoted their professional careers to the protection of the environment without seeking individual recognition. The author is most happy to have this opportunity of acknowledging the efforts of these members of the industry's staff, without which there would have been so little worthwhile to record for this Conference.

SIR JOHN HILL*

6   The nuclear fuel industry

SUMMARY

The use of atomic power as a means of producing electrical energy is now very well established, and the number of nuclear reactors is increasing rapidly. A supporting nuclear fuel industry has evolved which is capable of providing a whole range of services from conversion of imported uranium ore concentrate, through to transport and reprocessing of irradiated fuel elements. These operations are outlined, and a brief description is given of the more important reactor systems. Statistics are presented which demonstrate the economic and logistical advantages associated with nuclear energy.

The basic standards, which have been laid down by international bodies, setting maximum limits to the radiation dose to which the population can be exposed, are discussed; as is the legislation which has been enacted to control radioactive discharges from major nuclear sites in the UK, and to provide for the licensing of nuclear installations. Authorization to dispose of waste is issued jointly by the Department of the Environment and the Ministry of Agriculture, Fisheries and Food, in consultation with local authorities, and programmes of sampling and analysis are required. The Nuclear Installations Inspectorate was created to control, on behalf of the Secretary of State, the issue of licences for construction and operation of nuclear installations. The National Radiological Protection Board was created in 1970 with the responsibility of advising persons (including Government Departments) in relation to radiological protection, implementing research and training, and of representing UK interests in radiological protection internationally.

The operation of the magnox-uranium stations of the Electricity Generating Boards and the UKAEA reactors has taken place in the context of the controls referred to without significant interaction with the environment. New reactor systems, when introduced, are confidently expected to maintain this standard.

The operations of nuclear fuel reprocessing plants are described, and the means of controlling discharge of low-activity waste to sea, and of monitoring any possible impact on the environment, are discussed. Highly active liquid wastes are at present stored in high-integrity tanks and proposals to turn this waste into a solid form are outlined. Solid radioactive wastes have been disposed of at sea, and the precautions exercised are described.

Transport of radioactive materials is subject to regulatory standards laid down by the International Atomic Energy Agency and there is considerable experience in the UKAEA/BNFL of transporting these materials safely.

Finally, the UKAEA research and development programme is detailed. This is aimed at ensuring that the present very high standards of radioactive waste treatment, storage, etc., are more than maintained in the future.

*Chairman, United Kingdom Atomic Energy Authority
  Chairman, British Nuclear Fuels Ltd.

1. ATOMIC POWER AND THE NUCLEAR FUEL COMPLEX

Introduction

1.1. The term 'nuclear fuel cycle' refers to the processing of uranium from the mining and the milling stages, through conversion and fabrication, to the manufacture of elements for the core of a nuclear reactor and it includes irradiated fuel transport and reprocessing, the recovery and use of fissile uranium remaining in the fuel, and the plutonium formed during irradiation. A flow sheet for a typical fuel cycle is shown in Fig. 1, whilst Fig. 2 shows the location in the UK of the nuclear fuel plants and power stations.

1.2. Nuclear fuel cycle costs are about one-fifth to one-quarter of the total cost of generating power in a nuclear power station, or roughly one-tenth of the present cost

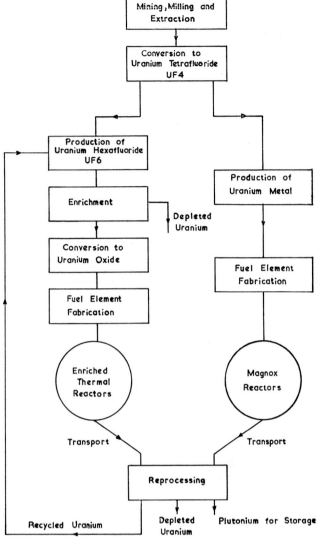

FIG. 1 Nuclear fuel cycle

is concentrated by a chemical process. The ore is first crushed and ground in wet rod and pebble mills to produce an aqueous slurry. Uranium is leached out of the ore pulp with acid or alkali, depending on its nature, and the uranium solution is separated from the pulp by filtration, or settling, and extracted from the clear liquor — either by an iron exchange technique, or by solvents. The uranium is stripped out of the resin, or solvent, and is precipitated and collected on a rotary vacuum filter, dried in a kiln and calcined in a furnace to form uranium ore concentrates which contain about 80 % uranium. Fig. 3 shows a typical flow sheet for the processing of uranium ore.

Conversion
1.5. Uranium for reactor use must be especially free from elements having appreciable neutron capture cross-section; and a purification process therefore follows concentration. The uranium ore concentrates are dissolved in nitric acid, insoluble impurities are filtered off and the filtrate passes into a purification plant in which separation from dissolved impurities is effected.

(a) To metal
1.6 Metallic natural uranium is obtained by decomposing concentrated uranyl nitrate liquid to uranium trioxide, and reacting this with hydrofluoric acid gas to form the tetrafluoride. This is subsequently reduced by magnesium to produce pure uranium billets which are remelted under vacuum, alloyed and cast into rods which, after heat-treatment, are machined to size.

of a unit of electricity in private consumption. The approximate contribution of each stage of the fuel cycle to the fuel cycle cost is shown in Table 1:

Mining, milling and extraction
1.3. The mining methods employed to extract uranium ore depend upon the characteristics of the ore deposit. Generally, deposits lying less than about 400 ft below the surface are mined by open pit methods. Underground mining is adopted when the ore deposit lies at some considerable depth and may be carried out using conventional techniques.

1.4. Most uranium ores contain between 2 to 6 lb of U_3O_8 per ton, i.e. roughly 0·1 to 0·3 % of uranium, which

TABLE 1 Fuel cycle stages and approx. contribution to the fuel cycle costs: (The figures refer to a typical enriched thermal reactor).

	%
Mining + milling (preparation of UOC)	40
Conversion to uranium hexafluoride and enrichment	40
Oxide conversion and fuel element fabrication	33
Transport and reprocessing of irradiated fuel	6
Credit attributable to plutonium and uranium	−19
	100

- ● MAGNOX POWER STATIONS OF THE CEGB AND SSEB
- ◪ AGR POWER STATIONS OF THE CEGB AND SSEB
- ▼ SITES OF BRITISH NUCLEAR FUELS LTD

FIG. 2 Location of nuclear fuel plants and power stations in the UK

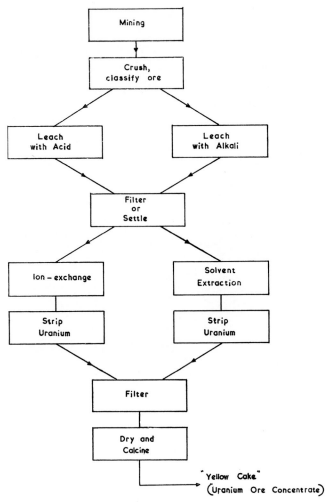

FIG. 3   Flow sheet for uranium ore processing

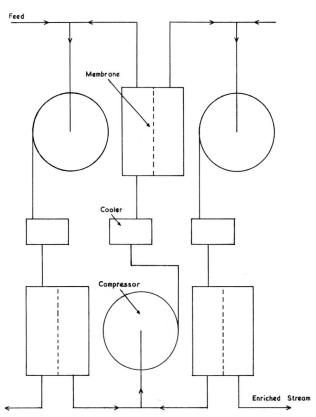

FIG. 4   Diffusion plant stages

so arranged that the slightly enriched fraction from one stage is fed to the next stage above in the cascade, whilst the slightly depleted fraction is passed to the stage below. A typical arrangement of diffusion plant stages is shown in Fig. 4. In addition to a product enriched in U235, a gaseous diffusion plant also produces 'waste' which is depleted in U235, typically, 0·25%–0·30% U235 which is stored for subsequent use, probably in fast reactors.

1.10. Currently, uranium is enriched mainly by the diffusion process, but a centrifugal method is being developed. It is anticipated that the cost of enriching

(b) To enriched hexafluoride

1.7. The fuels required for most thermal reactor types, e.g. the advanced gas-cooled reactor, the boiling water and the pressurized water systems, the SGHWR and the high temperature gas-cooled reactor, are enriched in the fissile U235 isotope above the level of 0·71% found in natural uranium.

Enrichment

1.8. The technology and industry of enrichment resulted from military needs and currently production is by the gaseous diffusion method. An alternative process under development, now at the prototype plant stage, is the gas centrifuge. Uranium hexaflouride, UF_6, is the most convenient uranium compound with which to carry out the enrichment process, since it sublimes at temperatures above 56·4°C at atmospheric pressure. This compound is made by reacting uranium tetra-fluoride with fluorine gas to form the hexafluoride.

1.9. The diffusion process depends upon the fact that when a gas passes through a tube or pore of diameter smaller than the mean free path of the molecules, the flow is inversely proportional to the square root of the molecular weight of the gas. If uranium hexafluoride is passed across such a porous material, or membrane, the gas passing through will be about 1·0043 times richer in U235 than that on a high pressure side of the membrane. To achieve useful enrichment levels, typically 1·5%-3·5% U235 for enriched thermal reactor use, the process must be repeated many times in a cascade of separation stages

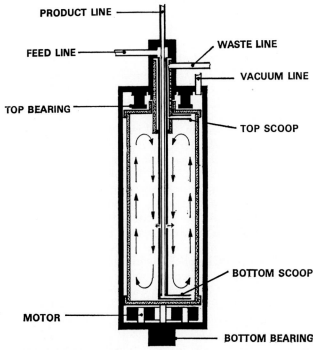

FIG. 5   A gas centrifuge

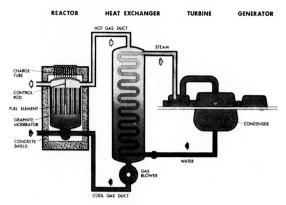

FIG. 6   Representation of a magnox reactor

uranium by the centrifuge process will be less than that of the gaseous diffusion process. Fig. 5 shows, diagrammatically, a section of a gas centrifuge. For useful output and enrichment levels they have to be operated in cascades, in the same manner as diffusion plant stages.

Fuel fabrication

(a) Oxide pellet fabrication

1.11. The greater part of enriched uranium fuel is in the form of uranium dioxide, which has the properties of high-temperature stability, good resistance to irradiation, ability to retain a large proportion of fission gases, and chemical inertness to hot water. Enriched uranium dioxide is produced at a single unit of plant in which uranium hexafluoride is successively reacted with steam to form uranyl fluoride, then reduced with steam and hydrogen to form uranium dioxide. The size and shape of the particles in the powder have a considerable influence on the properties of the final material. In the UK, the process stages comprise blending and micronizing of uranium dioxide powder, spray-drying of a slurry of the powder and binder to produce free-flowing granules which are fed to hydraulic pelleting presses. The pellets produced are continuously debonded, at approximately 800°C, sintered in hydrogen at about 1 650°C, diameter ground, washed and dried and then inspected to ensure compliance with specifications for dimensions and freedom from defects.

(b) Fuel element fabrication

1.12. Nuclear fuel must be contained in suitable cladding material for several reasons, including retention of fission products, structural support of the fuel during irradiation, protection of the fuel from the coolant, facilitation of insertion of fuel into, and removal out of the core, and accurate location of the fuel within the core. Consequently, fuel cladding materials must have good mechanical properties at working temperatures, and under prolonged irradiation, low neutron capture cross-sections, good thermal conductivity, chemical compatibility with fuel and coolant, and be amenable to production at a price compatible with good fuel cycle economics. For magnox type reactors, a magnesium-based alloy is used while stainless steel has been chosen for the advanced gas-cooled reactor. Zirconium-based alloys are used as cladding in water-moderated reactors, because these materials have excellent resistance to corrosion by high-temperature water, and have very low neutron capture cross-sections. The process of cladding consists of loading the fuel into tubes of the appropriate

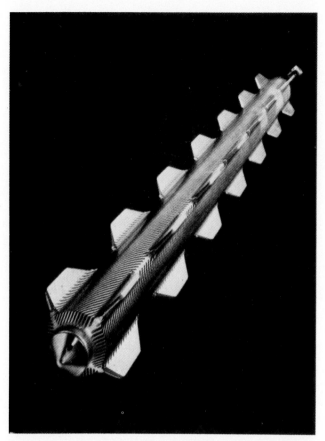

FIG. 7   Magnox fuel element for Sizewell reactor. This element contains a 42 in length of natural uranium bar weighing about 12 kg. It has a heat output of about 3 MW/tonne uranium with a burn-up of about 3 500 MWD/tonne uranium, equivalent to about 150 tons coal. Seven such elements occupy one channel in a reactor

material and dimensions, filling the loaded tubes with helium gas to improve heat transfer between the fuel and the cladding, and sealing the tubes by welding end-caps into place. The tubes are subsequently assembled into fuel elements by locating them, at each end, into support plates thereby forming parallel bundles in which accurate inter-tube gap size is maintained by spacers placed regularly along the length. In the case of magnox fuel elements only one tube is used and the surface is extended, by fins, to improve heat transfer.

Irradiation

(a) Thermal reactors

1.13. The several types of thermal reactor power stations can most conveniently be classified by their moderators. In the UK the graphite-moderated gas-cooled system has been developed and consists basically of a graphite core into which fuel elements are inserted, heat being transferred from the elements to boilers by means of circulating carbon dioxide gas under high pressure. The magnox stations utilize natural uranium metal clad in a magnesium alloy and these are restricted to relatively low operating temperatures. Fig. 6 is a diagrammatic representation of a magnox reactor whilst Fig. 7 shows a fuel element as used in the station at Sizewell.

1.14. The second generation of graphite moderated gas-cooled reactors, the AGRs, uses uranium oxide fuel enriched to about 2% U235, clad in stainless steel, shown in Fig. 8, and is capable of producing steam, comparable in quality to that of modern fossil-fuelled stations permitting advantage to be taken of recent

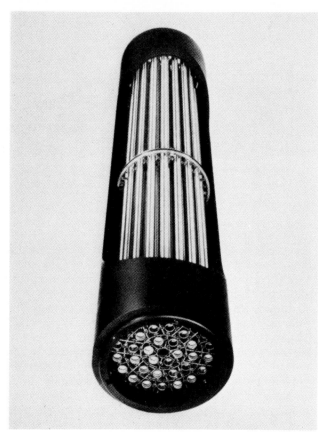

FIG. 8 AGR fuel element (sectioned through graphite sleeve). This element contains about 50 kg of low enriched uranium. It will have a heat output of about 12.5 MW/tonne uranium with a burn-up of about 18 000 MWD/tonne uranium, equivalent to the combustion of about 3 000 tons of coal. Eight such elements, mounted axially form a stringer which fills one channel in the reactor core. Dimensions of element: 41 in long × 7·5 in inside diameter of graphite sleeve. Each element contains 36 fuel rods

FIG. 9 SGHWR fuel element. This element contains about 200 kg of low enriched uranium. It has a heat output of about 14·3 MW/tonne uranium with a burn-up of 18 000 MWD/tonne uranium, equivalent to the consumption of about 12 000 tons of coal. Dimensions of element: 144 in long. Each element contains 36 fuel rods

advances in turbo-generator technology. The third generation of graphite-moderated reactors under development is the low-enrichment U235 Mark 3 system, using fuel consisting of small spheres of UO_2, about 800 μm diameter, coated in graphite and silicon carbide and bonded together in a graphite matrix. Power ratings several times as great as those for AGR can be achieved. Fuel for initial charges is enriched to between $4\frac{1}{2}\%$ and $5\frac{1}{2}\%$ U235, with replacement fuel being enriched to around 5–6% U235. The possible uses of this reactor system in conjunction with gas turbines, and as combined sources of heat and electricity for steel making, are actively being explored.

1.15. The use of heavy water as a moderator has resulted in the development of the SGHWR. The coolant used in the SGHWR is light water, which is allowed to boil. The steam produced drives the turbine generator on direct cycle. The fuel elements and light water are in pressure tubes which traverse the moderator tank, but are physically separate from it. The SGHWR system therefore lends itself well to modular factory-based construction and there is excellent operational experience from the 100 MW(E) station at Winfrith. Fig. 9 shows a fuel element being prepared for despatch from Springfields to the reactor.

1.16. In the USA the light water moderated reactors (LWR) have been extensively developed and two versions are being installed both in the USA and abroad. Both versions use enriched uranium dioxide fuel, typically

2–3% U235, clad in zircaloy, designed to operate at a higher rating than contemporary gas reactor fuel, and to achieve higher burn-ups. The pressurized water version operates with a coolant pressure of around 140 bars, which suppresses boiling, and incorporates intermediate heat exchangers between the primary (water) and secondary (steam) circuits. The boiling water version raises steam in the core at around 70 bars and this is fed direct to the turbines.

Transport and reprocessing

1.17. An important part of the nuclear fuel cycle is the reprocessing of irradiated fuel elements after irradiation and recovery of residual U235 and valuable plutonium which has been produced during irradiation. Typically, the plutonium content of spent fuel from enriched thermal reactor systems is between 5 and 9 kg/te of discharged fuel. Irradiated elements discharged from thermal reactor stations are first allowed to cool by storage under water. During this period the activity of the fuel elements is reduced to a level at which they can safely be moved in shielded transport flasks to the re-processing plant. Here the fuel is removed from the cladding, dissolved, and the uranium and plutonium separated from the fission products by a series of chemical processes. Uranium recovered by reprocessing fuel from enriched uranium reactors has a concentration of U235 which normally makes recycling an economic necessity. That recovered from natural uranium reactors is so

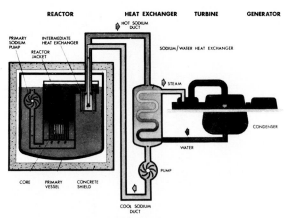

FIG. 11   A power reactor (PFR)

because of its good economics and high uranium utilization compared with other power-generating systems. A fast reactor core can be fuelled with mixed plutonium (natural and depleted) uranium oxides (typically a 20/80 mixture is envisaged for a commercial fast reactor) or with uranium oxide enriched to about 25% in U235. Blanket elements, consisting of uranium depleted in U235 (see paragraph 1.17 above) will surround the core and, by absorbing surplus neutrons in the U238 atoms, will help breed more plutonium than is consumed in the core. The most favoured coolant for the fast reactor is liquid sodium and the prototype reactors currently planned will utilize this medium, but gas-cooled fast reactors are being considered. The British prototype fast reactor, 250 MW(E) output, is nearing operation and incorporates many design features which will be used in the commercial fast reactors of the 1980s. Fig. 10 shows a model of a PFR fuel element assembly, cut away in parts to reveal its structure whilst Fig. 11 is a diagram of the reactor.

FIG. 10   Model of PFR fuel element. This element will be fuelled with approximately 50 kg of mixed plutonium/uranium oxide and will operate at a peak rating of 260 MW/tonne to $7\frac{1}{2}$% peak burn-up of heavy atoms. It will produce the same amount of heat as will the combustion of about 10 000 tons coal. Dimensions of element are: 150 in long x 5·6 in across flats of hexagon

depleted in U235 that, in the UK, it is stored for use in fast reactor blanket and core elements and it has little alternative commercial use.

Plutonium utilization
1.18. The plutonium recovered from thermal reactor fuel is stored against the time when it can be used for fast reactor fuel. The quantities of plutonium which it is envisaged will be available by the late 1970s (before demands for fast reactor fuel build up), are so great that its use for a while in thermal reactor fuel is probable, especially in LWRs in the USA. Plutonium used in this way will result in savings in uranium and in reduced demands for enrichment capacity.

(b) Fast reactors
1.19. Fast reactors have no moderator and utilize neutrons travelling at speeds much in excess of those typical of thermal reactors, hence their name. This system is actively being developed in the UK, France, Germany, Russia and the USA, and it is anticipated that it will be extensively installed once it is fully proven

2. PRODUCTION OF URANIUM FUEL AND THE NEED FOR NUCLEAR POWER
Our uranium fuel factories and their environment
2.1. The eight 2-reactor magnox-uranium fuelled CO_2 gas-cooled stations of the CEGB, the 2-reactor Hunterston A magnox-uranium station of SSEB, and the two 4-reactor magnox-uranium stations of BNFL at Calder Hall and Chapelcross, together feed about 4 000 MW(E), or nearly 10% of our total electricity supply, to the national grid. It can be shown that the annual fuel requirement of all these reactors can be satisfied by importing about 1 140 tonnes of uranium ore concentrate (see paragraph 1.4) each year. The five Mark 2 gas-cooled stations now nearing completion will supply a further 6 000 MW(E) to the national grid from 1976 onwards; because of their higher thermal efficiency (41% as against 30-33% from a magnox reactor) and their higher-burn-up (higher fraction of U235 consumed before the fuel element must be removed from the reactor), this 6 000 MW(E) can be supported by an annual importation of only about 800 tonnes of uranium ore concentrate. Thus, without going into further detail, we can say that 10 000 MW(E) or nearly 25% of our present-day electricity requirements can be met from 1976 onwards by the importation of about 1 900 tonnes of uranium ore concentrate annually. This material is processed at the Springfields Works of British Nuclear Fuels. (BNFL).
2.2 To date, many thousands of tonnes of uranium

ore concentrates have been imported and transported to Springfields Works, Lancashire, in 40-gallon mild steel drums of a type specified in transport regulations for radioactive materials of low specific activity. At Springfields the processes of purification, conversion to metal, metal fuel fabrication, manufacture of uranium hexafluoride for later enrichment at Capenhurst, uranium dioxide pellet manufacture and fabrication into oxide fuel elements (see paragraphs 1.5, 1.6, 1.7, 1.8, 1.11, 1.12) have been carried on for many years now without causing any environmental problems whatsoever. The possibility guarded against is that uranium, chemically toxic like many other heavy metals (the radiotoxicity of natural uranium is of secondary importance), might contaminate pastures near the works and affect the health of cows (it is not transferred to cows' milk). Checks made regularly verify that samples generally do not exceed a few percent of the limit imposed by Government officials. A Government authorization (the nature of these will be referred to again) permits Springfields Works to discharge up to 90 curies of alpha emitters and up to 3 000 curies of beta activity over any period of three consecutive calendar months, as a dilute aqueous solution by pipeline to tidal waters. As a matter of general principle (see paragraph 3.6), discharges are held as low as is practicable (i.e. currently in the region of 5% to 8% of the Government-authorized amount). The radiation levels measured, even near the pipeline end, are found to be negligibly small. Uranium in the form of hexafluoride is enriched in U235 at Capenhurst Works in Cheshire (paragraphs 1.8, 1.9, 1.10). This part of the fuel cycle has even less influence on the environment than the work at Springfields. The main requirement is to limit the average daily concentration of uranium in waste water discharged to Rivacre Brook nearby, and the measured values lie within 1% to 2% of the Authorized limit. Solid wastes of very low activity from Springfields and Capenhurst are buried at a site at Ulnes Walton, Lancashire, again under the terms of an Authorization; the maximum uranium concentration permitted is 0·5 per cent and, in practice, the authorized limit is not fully used.

Some comparisons with other fuels and other metals

2.3 About 10 000 MW(E) of nuclear capacity will be operating in the UK by 1976. This could be serviced by the annual consumption of the following quantities of fuel in power stations, operating at 75% load factor:

(*a*) 1 500 tonnes uranium refined from 1 900 tonnes of uranium ore concentrate; or

(*b*) 30 million tonnes of coal; or

(*c*) 15 million tonnes of oil; or

(*d*) 13 million tonnes of natural gas (600 000 million cubic feet)

The proportions of each to use in any country's economy is for technologists, economists and politicians to decide. But the nation's bargaining position, and hence its balance of payments, are obviously improved if this wide choice is available. The CEGB acknowledges that the cost of nuclear electricity is now lower than the cost of electricity from coal and oil. Our ability to utilize uranium promises to improve in the next decade from the operation of the AGRs, and in the decade after that will improve dramatically through development of the fast breeder reactor. This, through the medium of

plutonium 239, will permit utilization of U238 which comprises 99·3% of natural uranium purchased for thermal reactors, and which is available in quantity as waste from diffusion plants and depleted uranium. The economics of power production will also improve. This is a prize which all technically advanced nations are pursuing: a reduction in electricity generating cost coupled with greatly reduced call on reserves of fissile material, the scale of which will only become apparent after much research and development on fast reactors and their fuel cycles.

3. PRINCIPLES OF CONTROL OF NUCLEAR PLANT OPERATIONS IN RELATION TO THE ENVIRONMENT

Introduction

3.1. Interactions between industry and the environment need not, indeed cannot, be reduced to an absolute zero regardless of cost. But for the nuclear industry, such interaction can at reasonable cost be reduced to insignificance.

The natural background radiation

3.2. Natural background radiation in this country varies from about 0·10 to 0·15 rads/year depending on location. 0·023 rads/year comes from naturally radioactive isotopes present in our own bodies, mainly from potassium 40, a naturally occurring gamma ray emitting isotope created by nature and present in all potassium. Potassium is essential in the chemistry of life and without it we should die. About 0·030 rads/year is due to cosmic rays. The rest of the natural background radiation is gamma radiation from the ground we stand on and the materials of the houses we live in, and varies from place to place. Out of doors in Edinburgh, the gamma ray dose-rates vary from about 0·035 rads/year to 0·075 rads/year, while in Aberdeen the mean dose-rates for half-mile wide annular zones increase steadily from 0·075 rads/year at the periphery of the city ($2\frac{1}{2}$–3 miles) to 0·115 rads/year in the central zone. Typical figures for London are 0·020–0·030 rads/year. An interesting statistic just published about the Livermore area in California is that, in addition to lateral variations of dose-rate similar in size to the above, there are the usual vertical variations, and the dose-rate in downstairs bedrooms is 0·003 rads/year greater than in upstairs bedrooms. Natural background radiation at 5 000 m above sea level is double that at sea level. Because of the preponderant and increasing cosmic ray component, the natural background increases considerably at greater heights than these. With normal solar conditions in the temperate zone of the earth it is 2·3 rads/year at 11 000 m and 7·6 rads/year at 20 000 m above sea level. It has been estimated that Americans on average receive 0·001 rad/year through air travel. This is the dose received during one trans-Atlantic flight.

Background levels of radiation and radioactivity from world-wide weapons fall-out

3.3. Not all radioactivity found in the environment results from the operation of nuclear establishments; some occurs naturally, some as a result of world-wide fall-out from distant nuclear explosions past and present. It is usual before a nuclear establishment becomes fully operational to carry out an initial district survey of background radiation and radioactivity, both to establish what this background is and to give the monitoring team practice and familiarity with their district. Sometimes radioactivity from these other sources can be distinguished by its physical or chemical state but often this is not possible. In that event, distinctions can be made through comparisons with the results of countrywide fall-out surveys carried out and published periodically by the Letcombe Laboratory of the Agricultural Research Council.

Doses from medical diagnostic radiology
3.4. Radiation doses and dose limits (see paragraph 3.5 which follows) can be seen in perspective if we compare them with typical doses to patients from X-ray diagnostic examinations. Radiography of the chest delivers a dose of 0·2 rad to the patient, but many other procedures, e.g. abdominal examinations, deliver 1·4 rads, or more, to the patient. It has been estimated that the average annual genetically effective dose to the UK population as a whole from diagnostic radiology is about 0·014 rad per person, although with improved techniques all round this could be reduced to about 0·002 rad per person. The annual average genetic dose to the UK population resulting from radioactive waste disposal has been estimated to be of the order of 0·00001 rad per person (see paragraph 3.6).

**The International Commission on
Radiological Protection (ICRP)
Dose Limits**
3.5 Some recollection of the figures given in paragraphs 3.2 and 3.4 is necessary, if the dose limits promulgated by the International Commission on Radiological Protection (ICRP) for individual members of the public are to be seen in perspective. These dose limits are listed in Table 2. These dose limits refer to radiation exposure additional to normal levels of natural background (except that they do not apply to radiation exposures given for medical reasons).

The Radioactive Substances Act of 1960
3.6. The Radioactive Substances Act of 1960 controls the storage and disposal of radioactive wastes, and contamination of the environment is controlled by means of Authorizations issued under the Act. It is an offence to dispose of non-exempted radioactive wastes without authorization, and an offence to exceed an Authorisation. The following are the general principles concerning radioactive waste disposal, which had been set out prior to 1960 in the White Paper 'The Control of Radioactive

Wastes' (Cmnd 884) published in 1959 by a Panel meeting under the aegis of the Radioactive Substances Advisory Committee. The disposal of radioactive waste was to be carried out in accord with three general principles which still apply, namely:

(a) to ensure, irrespective of cost, that no member of the public shall be exposed to a radiation dose exceeding the ICRP dose limit (paragraph 3.5).
(b) to ensure, irrespective of cost, that the whole population of the country shall not receive an average dose of more than 1 rad per person in 30 years, and
(c) to do what is reasonably practicable, having regard to cost, convenience and the national importance of the subject, to reduce doses far below these levels.

The first principle needs no comment other than that ICRP is composed of the world's best experts in the field of radiological protection. The second principle controls the additional dose to the population from radioactive waste disposal; if there were no radioactive waste disposal the dose per person in 30 years would be rather more than 3 rads, from natural background radiation and from needfully incurred medical exposures. It has been estimated that the average genetic dose to the United Kingdom population from waste disposal is less than 1/1000th of the limited specified under (b) above. The genetic dose to the population is unlikely, therefore, ever to approach the limit prescribed because, as the total nuclear power installed increases, techniques for reducing the proportions of radioactive waste discharged tend to be improved so as to keep pace. The third principle is essentially the same as that propounded by ICRP in Paragraph 52 of its *Recommendations*.* Radiation doses resulting from waste disposal are now, and will continue to be low compared to the dose limits set out above, and to the other contributions to radiation dose in the environment described in paragraphs 3.2, 3.4.

**Radiological protection requirements
in the European Economic Community**
3.7. Now the United Kingdom is a member of the European Economic Community, and under Chapter III of the Euratom Treaty we will have to comply with the requirements concerning those matters which will, in time, become uniform throughout the Community. The present draft EEC directive concerned with radiological protection standards is, as one might expect, closely modelled on the 1966 Recommendations of ICRP and it is unlikely that any serious difficulties will present themselves so far as basic radiological protection standards are concerned. Under Article 37 of Chapter III of the Treaty each Member State of the Community must provide the European Commission with certain general data for the disposal of any kind of radioactive waste. These developments will clearly have important implications for the commercial operations of BNFL, and for the success of international companies which were recently created, such as United Reprocessors GmbH and Nuclear Transport Ltd. (concerned with reprocessing and transporting irradiated fuel), and URENCO and CENTEC which are concerned with the development and use of centrifuge enrichment.

TABLE 2   Dose limits (rems per year) for individual members of the public

Whole body, gonads, red bone marrow	0·5
Skin, bone, thyroid	3*
Hands and forearms; feet and ankles	7·5
Other single organs	1·5

*1·5 rems per year to the thyroid of children up to 16 years of age. (Note: for beta or gamma radiation, 1 rad = 1 rem: for alpha radiation 1 rad = 10 rems)

*Publication No. 9 of 1965-67.

The principle of controlling radioactive discharges to the environment by means of government authorization

3.8. Under the Radioactive Substances Act of 1960, discharges from major nuclear sites in England,* such as those of the United Kingdom Atomic Energy Authority, British Nuclear Fuels Ltd. and the Central Electricity Generating Board, are controlled by Authorizations issued jointly by the Department of the Environment and the Ministry of Agriculture, Fisheries and Food, after consultation with local authorities. The latter condition or custom is honoured in the observance and a notable instance occurred at Windscale in 1970. As a preamble it has to be understood that the amount which may be discharged over a given period of time must not only be below safe limits which are derived from ICRP dose limits, but must also be held at a yet lower level determined by the demonstrable need for the discharge to be made. At that time Windscale was finding difficulty in operating in a manner such that authorized discharges of alpha activity in dilute solution through the installed long pipeline to the Irish Sea were not exceeded. But the authorized amount was much less than ICRP dose limits would have allowed. This led the Atomic Energy Authority to apply for an increase in the authorized discharge of alpha activity. The local authority (Cumberland County Council) was consulted by the Department of the Environment. The County Council needing independent technical advice consulted the British Society for Social Responsibility in Science, a voluntary organization of which responsible scientists, including for example University staff, are members. After a visit to Windscale to examine the methods of effluent control in use there, the Society advised the Council that the Authority's proposal was reasonable, and that the proposed new discharge limit was safe.

District survey programmes for monitoring radioactivity introduced into the environment

3.9 Authorizations to dispose of waste generally require the UKAEA, BNFL or the Electricity Generating Board concerned to undertake sampling and analysis as required by the Controlling Ministries, and the results of this work are reported routinely to the Ministries and are also kept available at each site for examination by the Ministries' Inspectorates. As an independent check, samples of effluent are often collected by the inspectors who forward them to the Government Chemist and to other laboratories for independent analysis. Programmes of sampling and analysis are drawn up at individual sites. There are two kinds of programme. Firstly there are those designed to monitor directly the quantities and composition of the radioactive wastes before they are discharged, and thus to demonstrate that the discharges do not exceed the authorized limits. This type of monitoring is undertaken when it is liquid and solid radioactive wastes that are to be discharged. Quantitative limits are not specified for gaseous effluents. Their emission is so regulated that public health is not endangered, and the best practicable means are used for their reduction. There is a continuing dialogue with the inspectors responsible

for administering this part of the Act, to ensure that these aims are fulfilled.

3.10. The second kind of monitoring programme is one in which the environment around the nuclear site is monitored in order to check that the dispersal of radioactive waste, which has been discharged to the environment, does not lead to unacceptable exposure of the general public, either as individuals, or as a population group. For this purpose, independent monitoring programmes are conducted by staff from the nuclear establishment, and by staff of the Authorizing Departments. As an example of the latter, surveys of radioactivity in the sea, seashore, fish and seaweed are carried out by staff of the Fisheries Radiobiological Laboratory of the Ministry of Agriculture, Fisheries and Food. The principles which should govern such environmental monitoring programmes were set out in ICRP Publication 7.* Radioactive substance in the environment may give rise to an external radiation hazard, or may cause internal hazard if it is taken into the body by inhaling contaminated air, by drinking contaminated water or eating contaminated food. There are, thus, many and complex pathways by which releases of radioactivity may result in the irradiation of man, so that scientific studies in depth require examination in detail of the age structure, dietary, occupational and domestic habits and hobbies of the relevant population in order to establish the true level of exposure. Many cases, of course, are quite simple and easily encompassed. In any case it is necessary to collect information with the aim of identifying the most important pathways, the relevant nuclides, and the most highly exposed group of people within the whole of the population exposed. Specific examples of so-called pathways, and of significant nuclides, and population groups will be given in sections 4 and 5. ICRP does not specify permissible limits for the concentrations of the various radio-nuclides in the various foods which may be sampled but does so for air and water, and the basic dose limits are also available. Using local knowledge, for example, of how much locally-caught fish is eaten, and by reference to existing general data, one can deduce the appropriate 'derived working limit' (DWL) of radiation dose or activity concentration, and reach an agreed figure by discussion with the authorities. Compliance with the DWL, or better, will then ensure human exposure is within the basic ICRP limits.

The Nuclear Installations Act of 1965

3.11. In most of the chemical or metallurgical processes leading up to and including fuel fabrication, which were summarized in paragraphs 1.5 to 1.12, the possibilities for accidental release of quantities of material likely acutely to influence environmental conditions are small and the probability of such an occurrence is so vestigial as truly to merit the term 'incredible'. But in most of the processes (summarized at 1.13 to 1.19, i.e. those occurring during and after irradiation of the fuel in a nuclear reactor), risks of significance to public and environment must be said to exist to a lesser or greater degree according to the process considered. Principally it is the creation of very large quantities of radioactive fission products, and of radioactive transuranic elements such as plutonium, inside the fuel rods during irradiation by neutrons in the nuclear reactor, which is the source

---

*The arrangements are somewhat different for premises in Scotland, Wales or Northern Ireland.

*Published in 1966.

FIG. 12   Chapelcross Reactor Station

of risk. The nuclear industry has devoted a considerable part of its resources over the 30 or so years of its development to obtaining an understanding of its safety problems, and to the prevention of accidents. The Nuclear Installations Act of 1965 permits the construction or operation of nuclear reactors, and certain other nuclear installations under licences issued by the Secretary of State. The Atomic Energy Authority (though not British Nuclear Fuels Ltd.) and Government Departments are exempt from this licensing procedure, but in the operation of their own installations are directed to observe, as far as is practicable for a diversified R. & D. body, the safety requirements imposed on other licensees.

3.12. The Secretary of State for Trade and Industry and the Secretary of State for Scotland are aided and advised in the operation of their licensing procedure by the Nuclear Installations Inspectorate, which was created in 1959. On a new reactor, fuel may not be loaded, commissioning programmes may not be carried out, nor may power be raised, until approval has been given by the Nuclear Installations Inspectorate on behalf of the Secretary of State. The station can only be operated commercially after consent has been given, the nuclear site licence issued, and all the conditions of the nuclear licence met.

3.13. The Nuclear Installations Act of 1965 also

enabled the United Kingdom to ratify the International Conventions which deal with third-party liability in the event of a nuclear accident. Licensees and the Atomic Energy Authority are under an absolute duty to secure that no occurrences involving nuclear matter on their sites cause personal injury or damage to property, and are under a similar duty regarding ionizing radiations emitted on, or from, their sites from other material or equipment and as regards occurrences involving certain nuclear matter in the course of carriage outside their sites. Nuclear operators are liable to pay compensation under the Act within a limit of £5 million for any one occurrence, and Parliament is responsible for providing money to satisfy claims which exceed the funds available to licensees. The limitation period for claims is 30 years from the date of the occurrence. A singular feature of nuclear law, is that the operator of the installation is liable to pay damages, if an accident occurs, irrespective of whether he has been negligent; and the rule is similar if there is an accident while nuclear material is being transported.

The role of the National Radiological Protection Board

3.14. The National Radiological Protection Board was created by the Radiological Protection Act, 1970, and

Parliament gave the Board the following functions, namely:

(a) by means of research and otherwise, to advance the acquisition of knowledge about the protection of mankind from radiation hazards, and

(b) to provide information and advice to persons (including Government departments) with responsibilities in the United Kingdom in relation to the protection from radiation hazards either of the community as a whole or of particular sections of it.

Among other duties the Board is responsible for co-ordinating or representing United Kingdom interests in radiological protection internationally, and for co-ordinating national arrangements for incidents involving radioactivity. It therefore has an important and basic part to play in many of the activities which have been referred to in this section.

4. NUCLEAR REACTOR POWER STATIONS:

(a) The reactors in normal operation and the environment

Magnox-uranium gas-cooled reactors

4.1. The prototypes of the magnox-uranium gas-cooled reactors operated by the Electricity Boards are the four Calder Hall reactors adjacent to the Windscale site in Cumberland, and the four reactors on the Chapelcross site in Dumfriesshire. These were originally owned by the UKAEA; but are now part of BNFL. Each site generates about 200 MW(E) which is supplied to the grid. Fig. 12 is a photographic view of Chapelcross. It is relevant to discuss environmental aspects of this station, since it is equivalent to a commercial nuclear power station.

4.2. Gaseous effluent from the Chapelcross reactors, which have steel pressure vessels as do others of this period, consists of shield-cooling air which is drawn over the interior surfaces of the main concrete shield round each reactor, and discharged through the stacks at about 200 ft above ground level. The air discharged contains small quantities of the radioactive isotope Argon 41 created by neutron activation of the natural argon in air. (A small fraction of the Argon 41 discharged arises from leakage of the coolant gas, because carbon dioxide contains argon as a trace impurity). The dose-rate from Argon 41 to a person at the station fence is estimated as about 0·01 rad/year. In addition to the radiation from Argon 41, there is also direct radiation from the top ducts leading to the heat-exchangers. However, the radiation exposure just outside the fence is well below the ICRP dose-limit for members of the public (see paragraph 3.5). Samples of milk are collected regularly from 12 farms in two radial zones around the site, and analysed for iodine 131 and strontium 90. No iodine 131 has been detected in these samples; their strontium 90 content is the same as in samples taken in parts of the country where there are no nuclear installations, and is therefore the result of world-wide fall-out from nuclear weapons tests.

4.3. Liquid effluents from Chapelcross are discharged by pipeline to the Solway Firth during the two-hour periods following high tide. The Authorization restricts the discharge in 12 consecutive months to 150 curies of tritium and 700 curies of alpha plus beta activity (less tritium). Tritium is created by neutron activation of trace lithium impurity in the reactor's graphite moderator and is collected in the aqueous liquor produced by the

gas circuit driers. Another source of activity in the liquid effluent is water discharged via a detention tank from the 'pond' (a concrete tank), in which spent irradiated fuel is stored for a few weeks, to allow for some reduction of radioactive heat by natural decay before the fuel is transported by road to the reprocessing plant at Windscale. The third source of activity is the decontamination centre where tools and equipment are cleaned after they have been used in the reactors. The actual discharges usually run at 3% to 4% of the limit prescribed by the authorizing Government department. There is a marine monitoring programme associated with these discharges in the Solway Firth, in which white fish, salmon, shrimps and a non-edible seaweed (used simply as an indicator) are sampled. Measured activity in samples is less than 1% of the DWL. Surveys along the shores of the Solway Firth show that gamma radiation adds little to natural background.

4.4. Although based fundamentally on the original Calder/Chapelcross reactor concept, the considerably larger reactors of the Electricity Generating Boards differ in many details, and the later magnox-uranium stations, which have prestressed concrete pressure vessels, could be said to differ considerably. Considering the environmental effects, the latter type discharge less argon 41 because the shield cooling air is replaced by a water-cooling system. The most interesting point regarding liquid effluents is that aspects of the environmental survey which may be important at one site, are often of little consequence at another. The characteristics of each site depend greatly on local conditions, and each site requires a special study. At some sites, because of particularly favourable conditions afforded by vigorous tidal movement, and natural mixing of waters, the marine survey may yield so few positive results that only a small monitoring programme is needed.

Advanced gas-cooled reactors
(AGR or Mk. 2 gas-cooled reactors)

4.5. A number of Advanced Gas-cooled Reactors (described in paragraph 1.14) are nearing completion in the United Kingdom. Their influence on the environment can be predicted from theoretical design studies, the results of laboratory research, and from experience with the prototype AGR which has been in operation testing fuel and other components on the Windscale site for several years. Since the full-scale AGRs all have prestressed concrete pressure vessels, gaseous effluents will be limited to some argon 41, and possibly a little xenon and krypton when a leaking fuel can occurs. As in the case of the magnox-uranium reactors, leaking or defective fuel will not be allowed to remain in these reactors. Both types of reactor have their fuel loaded and unloaded while the reactor continues to operate at power, and faulty fuel elements can therefore be removed as soon as detected.

High temperature gas-cooled reactors
(Mk. 3 gas-cooled reactors)

4.6. Mk. 3 gas-cooled reactors may be built in this country. The coolant in these reactors will be helium, and the fuel will be of the type described in paragraph 1.14. UKAEA operational experience with the jointly owned OECD 20 MW(Th) experimental reactor (DRAGON) at Winfrith, evidence from other reactors of this type, and from fuel development studies, lead one to expect that

FIG. 13   Steam generating heavy water reactor at Winfrith Heath

gaseous effluents from a power reactor of this type will be insignificant. Liquid effluents from these reactors will be similar to those from other gas-cooled reactors.

Water-cooled reactors
4.7. The several water-cooled reactor systems available were mentioned in paragraphs 1.15, 1.16. The proto-type Steam Generating Heavy Water Reactor produces 100 MW(E) on the UKAEA's site at Winfrith Heath. Its gaseous effluent is of a kind characteristic of all direct cycle boiling water reactors. Non-condensable gases ejected by the main steam condenser vacuum pump contain small quantities of xenon and krypton isotopes if leaking fuel is present in the reactor. An adequate delay (three days or more for xenon isotopes) must be built in to the off-gas system so that isotopes of short half-life, (which constitute most of this gaseous activity) are removed by radioactive decay. This delay is usually provided by including a charcoal bed in the off-gas train. If this is done the dose-rate at the station fence arising from xenon and krypton will be of the order indicated in paragraph 4.2. for argon 41 from Chapelcross reactors, and the same remarks will apply. Experience with Winfrith SGHWR is that any iodine from leaking fuel is retained in the water and rapidly removed by the full-flow water purification system, so that no iodine escapes to the atmosphere. Back-up measurements in the Winfrith environment have confirmed that no iodine 131 is

detectable in the local milk. Fig. 13 shows the Winfrith SGHWR; and Fig. 14 shows members of the Winfrith environmental survey team at work in the field.
    4.8. Pressurized water reactors (PWRs) are being built and operated in many countries abroad. In these reactors, xenon and krypton which arise from any leaking fuel are stripped from the primary circuit water in a by-pass purification loop and stored before discharge. Gaseous and volatile fission products may also reach the atmosphere by leakage through the heat exchangers to the boiling water circuits and by leakage into the containment which must be vented from time to time. Data assembled in the United Nations report of 1972 on Ionizing Radiation Levels and Effects show that noble gas discharges from BWRs have, in the past, been larger than from PWRs with analogous discharges from gas-cooled reactors generally at an intermediate level. Tritium discharges in liquid effluents from BWRs have generally been less than from gas-cooled reactors, and much less than from PWRs. However, truly significant comparisons between reactor systems cannot be made on the basis of such figures because, in all cases, the activities discharged, and the estimated doses from them to the public, have been well within prescribed limits.

Sodium-cooled fast reactors
4.9. The nature and quantity of gaseous and liquid effluents which will arise when large sodium-cooled fast

neutron reactors come into operation, can be foreseen from the experience of operating the much smaller experimental fast reactors, and from the design studies now being made. Gaseous effluents will probably be so minute that the dose-rate at the station fence may be only 0·00005 rad/year (1/10 000th of the ICRP dose limit), a figure so low that it may as well be labelled zero. Liquid effluents will consist mainly of activity arising from the decontamination of mechanical components removed from the primary circuit for maintenance or repair. The small experimental fast reactors which have been built and operated so far, give confidence that effluents from the larger reactors now planned will be under good control. Valuable experience with a fast reactor of medium size will be obtained from the prototype fast reactor now nearly complete on the UKAEA's site at Dounreay. This is shown in Fig. 15.

Aesthetic aspects

4.10. Some people object to power stations and to overhead power lines in open countryside. Siting our first nuclear power stations in remoter areas of open seashore and countryside may have exacerbated these difficulties but as less remote areas are used this may become a less prominent issue. Although a nuclear station may be as large as a fossil-fuelled station, the space required for fuel and waste is smaller. Some of this advantage

FIG. 14 *Health physicists carrying out district survey sampling and monitoring*

FIG. 15 *Prototype fast reactor at Dounreay*

FIG. 16   Hunterston A reactor station

accrues from the fact that all the uranium used in our country is mined abroad. Nevertheless, if uranium mining should begin in Britain as a result of the continuing geological survey, if we take a figure from a recent USAEC report, we could expect that the disturbed area would amount to approximately 18 acres for the production of the annual fuel requirements for a 1 000 MW(E) reactor. According to the same report this is less than 10 per cent of the equivalent annual commitment of land to coal mining. Leaving aside questions of mines and waste, we introduce Fig. 16, a photograph of the Hunterston magnox-uranium reactors with the Firth of Clyde as background, partly for its pictorial quality and partly to show that such buildings can be designed so as to harmonize with the landscape.

Waste heat discharges from nuclear power stations
4.11. The rather low thermal efficiencies of the earliest prototype nuclear power reactors, chosen with appropriate caution by designers to fall in the range 25% to 30%, led some people to the conclusion that nuclear power reactors in general would discharge excessive amounts of waste heat to the environment. Others welcomed the prospect of added warmth to an often cool local climate. In fact, both houses have been to some extent confounded in their expectations. The thermal efficiencies of modern nuclear power stations cover the range 30% to 42% and compete on level terms with those of modern fossil-fuelled power stations. Thermal pollution, if the term has any meaning or application in Britain, is not an issue of a decisive nature betwixt nuclear and the other sources of electric power production.

A conclusion about reactors in normal operational conditions
4.12. Paragraphs 4.1 to 4.11 have shown the innocuous effects of the effluents discharged in a controlled manner from nuclear power stations in normal operation. If there is believed to be any risk from these effluents it is surely nominal and negligible. Considered rationally, is it not but the order of risk taken by anyone who moves safely from one place to another, thereby possibly submitting unawares to a small increase in the natural radiation background?

4.13. It used to be argued that coal and oil-fired power stations, burning fuels which contain traces of radium, were adding more radioactive 'hazard' to the atmosphere than nuclear power stations of equal electrical output. A more meaningful comparison is between the negligibly small risks from radioactive discharges from reactors and the tangible risks from noxious chemicals (oxides of sulphur and nitrogen, and smoke) discharged from conventional power stations.

(b) Reactor safety
Introduction

4.14. The nuclear industry has proved itself safe, but vigilance must be maintained. In Britain we know the Windscale accident of 1957 did not interfere with living apart from the temporary ban on local milk contaminated with iodine 131, and that there were no illnesses or casualties. Air-cooled reactors, as at Windscale then, are no longer used. Present-day power reactors use closed-cycle cooling. Great advances have been made in devising precautions intended to make a reactor accident incredible, and to contain an escape of fission products or other radioactivity should it nevertheless occur.

Gas-cooled reactors

4.15. In gas-cooled reactors the accident most guarded against is the rapid loss of gas pressure which could follow fracture of part of the gas circuit. However, provided the nuclear reaction can be shut down quickly, enough gas can be kept flowing at atmospheric pressure to cool the fuel, prevent damage to it and so prevent the escape of fission product activity. Very high reliability of shutdown is ensured by providing a large number of automatic shut-down devices (e.g. more control rods than needed to allow for random failure of a few), and of several different independent kinds (to avoid some unforeseen fault being common to all). In all reactors, ample emergency supplies of carbon dioxide can be fed to the reactor to prevent air entering the core. The greatest single advance in gas-cooled reactor safety has been the introduction of the prestressed concrete pressure vessel containing the reactor core and all the heat exchangers. Its main advantage is its inherent safety, the vessel strength being provided by a multiplicity of prestressing cables, no single one of which is essential, and all of which can be inspected, removed and replaced as desired. A second advantage is that because the gas circuit is now almost wholly enclosed in the vessel, a fracture of any part of it is now virtually inconceivable.

4.16. If, through a blockage of gas-flow in a channel, fuel should be damaged and fission products released to the gas circuit, the gas could be cleaned and discharged if necessary, using the installed system of particle filters and iodine-absorbing charcoal beds. Thus there need be no release of volatile or solid particulate fission products to the atmosphere.

Water-cooled reactors

4.17. If there were an accidental break in the pressure circuit of a water reactor, most of the water would flash off as steam. Because steam is a very poor coolant compared to liquid water, an emergency core cooling system must come quickly into action and project water on to the fuel. This is achieved in the Winfrith SGHWR by injecting the water through a perforated pipe which runs down the centre of each bundle of fuel elements. Full scale tests have verified that this system works, and all water reactors have an emergency core cooling system of some kind. The environmental safety of water reactors is also assured by the containment system, which is designed to accommodate the escaping steam, and to contain any gaseous and volatile fission products which might be released from damaged fuel. With proper design and care no significant escape to the atmosphere need occur.

Fast reactors

4.18. The first commercial fast breeder reactor in the UK will most probably be a pool-type reactor like the prototype fast reactor at Dounreay. In this the core and heat exchangers are immersed in a large pool of sodium. The large thermal capacity of the pool, and the excellent heat transfer properties of sodium, ensure that the core will not be damaged even by such highly unlikely failures as complete loss of supplies to all sodium pumps. Thus in many respects the fast reactor will be an exceedingly safe reactor.

4.19. The word 'fast' does not mean that the reactor will be tricky to control. As with thermal reactors, control is exercised through the delayed neutrons, and the fast reactor in normal operation is very well-behaved and controllable. Research and theoretical studies continue to postulate and explore a large variety of different accidents which could conceivably affect the reactor. These studies are expected to confirm existing evidence that even severe accidents would do no more than melt fuel. If this is the case, all plutonium and fission products would be retained in the sodium pool (except for the noble gases whose small volume would be easily accommodated in the enclosed cover gas above the pool). Whatever the result of these researches may be, the reactor and the sodium pool will be contained inside a reinforced concrete vault designed to withstand the energy release; and the operating area around and above will be designed to provide yet another barrier against the release of radioactive material to the atmosphere. There is, therefore, high confidence that fast breeders will be very safe reactors, which will not compromise the safety of the public.

Emergency control procedures

4.20 Sites presently selected for nuclear power reactors have only limited numbers of people living within two-thirds of a mile, and the Government exercises control over new developments within two miles which would tend to increase greatly the numbers of local population. The number of people who might be affected, if an accidental release of radioactivity should occur, should therefore be small. Local liaison committees are set up at each nuclear station to enable questions to be asked, and answered, on all matters affecting the environment and safety.

(c) Dismantling and demolition of obsolete nuclear power stations

4.21. Britain is heavily scarred with the abandoned debris of earlier industrial activity. It has been suggested that obsolete nuclear power stations will be added to this catalogue of desolate scenery. But need this necessarily be so? Most of the sites on which nuclear power stations are being built have been chosen for features which are strong incentives for the re-use of the same, or adjacent sites for future nuclear power stations. There

FIG. 17 Interior view of Windscale reprocessing plant

will be several methods for reclaiming the site. A considerable portion of the plant, for example in the turbine halls, is similar to that in conventional stations, and non-radioactive, so that dismantling is no special problem. Difficulties arise only in dealing with the radioactive central regions, the reactors themselves. The nuclear fuel having been removed in the usual way, the main nuclear problem is radiation from neutron activation products, iron 55 and cobalt 60 in reactor structural steel work. For many years continuous manual work will not be possible directly adjacent to the steel structure of the reactor core. Remote cutting and handling methods could, however, be employed if immediate removal and reclamation were essential. Work of this general kind has been done before in carrying out repairs to reactors. The UKAEA is putting in hand studies of how the techniques used then, together with other new methods, could be used to dismantle nuclear power stations when this becomes desirable in the future. If a delay of 30 to 50 years were acceptable before beginning the demolition job, it would be a simpler proposition. Carefully worked out explosive demolition techniques might then be feasible which would enable the site to be cleared, earthed over and planted with grass and trees for free access if desired.

4.21. Many reactors built today are constructed with the radioactive core just below ground level. When these reactors cease to operate, it will be possible to clear the works above ground quite quickly, whether the core is demolished or not.

5. NUCLEAR FUEL REPROCESSING PLANTS
(a) Reprocessing plant operations
5.1. The nature and purpose of nuclear fuel reprocessing were outlined in paragraphs 1.17, 1.18. Fuel which arrives at the BNFL reprocessing plant at Windscale is stored in the 'ponds' for about four months to complete the decay of iodine 131. A significant fraction of the low activity waste discharged from Windscale is water run off from settling tanks fed by the outflow from the ponds, but the greater part of this low activity discharge arises from the reprocessing plant itself. An interior view of the Windscale reprocessing plant is shown in Fig. 17. Here the first solvent extraction cycle separates about 99·96% of the fission products and this constitutes the highly active liquid waste, of which about $5m^3$ is produced per tonne of fuel processed. As described below this is concentrated by evaporation and stored in high integrity vessels on site, to await eventual solidification (paragraphs 5.7 and 5.8).

5.2. At Dounreay a plant for reprocessing highly enriched fuel from Materials Testing Reactors came into operation in 1958 and in 1961 a second plant began reprocessing high burn-up fuel from the experimental Dounreay Fast Reactor. Treatment of the wastes from these plants is broadly similar to the procedures at the much larger Windscale plant and will not be discussed in further detail here.

(b) Controlled discharge of low activity waste water to sea
The method of discharge and the official Authorization
5.3. The low-activity waste at Windscale is routed first to effluent-holding tanks where it is accurately sampled, analysed and measured. It is then discharged through the pipeline, which runs 2·5 kilometres along the sea-bed, during the two hour period after high tide which is the most favourable for dispersion of the effluent in the sea.

The Authorization (see paragraph 3.8) specifies the quantities of activity and of certain specific isotopes which must not be exceeded in the discharges totalled up over any period of three consecutive months, and certain subsidiary conditions. The most significant figure in the Authorization, in relation to possible exposure of the public, is that for ruthenium 106. The discharge is such that consequential exposure is well below the ICRP dose limit. For example, in 1970, the highest dose to the public (received in fact by only a few people) was only 6% of the ICRP dose limit.

Environmental monitoring in relation to sea discharges
5.4. The Cumberland coast and the neighbouring and more distant waters of the Irish Sea are surveyed and monitored by Windscale health physics staff, and independent measurements are made by officials of the Ministry of Agriculture, Fisheries and Food (Fisheries Radiobiology Laboratory). Radioactivity in an edible seaweed called Porphyra Umbilicalis is primarily what is studied. This weed is gathered on various parts of the British coast, and sent to South Wales where it is processed and cooked, to be sold as a food called 'laverbread'. In general about a quarter of the weed which arrives in South Wales has been gathered along the coasts near Windscale, from Barrow-in-Furness to near St. Bees Head. Samples gathered on the Cumberland coast are analysed at Windscale; the principal activity present is ruthenium 106, and this is the limiting radioactive isotope in regard to human exposure. Direct measurements on laverbread have shown that even for those few people who are said to eat exceptionally large amounts of this delicacy, the estimated dose is less than 6% of the ICRP dose limit. For the great majority of the laverbread-eating population, whose consumption is much lower, the estimated dose is of the order of one-hundredth of the ICRP dose limit (or, in fact, only one-hundredth part of a typical X-ray dose given for diagnostic purposes (see paragraphs 3.4, 3.5).

5.5. A minor aspect of the Windscale marine survey is the possible exposure of a few fishermen who work part of the time on fixed salmon nets in the Ravenglass Estuary, a few miles south of Windscale. The silt in the estuary contains traces of radioactivity. The greatest exposure these fishermen could receive is about 10% of the ICRP dose limit. Some trace caesium 137 contamination of plaice, caught off the Cumberland coast, has been identified, but the level is about the same as the naturally-occurring radioactive isotope potassium 40

in the fish flesh, and the population dose which may result is therefore negligible compared to natural background.

(c) The management of highly active liquid waste
Interim storage in the liquid state
5.6. At present it is the practice in all countries to store highly active liquid waste from fuel reprocessing in high integrity tanks. At Windscale the highly active liquid waste from reprocessing (see paragraph 5.1) is concentrated by evaporation and stored in the acidic state in stainless steel tanks, which are fitted with water-cooling coils for the removal of radioactive decay heat, and which are housed in stainless steel lined concrete vaults. The existing store comprises nine tanks, eight of which have a capacity of 70 m³ each, and the ninth and most recently completed, 150 m³. Two further tanks of the larger size are at present under construction. The present storage system has been in use for about 20 years, and no serious problems have been encountered. Surveillance is continuous, and spare tanks are always available to which waste could be transferred if a tank were ever to leak.

Solidification and vitrification of the waste
5.7. It is recognized that it will be necessary to solidify these liquid wastes, in order to make surveillance less vital and reduce or eliminate the need to replace tanks periodically over the longer term. Technical studies and development of a process called HARVEST are in progress and it is planned to begin solidifying wastes in the mid 1980s. This process is derived from the FINGAL vitrification process on which the original work was carried out at Harwell. Vitrification is to be preferred since simple calcination produces a mixture of friable oxides which could be easily dispersed by accident, and would have other undesirable properties. The vitrified material should, if possible, have good resistance to leaching by water, resistance to degradation by nuclear radiation and heat, a high thermal conductivity, high melting point and, finally, a good capacity for incorporating the waste oxides. The best material is a borosilicate glass with the following composition: waste oxide 25%. silica (SiO_2) 43% and borax ($Na_2B_4O_7$) 32% (by weight). The leachability of this glass is very low indeed; and, in addition, the glass will be stored in leak-proof containers made of stainless steel having a high corrosion resistance.

Storage of the vitrified waste
5.8. Ultimate or irrevocable disposal of the vitrified waste, say in some geological formation, is not at present contemplated. Such disposal might prove hazardous to man if the long-term properties predicted for the waste, or for the state of its environment did not come up to expectation. Instead, the waste containers will be stored under water in a pond on the reprocessing site. There the containers can be kept cool, and isolated from the atmosphere under minimum surveillance, and they can be inspected, retrieved and refurbished should this ever be necessary. (It is notable that such engineered surface storage has now been proposed in the United States, plans to use a Kansas salt-mine as a repository having been put in abeyance.) The total land area occupied by the highly active liquid waste store at Windscale is at present 900 m³ and it has been estimated this will increase to 1 500 m² in 1980. After the solidification programme

has come into operation, the total land area occupied by interim liquid storage tanks and solidified waste storage ponds in the year 2000 has been provisionally estimated as 7 000 m²—about the area of two football pitches, and quite trivial, if one considers that some three-quarters of our national electricity at that time will be generated by nuclear power stations.

(d) Gaseous effluents from reprocessing plants

5.9. When fuel enters the dissolvers, noble gases, tritium and iodine are liberated. The six-months between fuel discharge and reprocessing ensure that iodine 131 activity has been removed by radioactive decay. Three radioactive isotopes remain, which are at present discharged to atmosphere from the plant stack. In descending order of importance these are krypton 85, tritium and iodine 129. Much of the tritium (as tritiated water) and iodine stay with the aqueous streams in the plant, but some goes to atmosphere. Krypton 85 from reprocessing plants has been the subject of several national and international reports in recent years, stimulated by the knowledge that as national and world nuclear power programmes grow, these long-lived isotopes will assume greater significance in the local and world atmosphere. It is now widely agreed that krypton removal and storage will have to be included in future reprocessing plants. Further research is needed to decide the best of several possible methods available. If nuclear power programmes continue to increase, then tritium could be considered in turn. It may never be necessary to consider iodine 129, particularly if an incentive towards shorter cooling times before reprocessing fast reactor fuels encourages further advances in the efficiency of iodine 131 removal equipment.

(e) Solid wastes from reprocessing plants

5.10. The principal solid waste from reprocessing is the fuel canning material. This is stored on the Windscale site in concrete 'silos'. Other solid waste which has extremely low levels of activity is buried at Drigg, the licensed burial site near Windscale. The Drigg site also accepts solid waste of very low activity concentration from all parts of UKAEA and BNFL, and from the National Disposal Service. An Authorization governs the conditions under which such low activity solid wastes may be buried at Drigg. Solid wastes of intermediate activity which arise at Windscale are at present stored,

FIG. 18   Flask containing irradiated magnox fuel being loaded on a transporter

and it is planned to incinerate these to reduce their bulk (see paragraph 7.6).

(f) Deep sea disposal of solid wastes under international auspices

5.11. Specially constructed containers of solid radioactive waste have been disposed of in the great depths of the North Atlantic Ocean beyond the continental shelf, in co-ordinated international operations conducted under the auspices of the Nuclear Energy Agency of the Organization for Economic Co-operation and Development. The construction of the containers is such that those handling them are adequately shielded from radiation, and that they reach the bottom unbroken, e.g. water enters through a pressure-equalizing valve. The ocean depths provide isolation, and the enormous volume of water there ensures dilution as radioactivity slowly leaks from the containers. The wastes in the containers are solid and of relatively low activity. A thorough assessment of the safety of such operations has been carried out by specialist panels convened by NEA and IAEA. There is no question of highly-active wastes being disposed of in these operations.

6. SAFETY CONTROL IN THE TRANSPORT OF RADIOACTIVE MATERIALS

Introduction

6.1. The provision of safe means for transporting radio-active ores or concentrates from abroad to the nuclear industry in this country, for transporting radioactive fuel and other materials between the various nuclear plants in the United Kingdom, and to and from our customers in foreign countries, is an important aspect of the protection of the environment; as is also the safe transport of toxic or hazardous materials in other industries.

Regulations for the safe transport of radioactive materials

6.2. Whereas the control of hazards within nuclear establishments may, by virtue of the staff and equipment available, depend to a considerable extent on operational controls, such controls are not readily available during transport. Compliance with sound regulatory standards, which are based essentially on design requirements for packages, and transport 'flasks' (as the larger containers are called), is therefore necessary to ensure that the environment, and especially transport workers and the general public, are protected from the hazards of these materials. The development of regulatory standards for safe transport of radioactive materials was entrusted by the United Nations in 1959 to the International Atomic Energy Agency (IAEA), which has its headquarters in Vienna. Although the IAEA regulations are recommendations to Member States and International Transport Organizations, nevertheless they have been widely implemented largely because of the procedure used for their development. This has achieved a large measure of compatibility not only between Member States but within the various modes of transport, by road, rail, sea or air. Basically the regulations prescribe *what* must be achieved, rather than how, and this has encouraged designers and consignors to use new materials, improved constructional techniques and their ingenuity. Regulatory development has also kept pace with the rapid development of the nuclear fuel industry. United Kingdom policy has been to implement the regulations recommended by the IAEA and to give full support to their preparation.

6.3. The IAEA regulations include provisions to ensure the following:

(a) adequate containment of the radioactive material;
(b) protection against radiation emitted by the radioactive material;
(c) safe dissipation of heat generated by the radioactive material;
(d) prevention of criticality when the radioactive material is also fissile.

Resistance to the possibly damaging effects of transport accidents is demonstrated by prescribed tests. To help ensure compliance with certain specified regulatory requirements, the regulations embody the concept of approval by the competent authority, who is to signify approval by issue of a certificate before the movement of material can actually be made. In the United Kingdom, the Secretary of State for the Environment is the competent authority for road and rail transport, and the Secretary of State for Trade and Industry is the competent authority for sea and air transport. In practice, the Transport Radiological Adviser to the Department of the Environment (who is also Head of the Dangerous Goods Branch) acts on behalf of both competent authorities.

The transport experience of the British nuclear fuel industry

6.4. The large quantities of uranium ore concentrates safely shipped in, and transported to Springfields Works were mentioned in paragraph 2.1. One result of this and other activities has been that over 23 000 tonnes uranium in the form of new fuel have been safely transported by, or on behalf of, UKAEA/BNFL in the last 23 years.

6.5. Long mileages are covered mostly by rail, and of course by ship from overseas. A railway line is provided up to and inside the Windscale site, to handle the majority of the heavy traffic in irradiated fuel. Road transporters are used to bring irradiated fuel from Chapelcross to Windscale, and from Whitehaven docks to Windscale. Since the reprocessing of irradiated fuel began at Windscale about 20 years ago, over 15 000 tonnes of irradiated fuel have been transported to that establishment. This total includes some 600 tonnes from overseas. Since the

FIG. 20 *Flask and irradiated uranium dioxide fuel being transferred from ship to lorry*

reprocessing of highly enriched fuel began at Dounreay, about 860 kilograms uranium of this irradiated fuel have been transported to that establishment, about half coming from the Harwell reactors, and the rest from overseas. All these movements have been accomplished in safety and without mishap. Some appreciation of the transport flasks used in carrying out these movements may be obtained from Figs. 18 and 19 which show flasks and a road transporter used for irradiated magnox-uranium fuel, and Fig. 20 which shows a type of flask used for irradiated uranium dioxide fuel.

7. RESEARCH AND DEVELOPMENT
General objectives of the research and development programme

7.1. The programme seeks to provide information necessary to minimize discharges from sites, and to ensure the safety of stored wastes. It includes an examination of the concept of the 'zero discharge site,' such as is envisaged for future reprocessing plants, involving the total recycle of water used within the plant, and the essentially complete containment of radio-active gases and solid wastes. The relatively few environmental problems arising from nuclear power reactors are included in the programme directly, and will benefit indirectly from some of the techniques and processes developed for fuel reprocessing.

Gaseous effluents from reprocessing plants

7.2. It would be advantageous to liberate tritium from irradiated fuel before the latter is dissolved in nitric acid, since the tritium could then be trapped and stored. In spite of the inherent difficulties, research into ways of achieving this is being done in many countries. Studies are also being made of methods for the removal of tritium from aqueous process streams, and into its possible scientific and commercial uses. If, however, tritium is released ahead of the aqueous stages of the plant, at least part of the krypton is likely to be released at the same time. The stable xenon released simultaneously would have five to 10 times the volume of the krypton, so that its elimination would reduce the storage volume required for krypton and krypton 85 (see paragraph 5.9); however, it is not clear that the overall cost would be less. The xenon, being inactive,

FIG. 19 *Flask and transporter on road*

could be discharged safely, but if the price is low there are possible uses for both krypton and xenon which have to be trapped for other reasons. The problem of iodine 131 removal has been studied for several years (see paragraph 5.9) but will continue to receive attention in case short 'cooling' times are required in reprocessing fast reactor fuels.

Liquid effluents and liquid wastes from reprocessing plants and other nuclear establishments

7.3. New treatments for low activity liquid effluents are being studied in case more restrictive discharge limits are demanded in future. Research at Harwell is aimed at reduction of effluents. Many novel process techniques are being studied including some drawn from experience in the desalination field. Beside improved evaporation techniques, freezing, ion exchange and various membrane techniques are being examined. Membrane techniques include electrodialysis, which removes ionic species from water, ultrafiltration which removes colloids, and reverse osmosis which removes both. In all cases, membrane stability in a highly radioactive environment requires to be investigated. Membranes are expected to be particularly useful for large-volume, low-activity waters such as arise from fuel-storage ponds. They may, therefore, have applications at both reactor and reprocessing sites (see paragraphs 4.3 and 5.1). These lines of development will lead on naturally to consideration of the total re-cycling of water within nuclear plants. It should also be possible to remove specific toxic elements such as plutonium from waste streams, and put them aside for permanent storage.

7.4. Floc precipitation has been used at Windscale and Harwell to treat low-activity liquid effluents. We have many years of experience operating these techniques which have consistently removed activity to levels lower than in the intake from normal river water. The low-activity precipitates from these processes have to be disposed of eventually. Their volume can be reduced by using deflocculating chemicals and by subjecting the gelatinous solids to cycle freezing and thawing.

7.5. As mentioned in paragraph 5.7, a considerable research, development and design programme jointly between the UKAEA and BNFL is at an advanced stage concerning the solidification and vitrification of highly-active liquid waste arising from fuel reprocessing. A particularly favourable feature of the processes, which are now well-proven at Windscale, is the essentially 'salt-free' character of the highly-active liquid waste, since this permits very considerable reductions in the volumes of liquid and, ultimately, vitrified wastes to be stored. This principle has not been applied generally, however, and some lower activity wastes are therefore evaporable only to a limited extent. BNFL and UKAEA are working together on possible process improvements. Several solidification techniques including incorporation in bitumen, concrete and plastics have been developed to render these lower active wastes suitable for storage.

Solid wastes from reprocessing and other nuclear plants

7.6. A great variety of contaminated solid wastes arise in nuclear plants and laboratories. It is important to minimize such arisings in the first place, but even so there comes the time when the activity in such wastes must be assessed, specific radioactive species recovered (if this is desired and economical) and, finally, the bulk

of the residue reduced in order to minimize the storage space or disposal operations required. A large proportion of these wastes are combustible. Incineration is, therefore, a valuable method for reducing their bulk; but special precautions are necessary to contain any radioactive elements in the concentrate, and to remove them from the 'off-gases.' A development and design programme is being carried out by BNFL with UKAEA support. An experimental incinerator has been commissioned at Windscale, but further work is needed before a full-scale incinerator which can handle significant quantities of plutonium can be built.

Development of monitoring instruments

7.7. A wide range of instruments, usually electronic, has been developed to monitor radioactive elements discharged to the environment. Current research is directed particularly towards instruments for measuring the radioactive content, particularly of plutonium, in solid wastes.

International co-operation on research and development

7.8. The research programme being carried out by the UKAEA and BNFL has been selected to meet the needs of the nuclear industry in this country, but it is expected that it will, for the most part, be carried out in close collaboration with European laboratories and firms. Negotiations to this end are at an advanced stage between the French, British and German partners in the consortium, United Reprocessors GmbH.

8. CONCLUSIONS

8.1. Nuclear power has been developed in the UK to the stage where there exists a large-scale nuclear fuel complex and reactor-powered electricity-generating stations. This development has taken place with the most careful control over all aspects of the use of radioactive materials.

8.2. Limits upon radiation and dose-rates are strictly in accordance with international practices established by the world's leading experts in the field of radiological protection.

8.3. Information and advice for observing these requirements and the related research and training involved is provided by the National Radiological Protection Board. It can be fairly claimed as a result of all these precautions that the annual dose-rates received by members of the public from nuclear power is infinitesimal when compared with that experienced from natural sources.

8.4. Regular monitoring around nuclear power stations and fuel plants confirms the negligible levels of contamination of these areas by radioactive materials.

8.5. Whilst all aspects of radioactive waste storage or disposal from nuclear fuel reprocessing plants are carefully controlled and monitored to ensure compliance with the regulations drawn up by the controlling authorities (referred to in section 3) it is possible that, in time, even more stringent restrictions will be applied. With this in mind a research and development programme is in hand to devise improved methods of dealing with these problems, and it is envisaged that highly active wastes will ultimately be stored in glass form.

8.6. It is expected that the early 1980s will see the start of a rapid expansion in the use of nuclear power and, therefore, a corresponding increase in fission products and radioactive waste. Continuing progress in containment, treatment, storage, etc., will be more than sufficient to ensure that the present very high standards are maintained.

9. BIBLIOGRAPHY

1. ICRP Publication 9. 'Recommendations of the International Commission on Radiological Protection, adopted 17 September 1965'. Pergamon Press 1966. See also ICRP/72/G-4, 'Statement Issued by the International Commission on Radiological Protection after its meeting in November 1972, Nov. 1972.
2. ICRP Publication 7. Principles of environmental monitoring relating to the handling of radioactive materials, Pergamon Press 1966.
3. DUNSTER, H. J., KENNEY, A. W., NEIL, W. T. L., and PRESTON, A. The British approach to environmental monitoring, *Nuclear Safety*, **10**, No. 6, Nov.-Dec. 1969.
4. PRESTON, A., *et al*. UK experience of radioactive waste release to the environment and expected waste management in fuel cycles in the 1980's. Paper A/CONF. 49/P/512. United Nations, Geneva Conference on the Peaceful Uses of Atomic Energy, 1971.

5. Nuclear Energy Agency of OECD. Radioactive waste management practices in Western Europe, published 1971, available HMSO.
6. European Nuclear Energy Agency (now NEA of OECD, see above). Radioactive waste disposal operations into the Atlantic 1967 (published by OECD in 1968).
7. United Nations. Ionizing radiation levels and effects, a report of the United Nations Scientific Committee on the Effects of Atomic Radiation to the General Assembly, with annexes, published 1972.
8. NEA/IAEA Symposium on the Management of Radioactive Wastes from Fuel Reprocessing, Paris, November 1972. Proceedings: in particular the following papers:
MARLEY, W. G., NRPB. Atomic energy and the environment.
Paper 5 by BRYANT, P. M., and JONES, J. A., NRPB. The future implications of some long-lived fission product nuclides discharged to the environment in fuel reprocessing wastes.
Paper 9 by CLELLAND, D. W., BNFL. High level radioactive waste management in the UK.
Paper 22 by GROVER, J. R., UKAEA. Glasses for the fixation of high level radioactive wastes.
Paper 29 by SZULINSKI, M. J., and WARREN, J. H., Harvard, U.S.A. Engineered storage of radioactive waste.

J S S REAY, BSc, PhD, DIC, FRIC*

7 Monitoring of the environment

SUMMARY

Fuel combustion is the main source of man-made air pollution throughout the world. Global and regional monitoring are conducted under the aegis of international organizations. In the UK there is an extensive national monitoring network for sulphur dioxide and smoke. At a smaller number of points measurements are made of a range of other combustion-produced pollutants. Time-series data, showing levels and trends, are a guide to central and local government on possible needs for action and exhibit the results of any such actions.

1. INTRODUCTION

It has been decided that at this Conference the subjects of pollution, combustion and the environment should be dealt with under four headings. This paper deals with air quality monitoring and will stray as little as possible into the closely related fields of the fate of pollutants and their medical and ecological effects. The word monitoring will be used throughout to signify surveillance rather than measurement in relation to set standards or as part of a regulatory system.

We monitor our environment primarily to determine the levels and trends of various materials which might at certain concentrations have a harmful effect on human health or upon the natural systems upon which we depend. A minor part of the effort goes into the measurement of other items about which we may have no suspicions of adverse effects but which may be indicators of trends we do not understand. Outstanding amongst human activities emitting wastes into the atmosphere is the production of energy, mostly by combustion. Many of the by-products of combustion are materials which have harmful effects in certain amounts or at certain levels. These are therefore the subject of the bulk of our monitoring activity—globally, regionally and locally. Another object of regular monitoring, particularly detailed national monitoring, is to provide the basis for, and to observe the results of legislative or other measures affecting the emissions of pollutants of concern. This short paper seeks to outline the various scales of monitoring with particular reference to combustion or fuel products. Further attention is then devoted to monitoring within the United Kingdom, and to the information the results provide.

2. SCALES OF MONITORING

2.1. Global monitoring

The main global monitoring agency is the World Meteorological Organization (WMO). So far it has designated seven baseline stations at remote sites. A total of 12 stations are contemplated. These stations are encouraged to measure any constituent of the atmosphere. First priority is given to making those measurements that will permit documentation of long-term changes in atmospheric composition of particular significance to weather, and climate. In this class comes carbon dioxide, a rise in the level of which could cause surface heating. Particulate matter would operate in the opposite direction by diminishing the solar radiation reaching the earth. Turbidity is measured as an indicator of particulates in the atmosphere. The chemical composition of precipitation is also studied (SO_4^{--}, Cl^-, NH_4^+, NO_3^-, Ca, Mg, K, acidity or alkalinity, pH and conductivity). It seems likely that other pollutants of global significance will be added to the list.

A constant watch, largely under the auspices of IAEA is kept on isotopes related to nuclear reactions—mostly explosions. Over 150 stations measure the isotopes in precipitation.

2.2. Regional and national monitoring

These are concerned with the situation in which the pattern of concentrations of pollutants displays significant variations on a less than global scale, usually under the influence of variations in emission densities. The bulk of our monitoring is in this class.

The WMO is again active in this field with a growing network of regional monitoring stations. So far these are intended to measure much the same range of pollutants as the baseline stations, but extension to items of specific regional interest seems likely. The UK operates one site in this network and makes similar measurements at two

*Head, Air Pollution Division, Warren Spring Laboratory, Department of Trade and Industry, Stevenage, Herts.

H

other stations. AERE, Harwell, monitors the concentration of radio-activity in the air, and its deposition in rain. This is primarily a measurement of fallout from nuclear explosions. The drift of certain pollutants between countries calls for much more detailed studies, and the UK is at present taking part in an OECD project to elucidate the movement of sulphur dioxide and its transformation products between the countries of NW Europe. This will be dealt with in more detail in paper 8 (The fate of pollutants) by Dr M. W. Holdgate and Dr L. E. Reed.

At national level, the UK is one of the most intensely monitored countries in the world. Not all the studies have been co-ordinated, and there are moves to improve this situation. National surveillance at a large number of sites throughout the country seems justified only for hazardous pollutants which are widespread, and where the pattern of emissions and concentrations can be so dependent on local conditions that there may be significant differences between areas calling for local information as the basis for tackling the problem.

2.3. Local monitoring

Studies such as the surveillance of emissions or concentrations from specific emitters, often industrial processes, come into the category of *local monitoring*. The measurements of this sort in the UK are principally carried out by the producers and by HMACI or Local Authorities. Fly ash from power stations is a combustion product in this category.

3. MONITORING OF SPECIFIC COMBUSTION-PRODUCED POLLUTANTS

3.1. Carbon dioxide

This is the desired end-product of combustion of carbonaceous fuel. In terms of the normal criteria of health or amenity associated with pollution, carbon dioxide is not of serious concern. There is a large margin of safety between average levels and those causing adverse health effects. There is thus no requirement for intensive monitoring. It is however, studied on the global scale for the reason mentioned in 2.1.

3.2. Carbon monoxide

This product of incomplete combustion of carbonaceous fuels presents a sharp contrast. It is notably toxic and causes numerous fatalities when produced in confined conditions. We must, however, distinguish between such extreme local situations and the effect of human exposure to normal ambient concentrations. Background levels are not a cause for concern but vehicle emissions, in particular, are increasingly producing situations where steps may have to be taken to protect people from the effects of carbon monoxide. Attention must be paid to levels in tunnels and inside buildings where car engines are running. A study for over a year in six UK cities showed that in the 1960s, even in the busiest streets, the level of carbon monoxide rose only for a few minutes per month over the 50 ppm limit permitted for eight hour continuous daily exposure. Figures from other countries show that these levels can be reached for longer periods where suitable topographic or climatic conditions combine with very dense motor traffic. We have carried out studies inside vehicles with defective exhaust systems and will be measuring carbon monoxide at the kerbside in

our new Five Towns Survey (qv section 6). It must be remembered that many people further expose themselves to considerable concentrations of carbon monoxide through smoking.

3.3. Water

This end-product of combustion of the hydrogen component of fuels obviously is not a toxic product, but high local emissions (not so much from combustion itself as from the use of water as a coolant) do give rise to suggestions that it may exert a significant local climatic effect. There have been specific studies but there is no regular monitoring. Nor have we in the UK given more than local attention to the thermal pollution aspect of waste heat disposal either to air or water. In passing it is worth noting that clouds of condensed steam issuing from the chimneys of many process industries are frequently regarded, quite mistakenly, as visible evidence of substantial pollution. While these emissions may contain other pollutants, gaseous or particulate, their overall appearance can be a poor guide to their potential content of pollutants.

3.4. Oxides of nitrogen

During the combustion of organic fuels the high temperatures produced favour the oxidation of atmospheric nitrogen primarily to NO which further oxidizes to NO_2. The nitrogenous components of fuel may also be oxidized during combustion. Man's contribution to the input of oxides of nitrogen into the atmosphere is not insignificant. There is a lack of trend data on levels of nitrogen oxides but various indicators, such as nitrate in rain, would appear to reflect a substantial increase of aerial concentrations over this country. Furthermore, ground level concentrations in certain localities may be largely the product of vehicle emissions which in some Regions of the UK account for nearly half the total low-level emissions. In our Five Towns Survey we are making a start on detailed monitoring of these compounds in areas where the highest concentrations might be expected.

3.5. Fuel residues

Unburnt or partially burned fuel and the non-gaseous residues of combustion are very important materials from the point of view of pollution. We are all familiar with smoke, soot, smuts and ash. This has been a major area for activity in the UK both in terms of monitoring and abatement under the Clean Air Act. This will be described in the section on the National Survey of Air Pollution. There have been detailed local studies of specific carcinogenic components in smoke but the UK does not monitor this aspect of air pollution on a national scale. Gaseous and liquid hydrocarbons have provoked interest in relation to cars and aircraft. Warren Spring Laboratory has studied total hydrocarbon concentrations in typical airport conditions and hydrocarbons also feature in the Five Towns Survey.

3.6. Sulphur oxides

Many natural and processed fuels contain sulphur, from negligible amounts up to over 5%. Sulphur contributes to the energy produced on combustion as it is oxidized to sulphur dioxide. Together with smoke, sulphur dioxide has been the major pollutant identified with urban pollution in the UK for some decades and it has been intensively monitored in the National Survey. Particularly during high-pollution incidents interest has also been directed at sulphuric acid to which, with the addition of water, sulphur dioxide is transformed. Sulphuric acid droplets have formed a significant part of the total phenomenon in high-pollution incidents. Some measurements of sulphuric acid have been made at such times and more regular measurements have been proposed for the future. Measurement of sulphur oxides by the UK also features in the regional OECD project on long range transport of pollution which will be dealt with in the following paper (Holdgate and Reed: The fate of pollutants).

3.7. Lead

This is one of the most controversial environmental constituents at the present. The main difference, between it and most of the others listed, lies in our exposure to it through media other than the atmosphere, and in the complexity of human response to this significantly cumulative poison. Through our diet we ingest some 300 µg per day of which we retain only a small proportion. In addition we are exposed to lead in the air, especially in urban areas and particularly from the use of leaded petrol by motor vehicles. Some of this vehicle-emitted lead is transferred to surfaces and we are further potentially exposed to it through our contact with dust and contaminated articles including foodstuffs. The meaningful monitoring of this pollutant is thus a complex affair. Regular studies are made of foodstuffs; airborne lead levels have been measured for some years in London's Fleet Street; analyses have been made of city dusts and in some cases body levels of lead have been measured to assess how much of the environmental lead is in fact reaching human targets. It is not my purpose to evaluate the importance of lead as an environmental pollutant. Lead is also on the programme of the Five Towns Survey. In Birmingham, measurements are being made in the vicinity of the giant motorway interchange system. Other measurements have been made around industrial installations which process lead. Limited studies have been made of organic lead vapour from the use of leaded petrol.

3.8. Trace elements

Several elements, toxic if in sufficient quantities, are contained by a variety of fuels. Among these are germanium, selenium, vanadium and mercury. Their overall fate on combustion has not been examined in great detail. In this country the NERC has initiated a study by AERE, Harwell, of the aerial concentrations of some thirty elements at seven sites, predominantly rural. There are now proposals for the extension of multi-element measurements to a selection of sites ranging from rural to urban/industrial. The NERC study also covers rainwater and dry deposited matter.

3.9. Radiation contamination

While atomic fuel and atomic energy do not quite come into the same category as fossil fuels and their combustion, some consideration must be given to the monitoring of radioactive materials. Reference has been made to global monitoring of fallout from thermonuclear explosions but in this country we also monitor our atomic power plants. Solids are readily retained in filters but the volatile I^{131} is the greatest potential hazard from present atomic power plants. The level of antipollution

protection applied to the nuclear industry is very high but the potential risks from a large number of nuclear power plants, and the associated fuel processing installations, are not negligible. Against this must be weighed the advantage of the removal of most of the other forms of pollution associated with energy production from fossil fuels. As far as atomic energy pollution risks are concerned, attention should now be directed primarily to breeder reactors and the associated fuel processing. Heat dissipation is one of the major environmental problems. It is too early yet to look even further forward in pollution terms to fusion rather than to fission.

4. THE NATIONAL SURVEY OF AIR POLLUTION

4.1. While the preceding section summarizes the monitoring of a range of individual fuel-produced pollutants, it is appropriate to describe in greater detail some of the major air pollution monitoring activities in the UK, and what they reveal about the state of our environment. It has been mentioned that the Clean Air Act of 1956 brought legislation applying to the emission of smoke, while steps were also taken increasingly to minimize ground level concentrations of sulphur dioxide. It was decided that the existing survey measurements, which provided a general picture of air pollution in a number of towns, should be expanded to cover the country's urban areas on a logical basis and that there should be a standardization on appropriate procedures. Among the stated aims of the National Survey are the provision of guidance to central and local government in the application of clean air legislation and the assessment of improvements occurring as a result of such legislation or other causes. It is not my intention to describe the details of the Survey which is performed mainly by Local Authorities on a voluntary basis. It is timely, however, to look back over ten years of the Survey and see how it has met its various objectives and what it has told us.

4.2. Much of the most useful information that can be gleaned from the Survey is obtained by considering the pollution data in conjunction with figures for fuel consumption and for the resulting emissions of smoke and sulphur dioxide. Detailed statistics are collected for the Regions, covering fuel consumption by type of fuel and, in broad terms, the type of device in which it is burned. At the same time the mean sulphur contents are provided for different types of coal and for various grades of oil. Sulphur dioxide emission factors can be reasonably

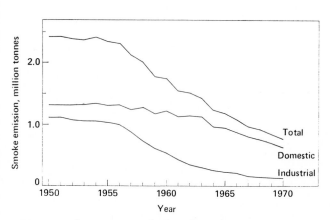

FIG. 1   Smoke emissions in the United Kingdom

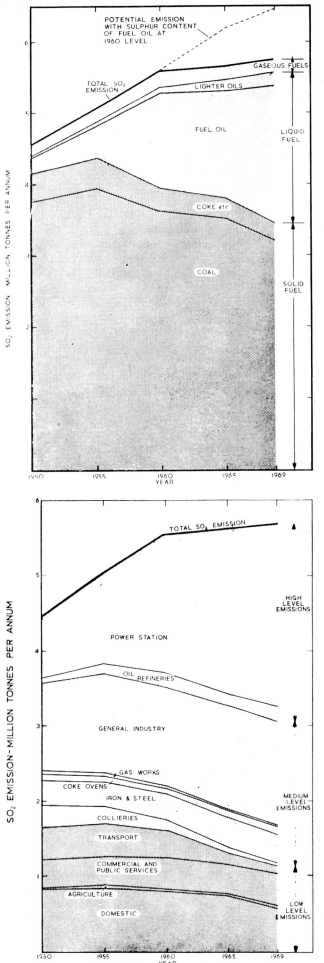

FIG. 2   SO₂ emissions in the United Kingdom by type of fuel

defined and finally smoke emissions from coal burning are estimated using a series of crude assumptions as to the weight of smoke emitted per unit weight of coal burned for a variety of purposes.

4.3. Figs. 1, 2 and 3 show the trends in smoke and sulphur dioxide emissions with details, for sulphur dioxide, of fuel type and height of emission.

4.4. In the first years of the Clean Air Act the main target in smoke reduction was industry where practically all smoke arose from the incomplete combustion of coal in inefficient boiler plant and furnaces. With the success of measures applied to industry, the bulk of remaining smoke emissions in recent years have been from domestic sources. Domestic smoke control, while supported financially by central Government, has been organized by Local Authorities. They have been able to use National Survey data to set priorities and to observe the results of introducing smoke control—and there have been striking successes. Fig. 4 shows the very close relationship between estimated urban smoke emissions, and the smoke pollution concentrations measured by the National Survey.

4.5. The Regional situations are shown by Tables 1 and 2. In considering coal consumption figures (Table 1) allowance must be made for the difference in the nature of coal used in some of the Regions. This is incorporated in calculated emissions used to produce Table 2.

4.6. Sulphur dioxide presents an interesting and more complex picture. While the total emissions of sulphur dioxide have been rising (Fig. 3), the urban ground level

TABLE 1   Regional distribution of smoke concentrations and of domestic coal consumption in the United Kingdom

Region	Domestic coal consumption per head, tonne 1970	Average smoke concentration µg/m³	
		1969–70	1970–71
North	0·56	95	88
North West	0·52	90	81
Yorkshire & Humberside	0·50	83	80
Northern Ireland	0·54	80	72
Scotland	0·38	79	69
East Midlands	0·44	64	60
West Midlands	0·34	54	50
East Anglia	0·29	46	46
London	0·04	46	42
South East, excl. London	0·14	34	31
Wales	0·54	32	33
South West	0·18	31	29

FIG. 3   SO₂ emissions in the United Kingdom from the combustion of fuels—1950–1969

TABLE 2 Regional trends in smoke emission and concentration in the United Kingdom

Region	Percentage decrease in smoke in the 6 years	
	1964–70 emissions	1964–65 to 1970–71 concentrations
North	34	34
North West	27	51
Yorkshire & Humberside	38	42
Northern Ireland	18	45
Scotland	38	58
East Midlands	42	43
West Midlands	36	54
East Anglia	32	48
London	67	49
South East, excl. London	35	40
Wales	29	24
South West	42	32

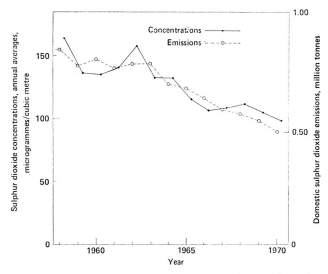

FIG. 5 Sulphur dioxide trends in urban areas in the United Kingdom

concentrations as shown by the National Survey have fallen by almost one-third between 1960 and 1970 (Fig. 5). This decrease follows the same pattern as the reduction in low level emissions of sulphur dioxide shown in the same figure. It is not the purpose of this paper to discuss the pros and cons of the high-stack policy adopted in this country but the National Survey monitoring results show the success of the policy in its declared intention to disperse large sulphur dioxide emissions, particularly from power stations, in such a way as to minimize their contribution to ground level concentrations of sulphur dioxide.

4.7. As with smoke, Regional figures for sulphur dioxide (Table 3) show central Government where overall priorities lie.

4.8. Local priorities can also be set where the monitoring information is adequate or where the pollution concentrations can be reasonably inferred. Within the London Region, which has the highest mean sulphur dioxide

concentration, the City of London has introduced legislation which will put a limit on the sulphur content of fuel used in all heating installations. This is particularly directed to the high density of commercial premises in the area. The monitoring results will chart the effect of this legislation in an area totally surrounded by the rest of London without such legislation.

4.9. Detailed epidemiological studies of human response to pollutant concentrations cannot be carried out nationwide. However, the relationships which are shown in such studies between symptoms, and the concentrations of smoke and sulphur dioxide, have provided a rough yardstick for authorities elsewhere in the country to relate their pollution monitoring results to the extent to which their populace is exposed to a real pollution hazard. The annual published tables of the National Survey highlight the numbers of days on which the daily mean figures for smoke and sulphur dioxide exceed various levels, 250, 500 µg/m³, etc.

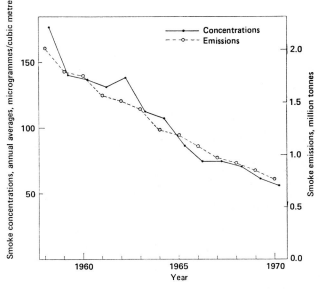

FIG. 4 Smoke trends in urban areas in the United Kingdom

TABLE 3 Regional distribution of sulphur dioxide concentrations in the United Kingdom

Region	Average concentration, µg/m³		Percentage decrease since 1963–64
	1969–70	1970–71	
London	143	132	29
Yorkshire & Humberside	140	121	23
North West	130	125	32
West Midlands	113	92	33
East Midlands	99	96	20
Northern Ireland	86	78	16
East Anglia	86	88	25
North	85	92	18
Scotland	85	85	23
South East, excl. London	79	74	23
South West	66	63	3
Wales	56	48	−3

4.10. It should be emphasized that the National Survey has been primarily concerned with urban pollution levels in relation to human health. It is unfair to castigate the Survey for not serving purposes for which it was not designed. It is also relatively easy to suggest with hindsight that it should have been designed to meet a number of other objectives without giving adequate thought to what the requirements would have been for instrumentation, for personnel and for finance.

4.11. Within the last few years increasing concern has been expressed about the level of sulphur dioxide pollution in rural areas and its possible effect upon vegetation rather than animal life. Accurate analytical methods specific for sulphur dioxide are employed at a small number of rural sites but the long sampling periods used are designed for the estimation of mean pollution levels. Significant effects on vegetation may be produced by short periods of high sulphur dioxide concentrations. This monitoring situation is under review.

5. DEPOSITED MATERIAL—GRIT AND DUST

5.1. Monthly measurements of grit and dust are reported from some 600 sites in the UK. By grit and dust is meant the insoluble portion of the particulate material which falls out of the air under its own weight and is collected by the standard deposit gauge (BS 1747, Part 1). Deposition rates are expressed in mg/m^2 per day. Some operators make selected chemical analyses on the deposited material.

5.2. Grit and dust are made up largely of ash, unburnt solid fuel, windborne dust from roads and industrial installations, including those handling solid materials in the open. Because the size of particles in this category tends to be well above the respirable range, their significance is primarily in terms of nuisance, or disamenity, rather than health hazard. Grit and dust are one of the forms of pollution with which the Clean Air Act of 1956 is concerned; this requires that grit and dust from existing solid fuel-fired industrial furnaces be minimized, and that new furnaces of this kind be fitted with arresters.

5.3. The UK measurements do not constitute a national survey. Many of the instruments have been sited in order to monitor the deposition from some particular industry. In contrast, particularly to sulphur dioxide measurement, deposition figures are valid only for a small area round a gauge and do not necessarily give much indication of average pollution in the rest of a town or region. The grit and dust monitoring reported by WSL thus constitutes a collection of highly local surveys as designated in section 3. The results for individual sites can be of considerable use in dealing with local situations. In addition to chemical analysis, mentioned above, microscopic examination of grit and dust can be a powerful aid to the diagnosis of the source of such deposits. A number of different forms of combustion-produced particulates can be clearly distinguished. In certain circumstances diagnosis of the source of particulate pollution can be assisted by the use of a standard directional deposit gauge (BS 1747 Part 5).

5.4. While certain industrial sources of grit and dust, including many combustion installations, have undoubtedly been dealt with, there has been no general downward trend in the median or peak deposition rates reported since 1963. The range of values found for yearly or seasonal averages is broad. It is not clear how far grit and dust deposition involves fresh material as against the cycling of previously deposited matter.

6. THE FIVE TOWNS SURVEY

Over a number of years UK surveys have been carried out of pollutants associated with motor vehicles; but, in contrast to sulphur dioxide and smoke, the approach has been to concentrate on situations in which the vehicle flow and street ventilation conditions are likely to lead to the highest levels of pollution. The carbon monoxide studies of the 1960s suggested that concentrations of this pollutant in our streets were not hazardous, but the seemingly inexorable rise in car density, and the development of interest in other aspects of vehicle pollution, has led the Government to initiate a more detailed study of a number of pollutants in the busiest streets of five towns in the UK for a period of five years. Levels of carbon monoxide, total hydrocarbons, lead and smoke, together with meteorological data are to be measured at all sites. In addition, at two sites in London, (one at the roadside and the other at a background station) nitric oxide, nitrogen dioxide, gaseous sulphur compounds and ozone are continuously measured. With the exception of lead and smoke, concentrations are measured at intervals of five minutes and the data, logged on punched tape, are analysed off-line by computer. It will be possible to register trends in pollution levels by varying traffic load or as a result of legislation on engine design, exhaust systems, or fuel. At the same time, the detailed nature of the results obtained should be of value to medical researchers concerned with short-term peaks in various pollutants which may also exhibit co-operative or synergetic effects. Furthermore it has proved possible to study such items as the contribution of photochemistry to the production of certain pollutants. The survey information will be one of the bases for deciding on the possible need for further legislation.

7. CONCLUSIONS

With the growing energy requirements of the world continuing (for some time, at least) to be met primarily from fossil fuel combustion, more thought must be given to the air pollution aspect of this activity. Air quality monitoring is essential as a guide to conditions in areas of concentrated and, particularly, low level emissions. There are substantial underlying movements in the patterns of fuel combustion by various sectors, domestic, industrial, transport and power production. These have significant pollution implications. Carefully analysed pollution trend data can be an important guide to policy for the optimum use of various types of fuel and the possible need to make changes in fuels, or in combustion practice.

M W HOLDGATE, MA, PhD, FIBiol* and L E REED, MSc, PhD, MIMechE, MInstF†

8 The fate of pollutants

SUMMARY

Pollutants are here considered in four main categories: those where local concentrations are of importance; those which react in the atmosphere to form products with less desirable characteristics; those which are persistent but deposit out of the air to swell the background concentration in soil and water; and last those which are persistent and whose concentration may be increasing in the atmosphere.

The local problems of sulphur dioxide and carbon monoxide still require most attention; but the question of the transport in the air of sulphur compounds between countries is now being studied intensively. Pollutants produced by photochemical reactions have been detected in the air in Southern England, but the risk of severe damage as a result is small. Carbon dioxide released from combustion processes is adding to the background concentrations already present: dire consequences have been forecast if the trend continues; but these extreme views do not commend acceptance. Nevertheless, carbon dioxide is a potential climate modifier and, as such, demands serious study.

1. INTRODUCTION

Fuel combustion is still the main source of energy, derived initially as heat, and then converted to other forms appropriate to particular purposes. We are all familiar with the steep, rising curve of energy demand, and with forecasts of a doubling or trebling of *per capita* demand in the United States and other developed countries between a base date in the 1960s and 2 000. Various figures are, for example, given in SCEP[1] and Kneese, Rolfe and Harned.[2] Although steady improvements in thermal efficiency are reported by these authors, all the projections are for a parallel steep rise in fuel consumption, and in the pollutants that must inevitably be generated in consequence.

Since man first discovered fire, the atmosphere has

*Director, Central Unit on Environmental Pollution, Department of the Environment.
†Deputy Director, Central Unit on Environmental Pollution, Department of the Environment.

been the sink into which our airborne wastes have been emptied. Until recent years this has been of little concern on a global scale. However, the rapid increase in human population, coupled with rising standards of living and increasing industrial development, has led to a present situation in which the contribution by man to the levels of some pollutants in the air is approaching, or even overtaking, the contribution from natural sources.

Although dispassionate studies (eg SCEP[1], SMIC[3], Royal Commission[4], suggest that air pollution is still not causing changes in climate or atmospheric properties on a world-wide scale, the possibility that it might do so, and the need for research in order that unwelcome effects can be foreseen and forestalled, are admitted. At local level, moreover, there is no question that there has been severe damage, sometimes over regions of hundreds of square miles in extent, attributable to atmospheric pollution: Fig. 1 is a familiar example.

So far, technology has only provided means for removing a proportion of the obnoxious or potentially dangerous pollutants from emissions. Dispersion of the remainder into the atmosphere by means of chimneys has become accepted practice. Underlying this practice there has (at least since the enactment of the first Alkali Act of 1863) always been the principle that the ensuing concentrations at ground level should not cause unacceptable hazard or damage. Our standards of acceptability have risen over recent decades, and any demonstrable rise in human mortality or illness, significant damage to agriculture, forestry or horticulture, or ecological changes over substantial areas would now be regarded in most developed countries as 'unacceptable'.

The need for predicting and preventing unacceptable effects has led to research, especially on the patterns of dispersion of pollutants, and the relationship between concentration and effect. Theories of dispersion are well developed, and it is as a result of these that we have confidently accepted the discharge of large volumes of polluted air through tall chimneys.

However, although present practices may minimize the concentration of pollutants locally at ground level, we are now being forced to recognize that what happens to them in the atmosphere at large can no longer be ignored. They may persist or accumulate; they may be washed out by rain, they may be deposited on the soil,

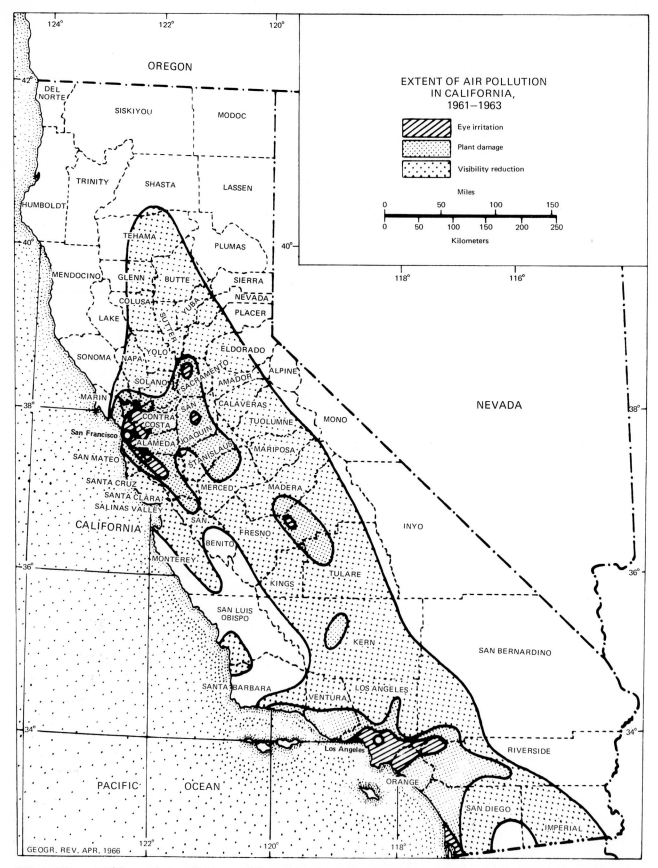

FIG. 1   The extent of general air pollution in California, 1961–1963. The plant-damage areas are specific, but the eye irritation and visibility reduction may be due in part to forms of general pollution other than photochemical. Sources: for plant damage, J. T. Middleton; California against Air Pollution (California Department of Public Health, Sacramento, 1961); for eye irritation and visibility reduction, local reports and personal observations up to December 1963. After Leighton[36]

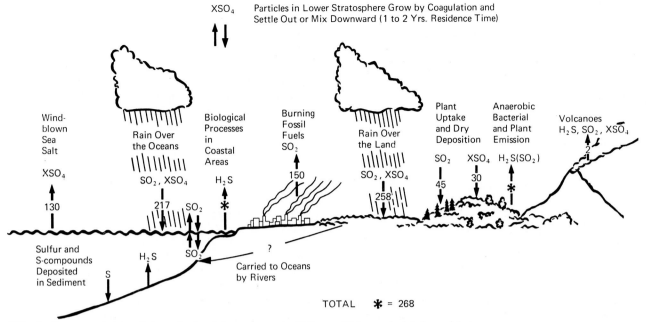

FIG. 2 *Sources and sinks of atmospheric sulphur compounds. Units are 10^6 tons calculated as sulphate per year*

water, vegetation or structures; or they may be transformed in the air into other substances which, in some cases, may be more obnoxious than the original ones.

At the outset, there are certain basic relationships to define. Pollutants are of significance because of their effects, and the magnitude of an effect is generally proportional to pollutant concentration. This depends on the balance between the rate of generation of a substance and its rate of removal. Removal and generation can both involve several pathways. It is particularly important to discriminate between natural and man-made sources, and between removal involving the actual degradation or combination of the pollutant molecule, and that caused by physical transport from the point of origin and dilution. In this paper we propose to look at these broad questions for a selected group of atmospheric pollutants. Other authors have dealt in detail with the chemistry of the atmosphere, and of the vast number of pollutants that can be found in the air.

2. POLLUTANTS IN THE AIR

An enormous number of substances can contaminate the air. One recent American Handbook[5] lists 260 'potential atmospheric contaminants' in a list that is certainly not exhaustive. But many of these substances are released as a result of particular industrial processes, and on a small scale. The atmospheric pollutants that cause most

concern are those emitted in large quantities, especially as a result of fuel combustion. The chief of these are carbon dioxide, carbon monoxide, hydrocarbons, oxides of nitrogen, oxides of sulphur and particulates (including various metals). Table 1, based on data in SCEP indicates the quantities contributed by various sources in the United States in 1968. It is clear that fuel combustion in stationary sources contributed most of the sulphur oxides, and a substantial proportion of the particulates and oxides of nitrogen; while motor vehicles were the main sources of hydrocarbons and carbon monoxide.

Naturally, the priorities for control in various countries relate to value judgements about the seriousness of these different pollutants, and in this context a recent OECD survey[6] is illuminating. It shows that in most countries (including Germany, France, Italy and Japan) sulphur oxides were accorded first priority. In contrast, in the United States sulphur oxides were rated eighth in order of priority, with nitrogen oxides and hydrocarbons coming first and second, and carbon monoxide fourth. This reflects the American preoccupation with motor vehicle emissions and oxidant smog. In the remainder of this paper we look in more detail at some of these pollutants, assess their origin and fate in the environment, and point to ways in which our judgement of the scale of the problems they pose can be sharpened.

2.1. Sulphur

Sulphur is an element essential to man and to animal and plant life, and a component of some protein molecules. In nature, it is taken up by green plants from the soil as sulphate, and obtained by animals from their food. Sulphur in dead plant and animal tissue is released by bacteria as hydrogen sulphide which is then (under aerobic conditions) oxidized by other sulphur bacteria to sulphur dioxide, and in some instances to sulphuric acid. This acid reacts in the soil with a variety of bases, in some cases forming salts such as calcium sulphate which are directly available to plants. Some of the H_2S and SO_2 produced by decomposition is released to the atmosphere, where SO_2 of volcanic origin also occurs. Lovelock et al[7] have suggested that dimethyl sulphide

TABLE 1   Major air pollutants and their sources in the United States (1968).

| Source | Millions of tons | | | | | |
	CO	Hydro-carbons	NO_x	SO_x	Parti-culates	Total
Fuel combustion, stationary sources	1·9	0·7	10·0	24·4	8·9	45·9
Transportation	63·8	16·6	8·1	0·8	1·2	90·5
Industrial processes	9·7	4·6	0·2	7·3	7·5	29·3
Garbage incineration	7·8	1·6	0·6	0·1	1·1	10·2
Miscellaneous*	16·9	8·5	1·7	0·6	9·6	37·3
TOTALS	101·1	32·0	20·6	33·2	28·3	215·2

*includes emissions from forest fires, structural fires, coal refuse, agriculture, solvent evaporation, gasoline marketing.

TABLE 2 Oxidation of SO₂ in the atmosphere (after Junge[14])

	Reactions	Rate % h⁻³	SO₂ Life time
	Gerhard and Johnstone 1958 Cox 1970	0·03–0·2	140–21d
Photochemical oxidation under Normal sunshine	in presence of NO: 0·3–0·7 ppm Olefins: 0·1–1·0 ppm Cox and Penkett 1971a	2–11	50–9h
Non-photochemical homogeneous gas gas reactions	SO₂: 0·1 ppm O₃: 0·05 ppm Olefins: 0·05 ppm Cox and Penkett 1971b no NH₃ present 50 μg/m³ Mn in Aerosol Cheng et al. 1971	0·4–3 ~2	10–1d ~2d
Liquid phase reactions in clouds and fog	with NH₃ present depending on NH3 and catalyst present e.g. Scott and Hobbs 1967	10–100	10–1h

may also play an important part in the transfer of sulphur between air and the surfaces of land and sea. Any appraisal of the significance of emissions from human activities must take into account this natural background.

Estimates of the scale of natural production of hydrogen sulphide and sulphur oxides inevitably vary, and contain a large amount of guesswork. Kellogg *et al*[8] have drawn on the estimates of Junge[9], Katz[10], SCEP and Robinson and Robbins[11] to arrive at a global sulphur balance which is reproduced as Fig. 2. It will be noted that the figure of 268 million tonnes yr.⁻¹ sulphate cited for natural releases is that required to balance the other emissions and depositions for which estimates can be

made with less dubiety. A Swedish study[12] suggests a range of 100-250 million tonnes yr⁻¹ sulphur from biological decay. In North-West Europe a natural emission of 62 million tonnes is estimated.

Various authors have estimated the total emission of sulphur due to man's activities. SCEP calculated a global figure of 93 million tonnes yr⁻¹ SO₂: Robinson and Robbins had earlier suggested 146 million tonnes yr⁻¹. The Swedish study referred to above calculated a lower value around 35-55 million tons. Kellog (Fig. 2) adopts an emission figure of 150 million tonnes yr⁻¹, as sulphate. Of this some 93·5% is produced in the northern hemisphere; in 1968 16·7 million tonnes yr⁻¹ are calculated to have been released in NW Europe alone. Kellog attempted to forecast the rate of increase in SO₂ emission, deriving a figure of 275 million tonnes yr⁻¹ by the end of the century: as he points out, this prediction is based on estimates of the future use of fossil fuels and is, inevitably, uncertain. OECD[13] estimated that European emissions would rise to 27·4 million tonnes yr⁻¹ by 1980, and that some 60% would come from power stations. The same statistics suggest that the UK emission, at present some 6 million tonnes, is more than that from any other European country. It is clearly important to know what happens to these emissions.

There is no evidence that sulphur dioxide is accumulating in the atmosphere. The increasing quantities discharged by man must therefore be returned in due course to land or sea. Because of the ease with which sulphur dioxide is converted into sulphates and sulphuric acid in the air, its lifetime there is short. The precise rate of this transformation, which can involve a variety of processes, is less easy to calculate. Table 2, taken from

FIG. 3 Computer calculations of sulphur dioxide concentrations for Nashville, Tennessee. Solutions of a multiple-source meteorological diffusion model compare favourably with measured values. Abscissa and ordinate are city dimensions (miles); isopleths are concentration values in parts per 100 million (pphm). After Pack[36]

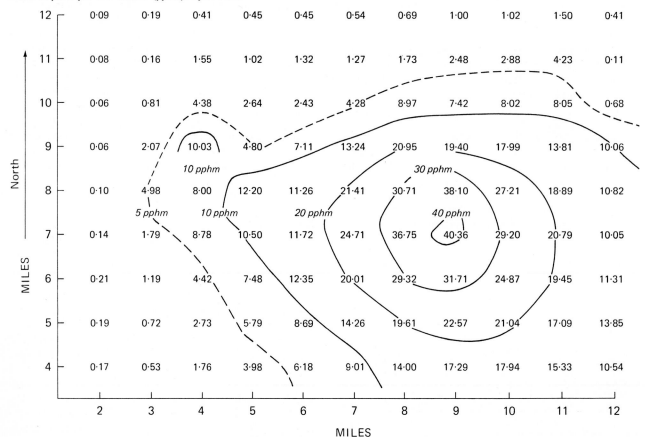

Junge[14] shows that the lifetime of SO_2 in the atmosphere could vary from 1 hour to as much as a few weeks, depending on which of the alternative reactions occurs. Where humidity is high, there is rapid formation of sulphate, which may exist as sulphuric acid or as ammonium sulphate if sufficient ammonia is present. More recent work by Atkins et al[15] has described how photochemical processes may enhance the oxidation of sulphur dioxide to sulphate. The increase in particulate formation could be an important factor in reducing visibility in urban areas under calm summer conditions and Tokyo, New York and other cities have already experienced severe oxidant conditions with reduced visibility. No such episodes have been recorded in the UK, but their occurrence cannot be precluded, and might be encouraged by the increasing emission of pollutants from motor vehicles.

Calculations of the rates of these transformations under various conditions are critical in the estimation of the distance that sulphur oxides can travel in an air mass. This, in turn, is vital to the assessment of the pattern of effects which a source may create, and to vexed issues such as that of trans-frontier pollution. Generally speaking, sulphur oxides generated by man predominate in cities, and there is a fairly steep gradient of concentration to rural areas. Such patterns of concentration are predictable, with the aid of mathematical models (Fig. 3). Patterns of effect of air pollution can likewise be correlated with the distribution of urbanization.

This pattern of concentration has been known for many years. In the United Kingdom it has been policy to reduce patches of high ground-level concentration, and ensure rapid dilution of emissions, using tall chimneys. It has been assumed that this practice will keep levels generally below the threshold at which human health or agricultural productivity is endangered; and, given the short residence time of sulphur dioxide in the air, it is considered by many that there is little prospect of serious cumulative effect. Indeed, in places where natural sulphur is deficient, airborne sulphate could be beneficial (Cowling and Jones,[16] Ross[17]).

These assumptions have recently been questioned by studies that suggest that, under certain meteorological conditions, airborne sulphur compounds can be transported substantial distances, and be precipitated as acid rain which can cause significant damage in places where the soils are of low alkalinity and the lakes oligotrophic. In Sweden the acidity of soils is said to be increasing in a manner that threatens to reduce forest productivity by some 10-15% over the next 30 years; and the acidity of many rivers and lakes is increasing to the point (below about pH 4·0) at which game-fish stocks decline.

This hypothesis of long-range transport is now being tested by a collaborative study under the auspices of the Organization for Economic Co-operation and Development (OECD). Measurements at about 40 strategically-sited land stations in the ten participating NW European nations will show the distribution of pollutants and how this responds to changing weather conditions. When high concentrations occur they will be tracked back to their area of origin by analysis of meteorological data.

There are six sites in the UK (Fig. 4). At each site arrangements have been made to measure rainfall and its acidity, airborne sulphur dioxide, and total particulate sulphur. Routine measurements are made on a 24-hour basis at Cottered and Eskdalemuir; but when the

weather pattern suggests that dispersion is likely to be poor and that there may be increased pollution in one of the participating countries, measurements are stepped up to a 6-hour frequency, and the other 4 stations start to make daily measurements. All results are sent to the Norwegian Centre for Air Research, which is co-ordinating the project, and analysing the data.

Sulphur dioxide is a reactive substance, and measurements of its concentration alone in the atmosphere are not sufficient to establish the history or lifetime of related sulphur compounds. Some information has been obtained by using aircraft to collect samples of sulphur compounds: with a flight path parallel to the east coast of England, at a distance of a few kilometres out to sea, it is possible to estimate the flux of sulphur compounds transported out of the area under study. By estimating the source strength, the percentage of sulphur deposited before reaching the coast can be calculated. Repeating the procedure near the Danish-Norwegian coastline gives an additional estimate of the losses over the North Sea. Both the Warren Spring Laboratory and the Meteorological Office are engaged in this sampling, and their preliminary findings* indicate that between 15 and 60% of the sulphur dioxide emitted can leave the English Coast in an easterly direction. Other forms of sulphur have not yet been studied so that a total balance cannot be attempted; but on subsequent flights there will be sufficient measurements to allow the construction of a sulphur budget.

Direct uptake by the ground (dry deposition) is

FIG. 4 Location of UK measuring sites

* Currently at press.

probably one of the most severely under-estimated ways by which sulphur is removed from the atmosphere. Meetham[18] calculated that only some 20% of the SO_2 emitted was blown out to sea, the remainder being either washed out by rain or deposited directly on soil or vegetation. Current work at AERE Harwell[19]* has confirmed that dry deposition may be a significant 'sink' for airborne sulphur. Direct measurement of the SO_2 gradient near the ground has shown that grassland will take up appreciable quantities of it. Extrapolation of the result from the Harwell site to the whole country suggested a total deposition of the order of 1·2 million tons annually (some 20% of the total UK emission). This is considerably less than the figure suggested by Meetham, but it is far from clear how representative the grass surface at the Harwell site is of the country as a whole.

3. CARBON MONOXIDE
Carbon monoxide as a pollutant is formed by the incomplete combustion of the carbonaceous matter of fuel, and is a sign of inefficiency. Petrol-engined motor vehicles are the principal source.

There is not the same wealth of information on carbon monoxide as there is on sulphur dioxide, and it is only in recent years that research has started to uncover its cycle in the atmosphere. It has been known for some time that carbon monoxide does not appear to be accumulating in the atmosphere, but the means of its elimination were not understood. The quantity of carbon monoxide in the troposphere has been put at 530×10^6 tons by Maugh[20] with an annual contribution by man of about 270×10^6 tons. Junge[14] quotes a figure of about 200×10^6 tons for man's emissions, (85 per cent of which comes from motor vehicles) and reviews the atmospheric cycle of carbon monoxide and its possible sinks. The oceans that were thought to provide a capacity to absorb carbon monoxide in fact play the opposite role. Their surface waters appear to be supersaturated in respect of the solution equilibrium with ocean surface air. The carbon monoxide concentration is highest in the surface layers and decreases with depth. Junge therefore points to the oceans as being a source, and not a sink, for carbon monoxide. Crude estimates indicate that the total emission would not be dissimilar to that produced by man.

The importance of natural emissions was confirmed by Stevens et al[21] of the Argonne National Laboratory who examined the isotopic composition of samples collected at various locations and times, and traced the origin of several different species of carbon monoxide. By comparing the average isotopic composition of carbon monoxide with that from a species whose production rate was known, they were able to estimate that natural production, mostly from the oxidation of methane and the degradation of chlorophyll, was about $3\cdot5 \times 10^9$ tons yr^{-1} in the Northern Hemisphere alone. This is about 10 times the total emission by man.

The confirmation of this massive natural emission at the same time implies that sinks of an equivalent capacity must be available, otherwise accumulation in the atmosphere would be evident. If the oceans have been discounted as sinks the only remaining ones are the atmosphere itself, and the ground.

Measurements of the CO concentration in the atmos-

phere made by Seiler and Warneck[22] from an aircraft showed that there was a marked reduction at heights above 27 000 ft. This suggests very strongly that the stratosphere is a powerful sink for carbon monoxide; they found, however, a difference between northern and southern hemispheres, and with appreciably more carbon monoxide present in the atmosphere of the former. They attributed this difference to the larger volume of human emissions in the northern hemisphere; but it is equally possible that natural emission rates will differ, in association with differences in land and sea areas, and in the rates of CO generation by terrestrial and aquatic vegetation.

Recent research has begun to clarify the role played by soil in the carbon monoxide cycle, but the evidence to date is far from conclusive. It appears that soils can be either a source or sink of carbon monoxide depending upon the microflora or microbial activity present[23]. Present estimates indicate, however, that the soils alone would provide a sink of sufficient capacity to absorb both the natural and man-made sources of carbon monoxide.

There is no evidence that the carbon monoxide discharged as a result of man's activities is of any global significance. The only adverse effects known occur in urban areas (especially road tunnels and confined spaces with heavy traffic) where levels can rise sufficiently to block a small proportion (rarely more than 2–3% of the oxygen-carrying capacity of the blood. Provided that discharges are arranged in such a way that their local concentrations do not exceed those at which a health risk might be incurred, there seems no need to adopt heroic and expensive measures to control emissions.

4. CARBON DIOXIDE
Carbon dioxide is the final oxidation product of carbonaceous fuels; it is also an abundant compound, intimately involved in the natural cycle and essential to the maintenance of life. It exists in the ambient air at a concentration of around 300 ppm, and it is only if man's activities enhance this value, so as to interfere adversely with natural processes, that carbon dioxide could be considered a pollutant.

Early measurements of the concentrations of carbon dioxide in the atmosphere lack the accuracy of those made today, but the evidence points to a concentration of about 290 ppm at the beginning of the industrial revolution. Precise data are now available from measurements made at a number of sites, of which the most well known is that at Mauna Loa, Hawaii, where an observatory is sited at a height of 11 000 ft free from the influence of local sources. Records of the carbon dioxide content of the air have been maintained continuously from 1958 onwards. The trend is shown in Fig. 5. The pronounced seasonal variation is evident: the summertime decrease being due to photosynthetic uptake during the growing season. The net annual increase between 1958 and 1968 was about 0·7 ppm, but there is evidence that accumulation proceeded faster in the later part of this period, and by 1971 the rate of increase had risen to around 1 ppm yr^{-1}. Other stations confirm a current annual increase of around 1 ppm; but at the South Pole the earlier rate of 0·7 ppm still appears to prevail. In 1971 the CO_2 content of the atmosphere was about 323 ppm.

Machta[24] discusses the variations in the rate of annual increase, and points out that the accumulation of 0·7 ppm

---

*Currently at press.

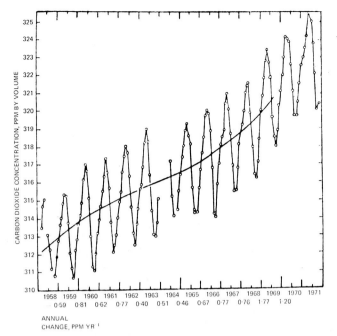

FIG. 5 Mean monthly atmospheric carbon dioxide concentrations at Mauna Loa. After Machta

yr⁻¹ between 1958 and 1968 represented on a global basis only about half the total carbon dioxide discharged from the combustion of fossil fuel. Plant photosynthesis and absorption in the oceans are believed to be the main 'sinks' for CO_2, and the variations in annual increase are best accounted for by variations in uptake by plants and changes in sea temperature. Machta has developed a model for total CO_2 movement in the environment using carbon-14 as a tracer. This predicts a mean atmospheric level of 380 ppm by the year 2 000. Such predictions depend on assumptions about the interchange of CO_2 between air and ocean. The total quantity of carbon dioxide in the air is equivalent to only about one-sixtieth of that in the ocean. Junge[14] has reviewed the interchange of CO_2 and oceanic carbonates and bicarbonates on a geological time scale and, against this, it is almost impossible to distinguish the short-term perturbations caused by man. Machta has speculated that if atmospheric temperature and CO_2 levels rise, this might reduce the oceanic capacity to absorb the gas and lead to an acceleration of atmospheric warming through the postulated greenhouse effect. However, the full understanding of these interactions demands a knowledge of how far, and how fast the deep ocean takes up carbonates from the surface layers, acting as a 'sink' for CO_2.

The carbon dioxide in the atmosphere permits incident solar radiation to pass through it and reach the earth. The resulting re-radiation is of longer wave length, and more of it is absorbed by the CO_2 layer. There has been much conjecture over the extent to which rising CO_2 levels might increase this greenhouse effect, causing general warming of the climate; various dire consequences, including melting of the polar ice caps, have been threatened.

On a more reassuring note, Manabe and Wetherald[25] suggest that if the mean atmospheric CO_2 level rises to about 380 ppm by the year 2 000, the mean temperature might rise by about 0·5°C in the lower atmosphere. There has been equal conjecture over how far this warming might be offset by increases in atmospheric turbidity, screening out radiation. A further twist to the argument

is provided by the suggestion that climatic warming might increase cloud cover, and that an increased mean cloudiness of only 0·6% could cancel out the entire warming attributable to the rise in CO_2 levels this century. The speculative nature of these arguments precludes any one of them being taken very seriously but CO_2 remains the chief pollutant which can at present be plausibly suggested as a potentially significant modifier of climate. For this reason, the factors controlling its atmospheric concentration and effects demand further study (Royal Commission, 1971)[4].

5. PARTICULATES

Smoke, produced by fuel-burning in the home and by industry, has been incriminated as the most damaging air pollutant in Britain. With sulphur oxides, it was the dominant component of the London smog of 1952/53, when 3 000-4 000 excess deaths were recorded. It has been the main target for control under the Clean Air Acts which, in parallel with changing patterns of fuel consumption, has led to a reduction of some 60% in smoke emissions in the UK and the abolition of the 'London particular.'

Smoke and fine particles are not solely man-made. Natural forest fires, volcanic eruptions and dust storms can inject substantial amounts into the air. It has been calculated that volcanic eruptions such as Krakatau or Hekla have put more such material into the atmosphere than man has done in the whole of recorded history. Violently explosive eruptions of this kind eject very fine dust into the stratosphere, where its residence time can be measured in years before it ultimately settles to earth. The 1963 eruption of Mt. Agung, on Bali, produced a dramatic increase, persisting for three or four years, in the atmospheric turbidity recorded at Mauna Loa, Hawaii—where levels have otherwise remained more or less constant since 1958.

Over some land areas in the northern hemisphere there is, however, evidence that levels of particulates in the air have risen: in parts of the USSR, the Swiss Alps and North America the amount of radiation reaching the ground has declined by up to 10% over the past 50 years[24]. But over tropical and southern hemisphere areas there is no indication of any recent trend. It seems likely, therefore, that particulates are largely significant as pollutants on a local, or zonal scale. Nonetheless, because atmospheric turbidity could affect the climate in a manner antagonistic to CO_2, further study of particulates and their residence time in the air is needed.

6. LEAD AND OTHER METALS

Various metals, including lead and mercury, occur as contaminants in coals and oils, and are released in a varying degree by combustion, often in complex particulates. These metals are generally well-dispersed and pose no (or only local) pollution problems. But lead is a special case deserving some attention.

Lead is only a minor constituent of fossil fuels and its evolution during combustion is of little significance. Lead, in the form of tetra-ethyl or tetra-methyl lead, however, is added to petrol to improve its octane rating. Most of this is emitted as inorganic lead with the exhaust products. Compared with the total tonnage of other pollutants emitted, that of lead compounds is minute; but lead is not a natural airborne contaminant (except for trace quantities), and it is highly persistent. Added to this is the high toxicity of the metal and its compounds, which may accumulate in vegetation, animals or humans.

The consumption of lead as additives to petrol was estimated to be 350 000 tons in 1970, the UK using about 10 600 tons in 1971. Approximately 70% of the lead is emitted as an aerosol from the exhaust pipe mostly in combination with bromine, chlorine or phosphorus. Recent work by Lawther et al[26] has indicated that the lead in the exhaust of petrol engines exists as small dense particles contained in an aggregate of carbon particles. The mean size of the aggregate is usually less than one micrometre. It is postulated that the inclusion of the lead in an inert carbon shield effectively denies the lead as a source that can be easily mobilized by man.

Although the medical significance of the availability of the lead from motor vehicles may be in question, its presence throughout the environment is not in doubt. The dust in most urban streets carrying a reasonable burden of traffic, contains lead in concentrations of 0·1 to 0·3% (1 000 to 3 000 ppm). (Concentrations ten-fold higher may occur around some industrial sources). Many studies have shown that the concentration of lead in dust, soil, or on vegetation declines sharply with increasing distance from a road. Much of the lead emitted (perhaps 50%) therefore will be removed from the atmosphere within a very short period. But Ruhling and Tyler[27] calculate that a comparable total quantity remains in the air for sufficient time for it to be widely distributed. This is illustrated by the data[28] shown in Fig. 6.

The quantity of lead deposited in Greenland snow can be followed back to about the year 800 BC. The values corresponding to the period around 1750 show the increase in emissions resulting from the beginning of the industrial revolution. The upsurge around 1940 reflects the increase in the use of lead as an additive to petrol. Similar increases are not found in snow in the Antarctic to the same extent as those in the North, first because emissions in the Southern Hemisphere have not increased to anything like the same extent; and, secondly, because the natural barriers to mixing between the atmosphere of the northern and southern hemispheres have inhibited

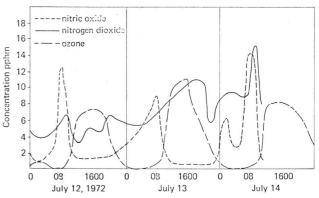

FIG. 7 Diurnal variation of pollutants in London. After Derwent and Stewart

transfer of lead aerosol to the Antarctic. It is likely that the global use of lead additives has reached a peak as many countries are beginning to introduce measures to prevent a continued increase in lead additives and, where possible, to reduce their use. It will, however, be some years before the effectiveness of these actions will be reflected in a drop in the concentrations found in the environment.

7. OXIDES OF NITROGEN, HYDROCARBONS AND OXIDANTS

Recently, these atmospheric pollutants have caused concern not because hydrocarbons or nitrogen oxides are particularly damaging on their own (although some of the former are smelly) but because they can interact, in still air and bright sunshine, to form ozone and other oxidants—viz. the notorious Los Angeles smog.

Yet neither oxides of nitrogen nor hydrocarbons, are generated solely by fuel combustion. Calculations indicate[11] that natural sources emit $1·5 \times 10^9$ tons of nitrogen oxides per year, compared with man's total emission of 50×10^6 tons; while forests release 175×10^6 tons yr[-1] of reactive hydrocarbons compared with man's 27×10^6 tons. The significance of man's emissions is that they are locally concentrated and their oxidant deriva-tives are damaging to plants (citrus-growing is said to be uneconomic within some 80 km of Los Angeles). They are the main cause of the damage pattern in Fig. 1.

Such damage is unknown in the UK but signs of photochemical reactions in the urban atmosphere have been detected. Atkins et al[15] have recently demonstrated that concentrations of ozone found in the ambient air at Harwell on certain days in summer were appreciably higher than those likely to be present from natural sources. They attributed these enhanced values to photochemical reactions. They also produced evidence that oxidation of sulphur dioxide to sulphuric acid droplets, or sulphate, occurred under the same conditions. It is this latter process which could be responsible for the presence of a visible white haze which reduces visibility on those infrequent occasions in summer when the intensity of sunshine is high and the wind speed low.

Derwent and Stewart[29] reported on measurements made in London in 1972 which show the diurnal variation of pollutants on days when photochemistry plays a significant role. Fig. 7 illustrates how a drop in the concentration of nitric oxide precedes an increase in the concentration of both nitrogen dioxide and ozone. There

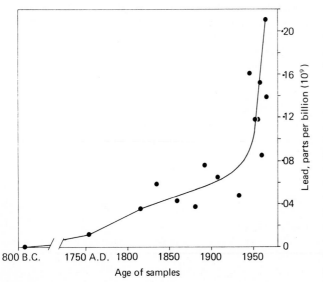

FIG. 6 Increase of lead in snow at Camp Century, Greenland, since 800 B.C. After Murozumi et al

have been no reports of eye-irritation or plant damage which might be associated with elevated levels of ozone and more complex compounds formed in the process. It is certainly premature to say that photochemical smog has occurred in England, but now that positive evidence of photochemical reaction exists, episodes on a minor scale cannot be ruled out in the future if the right climatic conditions prevail. Fortunately, under the turbulent atmospheric conditions and low sunlight commonly found in Britain it is highly improbable that oxidant levels will build up to damaging proportions.

8. CONCLUSIONS

Space has not permitted us to consider more than a handful of the pollutants for which the air is an important pathway. But the examples we have chosen illustrate some of the main kinds of substance with which we must be concerned. Carbon dioxide is the only pollutant so far known which is accumulating in the atmosphere in a way which may have a global effect, albeit to a marginal degree in the foreseeable future. Oxides of nitrogen, hydrocarbons, and sulphur dioxide exemplify pollutants whose interactions in the air can produce other substances with distinct, and less welcome, effects than those of the original substances. Lead is an example of the kind of pollutant which is dispersed widely in the air, settling out to swell background levels in soil and water far away. Carbon monoxide is an example of a kind of pollutant which does not persist for long in the environment, and causes concern only where its local concentrations build up to high levels.

For the fuel technologist the first conclusion must be that, at present, pollution problems are largely localized. Damage is concentrated around 'hot spots' near to major sources of emission. At the design stage, a balance of good siting and the provision of an adequate chimney to disperse pollutants and ensure good dilution is the best contribution fuel technologists can make to minimizing pollution of this kind.

But it is also clear that some pollutants can disperse very widely in the air and accumulate there. This paper has demonstrated that even for an intensively studied and well-known pollutant like sulphur dioxide, we lack essential data on rates and pathways of dispersion, and the transformations and interactions it may undergo. These aspects need study. Where a pollutant, like carbon dioxide or fine particulates, may accumulate in the atmosphere and affect climate, we need basic research so that we can predict, and if necessary shape our policies to forestall, unacceptable effects. These areas of research at national and international level, should have the backing of fuel technologists, even if others will undertake most of the actual studies.

Hitherto, British policies for pollution control have been developed to meet our national priorities, largely in isolation from the policies of other nations. It is clear that this will not be the case in future, for both scientific and political reasons. Within the European Community, the development of a common environmental policy brings with it demands for harmonized standards for widely traded products that may cause pollution, and for basic environmental quality objectives set so that human health is safeguarded adequately everywhere.

The trend towards increasingly uniform standards and practices is apparent. As knowledge of the dispersion and fate of pollutants grows, the possibility that some will be shown to have significant effects far from their point of origin must be recognized. and research to determine whether, and when this is the case, must be maintained. It is possible that the traditional British practice of reliance on dispersion and dilution to dispose of many of the products of fuel combustion will need to be supplemented in the future by additional control processes, or by the use of cleaner fuels. The development of economic and efficient control processes is a field where the fuel technologist can play a major part.

9. REFERENCES

1. Man's impact on the global environment. Report of the study of critical environmental problems, 1970. Massachusetts Institute of Technology.
2. KNEESE, A. V., ROLFE, S.E., and HARNED, J. W. Managing the environment 1971. Praegers Publishers.
3. Inadvertent climate modification. Report of the study of man's impact on climate, 1971, MIT Press.
4. Royal Commission on Environmental Pollution. First Report February 1971. HMSO.
5. SHEEHY, J. P., ACHINGER, W. C., and SIMON, R. A., Handbook of air pollution. US Government Printing Office, Washington. 1969.
6. Air management problems and related technical studies. OECD Paris 1972.
7. LOVELOCK, J. E., MAGGS, R. J., and RASMUSSEN, R. A. Atmospheric dimethyl sulphide and the natural sulphur cycle, Nature (1972), 237, 452–453.
8. KELLOGG, W. W., CADLE, R. D., ALLEN, E. R., LAZRUS, A. L., and MARTELL, E. A. The sulphur cycle. Science (February 1972), 175, 4022, 587–596.
9. JUNGE, C. Air chemistry and radioactivity 1963. Academic Press, New York.
10. KATZ, M. Air pollution handbook 1956. McGraw-Hill, New York.
11. ROBINSON, E., and ROBBINS, R. C. Sources, abundance and fate of gaseous atmospheric pollutants. Stanford Research Institute Report PR–6755, February 1968.
12. Air pollution across national boundaries. The impact of sulphur in air and precipitation. Sweden's case study for the UN conference on the human environment, 1972.
13. Report of joint ad hoc group on air pollution from fuel combustion in stationary sources. OECD Paris, 1973.
14. JUNGE, C. The cycle of atmospheric gases—natural and man-made. Q. J. Roy. Met. Soc. (1972), 98, 418, 711–729.
15. ATKINS, D. H. F., COX, R. A., and EGGLETON, A. E. J. Photochemical ozone and sulphuric acid formation in the atmosphere over Southern England. Nature (1972), 235, 372–376.
16. COWLING, D. W., and JONES, J. H. P. Soil Science (1970), 110, 346.
17. ROSS, F. F. Air pollution sense and nonsense. Env. Poll. Management (1972), 2, 44–51.
18. MEETHAM, A. R. Natural removal of pollution from the atmosphere. Q. J. Roy. Met. Soc., October (1950), 76, 330, 359–371.
19. GARLAND, J. Deposition of gaseous sulphur dioxide to the ground. (In press.)
20. MAUGH, T. H. Carbon monoxide, natural sources dwarf man's output. Science (1972), 177, 338–9.
21. STEVENS, C. M., KROUT, L., WALLING, D., VENTERS, A., ROSS, L., and ENGELKEMEIR, A. The isotopic composition of atmospheric carbon monoxide. Paper to meeting of Amer. Geoph. Union and Amer. Met. Soc. August 1972.
22. SEILER, W., and WARNECK, P. Decrease of CO mixing rates at the tropopause J. Geophys. Res. (in press).
23. INMAN, R., INGERSOLL, R. B., and LEVY, E. A. Soil; a natural sink for carbon monoxide. Science (1971), 172, 1229–1231.
24. MACHTA, L. Mauna Loa and global trends in air quality. Bull. Amer. Met. Soc. (1972), 53, 5, 402–420.
25. MANABE, S., and WETHERALD, R. T. Thermal equilibrium of the atmosphere with a given distribution of relative humidity. J. Atmos. Sci. (1967), 24, 241–259.
26. LAWTHER, P. J., COMMINS, B. T., ELLISON, J. MCK., and BILES, B. Airborne lead and its uptake by inhalation. Inst. Pet. Conf. 1972 'Lead in the Environment.'
27. RUHLING, A., and TYLER, G. An ecological approach to the lead problem. Bot. Notiser (1968), 121, 321–342.
28. MUROZUMI, M., CHOW, T. J., and PATTERSON, C. Chemical concentrations of pollutant lead aerosols, terrestrial dusts and sea salts in Greenland Antarctic snow strata. Geochim. Cosmochim Acta (1969), 33, 1247–1294.
29. DERWENT, R. G., and STEWART, H. N. M. Nature (1973), 241, 342.
30. GERHARD, E. R., and JOHNSTONE, H. F. Photochemical oxidation of sulphur dioxide in air. Indust. Eng. Chem. (1958), 47, 972–976.
31. COX, R. A. Quantum yields for the photo-oxidation of sulphur

I

dioxide in the first allowed absorption region. Harwell Report AERE–R6570, 1970.

32. COX, R. A., and PENKETT, S. A. Photo-oxidation of atmospheric SO₂. *Nature* (1971), **229**, 486–488.

33. COX, R. A., and PENKETT, S. A. Oxidation of atmospheric SO₂ by-products of the ozone-olefin reaction. *Nature* (1971), **230**, 321–322.

34. CHENG, R. T., CORN, M., and FROHLIGER, J. O. Contribution to the reaction kinetics of water soluble aerosols and SO₂ in air at ppm concentrations. *Atmos. Envir.* (1971), **5**, 987–1008.

35. SCOTT, W. D., and HOBBS, P. V. The formation of sulphate in water droplets. *J. Atmos. Sci.* (1967), **24**, 54–57.

36 Man's impact on environment (Ed T R. Detwyler). McGraw-Hill 1971.

Professor P J LAWTHER, MB, DSc, FRCP*

9   Medical effects of air pollution

SUMMARY

The combustion of any fuel, which may or may not be contaminated by impurities, produces pollutants that are potentially toxic to man; their toxicity depends on their chemical and physical nature, their concentration in the ambient air, and the duration of exposure of the subject. The modes of action of air pollutants derived from combustion vary greatly and may range from barely measurable impairment of perception to the production of cancer; in many cases the effect produced will depend more upon the state of the subject than on the concentration of the pollutant.

Many methods are used in the study of the clinical relevance of pollutants from fuel; in this short paper the results of some experimental and epidemiological studies are briefly reported.

1. INTRODUCTION

Air may be polluted by a host of substances, gaseous and particulate, natural and man-made; but it would seem appropriate here to consider the pollutants derived from the combustion of fuel. In the space available no formal review of this vast subject will be attempted; there is an immense 'literature' on the effects of air pollution on man; a useful introduction to key references is the WHO Technical Report 506, 'Air Quality Criteria and Guides for Urban Air Pollutants.'

Ideally the perfect combustion of a pure carbon-containing fuel would give rise only to carbon dioxide and water. Oxides of nitrogen may be formed by fixation of atmospheric nitrogen if the flame temperatures and pressure are adequate, and if the products of combustion are removed quickly. These compounds may also be derived from the combustion of any nitrogenous compounds in the fuel.

Carbon monoxide is produced if the fuel is burned in an inadequate supply of oxygen. Likewise, smoke, which may be simply carbon or complex organic compounds, may be formed in inefficient combustion; sometimes it is merely the result of distillation, simple or destructive, of volatile components of the fuel; some pyrolysis products are carcinogenic.

Pollutants derived from impurities in the fuel are legion. They include sulphur oxides from burning pyrites in coal or organic sulphur compounds in heavy oil, fluorides and trace metals, and such elements as lead and barium from additives in petroleum products.

The physical and chemical forms in which many pollutants are found in the air may be very complex. Moreover, not all are stable, and many take part in chemical reactions in the ambient air and even change their physical form. All these facts must be considered in detail when assessing the effects that pollutants may have on health.

One of the greatest difficulties encountered in giving a short account of the possible effects of so many air pollutants is due to the fact that many of them co-exist and their effects may be additive; again, their effects are rarely linearly related to concentrations or dose, and the population exposed to them is far from homogeneous; young, old, healthy and sick have to breathe the same air.

For the sake of brevity, and risking over-simplification, possible effects will be discussed under the headings of the various pollutants mentioned above.

2. CARBON DIOXIDE AND WATER

These two substances, usually regarded as the 'ideal' products of combustion, cause little if any trouble as pollutants of the ambient air, although carbon dioxide can be a serious hazard in industrial situations. There has, however, been much discussion regarding their

---

*Professor of Environmental Medicine at St. Bartholomew's Hospital Medical College; Physician-in-Charge, Department of Preventive and Environmental Medicine at St. Bartholomew's Hospital, and Director, Air Pollution Unit, Medical Research Council.

significance as 'global' pollutants derived both from fossil fuels burned on earth, and from petroleum burned in high flying aircraft from which they are emitted into spaces from which they are not readily removed. Carbon dioxide is transparent to ultra-violet light but relatively opaque to infra-red radiation from the earth: the 'green-house effect' could lead to a rise in terrestrial temperature that could have far-reaching ecological effects. Water vapour (and fine smoke particles) absorb radiant energy and have been thought to pose a threat by cooling the earth. Whatever the balance might be, the effects on man will not be immediate but, nevertheless, the accumulation of these pollutants merit careful study.

3. CARBON MONOXIDE

This colourless and odourless gas is emitted wherever combustion is incomplete; but the main source contributing to ground level concentrations in towns is the petrol engine, burning a 'rich' mixture. Dangerous levels may be found indoors where flues or heating appliances are faulty. Main stream smoke from tobacco-burning can contain as much as 4% carbon monoxide. Relatively high concentrations of this gas are commonly found in the blood of smokers who inhale.

Carbon monoxide enters the blood only through the lungs. It diffuses easily through the alveolar and capillary membranes to become attached to the haemoglobin in the red blood corpuscles, the main function of which is to carry oxygen to the cells of the body. The affinity of carbon monoxide for haemoglobin is remarkable and accounts for its toxicity since, in comparatively small concentrations in inhaled air, it impairs the oxygen-carrying capacity of the blood. The affinity of carbon monoxide for human haemoglobin is about 240 times that of oxygen; the combination is reversible, and the elimination of the gas from the body is enhanced by increasing the partial pressure of oxygen in the inhaled air, and by hyperventilation. The clinical relevance of carbon monoxide as an air pollutant is not solely related to its concentrations in the ambient air. The concentrations in the blood which would be present at equilibrium with specified concentrations in inhaled air can be readily calculated but, short of equilibrium conditions, the concentrations of carboxyhaemoglobin found in the bloodstream depend not only on the concentration of carbon monoxide in the inspired air, but on the initial concentration present in the blood, the rate of pulmonary ventilation and on the time of exposure. Obviously, in survey work, measurements of carboxyhaemoglobin in exposed subjects are preferable to determination of the gas in air—however valuable the latter is as a screening, or control measure.

It is not generally realized, though implicit, that when the concentration of carbon monoxide in the ambient air is below that which would be in equilibrium with that in the blood, the subject exhales carbon monoxide; thus inspired gas does not necessarily have an additive effect with that already present in the body. A smoker may actually be losing carbon monoxide in traffic polluted air; for example, if a subject is exposed continuously to 25 ppm carbon monoxide his blood will eventually become 4% saturated with carboxyhaemoglobin whatever his initial saturation: if this had been less than 4% he would absorb the gas, but if he had already more than 4%, due say to smoking, he would exhale carbon

monoxide until the carboxyhaemoglobin concentration fell to 4%.

The relationship between ambient CO concentrations, exposure time and levels of carboxyhaemoglobin (assuming initial values are basal, i.e. about 0·8%) is shown in the following table.

TABLE 1 Relationships between ambient CO concentrations, exposure time, and levels of carboxyhaemoglobin*

Ambient CO		Carboxyhaemoglobin level (%) at		
mg/m³	ppm	1 hour	8 hours	equilibrium
117	100	3·6	12·9	16·0
70	60	2·5	8·7	10·0
35	30	1·3	4·0	5·0
23	20	0·8	2·8	3·3
12	10	0·4	1·4	1·7

*Assumes an average individual engaging in light activity and with an initial 'basal' value.

Many of the effects of carbon monoxide are subtle, and quite beyond the scope of this note (as well as being in all probability irrelevant in practice).

In high concentrations, such as are encountered in forensic or industrial practice, the anoxia produced can of course be fatal. It is accepted by all workers that 20% carboxyhaemoglobin can cause symptoms and impair performance but healthy subjects rarely complain of any symptoms with saturations less than 10%. Much attention has been given to the possible effects of saturations such as could be produced in practice by inhaling polluted street air. In much of the work it is claimed that saturations of less than 5% had technical defects (it is essential, for example, in experiments assessing effects on perception or performance that the administration of the gas is by 'double blind' technique in which neither the experimenter, nor the subject, is aware of what is being inhaled). There is little convincing evidence that concentrations of gas giving less than about 8% saturation has any effect on perception on 'intellectual' performance, and there is evidence that EEG activity remains unaffected until saturations approach 33%.

But these results, though reassuring in some ways, do not exonerate carbon monoxide as a potentially dangerous pollutant of the ambient air. Rightly, the effects of the gas on the central nervous system were studied first since it is well known that this system in normal people is most susceptible to oxygen deprivation. But it may be that when considering effects of pollution on the general population, which includes sick people, its effects on the cardiovascular system may be more serious; the results of sound work (on normal subjects) have shown that saturations of about 5% have measurable effects on cardiac function. Obviously in the general population there will be patients with already impaired myocardial function, and with severe respiratory disease, to whom any further lowering of oxygen supply would be harmful. The whole question of protection of populations from any risk is complicated; in the case of the risks afforded by carbon monoxide the picture is further complicated by the fact that many people have already inflicted on themselves, by smoking, considerable damage by carbon monoxide. But these problems must be faced if criteria are ever to be adopted.

4. SULPHUR OXIDES AND PARTICULATE MATTER

The problems posed by carbon monoxide, which may be considered to be a no-threshold pollutant for reasons given above, are simple compared with those set by sulphur oxides and particulate matter—if only because, in many environments, these two pollutants occur together, and their separate effects are not discernible by epidemiological techniques. Probably more research has been done on these two pollutants than on any other, but there is still much doubt about their relevance in the concentrations commonly found these days.

Before considering their possible effects some brief account of their chemical and physical form is needed. Sulphur dioxide is emitted as a gas when fuel contaminated with sulphur compounds is burned; some sulphuric acid may be formed at the same time, and yet more may result from subsequent oxidation after emission from the stack. In coal-fired furnaces much of the sulphur in the fuel is retained in the ash. Some of the sulphuric acid in the air may be neutralized to form sulphate (there is good evidence that photochemical reactions can take place in this country in which SO_2 forms an aerosol of minute crystals of sulphuric acid when acted on by ozone in the presence of olefines). The particle size of any aerosol is relevant to the fate on inhalation: its site of deposition and the amount retained; and, furthermore, it is important to remember that hygroscopic particles may grow rapidly in the warm moist air in the respiratory tract, thus altering their penetration and retention characteristics. Sulphur dioxide is, of course, a soluble gas which is rapidly scrubbed out in the upper respiratory tract; obviously its fate will be quite different if it is adsorbed on particles, even if the latter are inert. The most dramatic demonstration that air pollution is harmful to health was the increase in mortality and morbidity in the Great London 'smog' of 1952 which has been so well documented by other authors. Following this episode work designed to identify the causal pollutant, or pollutants, tended to be based on over-simple hypotheses in which the blame was allotted to smoke or sulphur dioxide, or a mixture of these with fog droplets added. Insufficient notice was taken of the enormous complexity, chemical and physical, of the miasma.

Two main methods of investigation were employed in order to find the cause of the acute effects (felt mainly by sick or feeble people and manifest by impairment, sometimes fatal, of the cardiac or respiratory systems). Suspect pollutants were selected and administered, sometimes in silly doses, to experimental animals and normal, fit humans. All too often false extrapolations were made from unrealistic experiments (on animals and on man), to forecast or explain the effects on populations. Much unsound work was published and, unfortunately, some legislation has been based on such inadequate evidence.

But what can be said, from a study of a colossal amount of experimental data, good and not so good, is that administered by itself 1 ppm sulphur dioxide can produce increases in airway resistance in sensitive people which, though innocuous to them, may be assumed to be highly undesirable in patients whose ventilation/perfusion ratios are already disturbed. Furthermore, experiments with animals have shown that the effects of sulphur dioxide can be enhanced by the addition of particles, albeit in relatively massive amounts.

There are, however, remarkably few convincing accounts of realistic concentrations of acid particles producing derangement of lung function, but it would be prudent to assume that this lack of data is due rather to inadequacies of the experimental method. A sad, but insurmountable limitation of the experimental technique is the need to confine one's exposures to fit people, whereas one knows that it is patients with lungs already damaged who are especially at risk.

Because of these inadequacies, epidemiological techniques must be applied to supplement the experiments. The simplest are those which relate such gross events as the so-called 'smog' with increases in mortality and morbidity. The techniques can be refined so that lesser daily variations in pollution, and other environmental factors, can be studied if the populations exposed are large enough to contain an adequate number of susceptible individuals (often workers seem to forget that a response to a given stimulus is related not only to the strength of the stimulus, but also to the susceptibility of the stimulated). These techniques have been applied to London since the 1952 episode and now, happily, any relationship between mortality and morbidity, and the greatly reduced pollution is barely, if at all, discernible.

A further refinement of more orthodox epidemiological techniques has proved invaluable: patients with fairly severe chest disease have been studied in detail by analysing their own daily assessment of their well-being recorded in diaries. Very close correlation was shown to exist between smoke and sulphur dioxide concentrations and their health in the early days of the experiments when episodes of high pollution were all too common; but as the years have gone by, with the implementation of the Clean Air Act pollution by both smoke and (rather unexpectedly but explicably) by sulphur dioxide, has fallen dramatically until virtually no response is seen, even in highly selected patients. This technique has failed to enable us to discriminate between the separate effects (if indeed there be any) of smoke and sulphur dioxide, but it has enabled us to discern what we think is a 'threshold' dose below which pollution by the sulphur dioxide/particulates complex seems to have little effects on groups of patients (individuals may, however, be affected although we have been unable to demonstrate this). Patients would appear to be more sensitive to the effects of pollution at the beginning of the winter and throughout our series of quinquennial experiments the minimum pollution seen to be associated with worsening to any degree is that indicated by about 500 μg/m³ *together with* about 250 μg/m³ smoke each representing 24-hour averages at a group of seven sites in Inner London. There is no evidence to suggest that either of these pollutants in these concentrations would by itself produce the same response; our efforts to continue to use this technique, to identify the responsible pollutant is, happily, being frustrated by the rarity of any episodes of high pollution.

So far the sulphur oxides/particulate complex has been discussed only in relation to the exacerbations of respiratory disease with which it is associated. But it is rightly thought to play an important part in the causation of chronic bronchitis (though happily this might yet be seen to be of historic interest only). Acute episodes of pollution, such as the 1952 'smog', were reported by patients to have caused their breakdown in health and the development of clinically-manifest chronic bronchitis;

the mechanisms by which these changes may be produced are beyond the scope of an essay of this brevity; but before leaving this topic to discuss the role of air pollution in the production of disease one may, with profit, remember that the population on whom air pollution acts is altering no less than the pollution. The introduction of antibiotics, diuretics, antispasmodics, oxygen and other drugs into the treatment of cardio-respiratory disease has produced a new population of patients of greater susceptibility to non-specific stress. We refuse to allow broncho-pneumonia to kill our old men with emphysema; and they must wait till their lungs disappear, or till they get cerebral anoxia or carbon dioxide narcosis; and even then, sometimes, we revive them by intensive care when a bronchial tube to a remaining piece of lung gets blocked with mucus.

Again, the response of a population to a given dose of pollution will depend on antecedent or concurrent epidemics of infection; an epidemic of influenza, while removing many patients who might have succumbed to an episode in which air pollution reached high levels, creates a population of convalescents who, having survived one stress, cannot in their weakened state withstand more.

The urban excess of chronic bronchitis is well recognized and has been attributed, in part, at least to exposure to air pollution although many other factors play a major part. The simple hypothesis that bronchitis is caused by simple chemical irritation by some substance, such as sulphur dioxide in the concentrations found in urban air, is even less tenable than the idea that this gas alone is responsible for the observed effects of so-called 'smog'. There have been many studies of morbidity and of lung function among populations who, by virtue of their occupations, have been exposed to very high concentrations of simple irritants without developing bronchitis; but one must not lose sight of the fact that, since not all who smoke heavily and live in polluted cities get bronchitis, an individual factor may be involved and this might be the hyper-reactivity of the bronchial tree favoured by the Dutch school.

There have been several recent advances which have helped in our understanding of the mode by which this disease is produced. Among the most promising techniques currently used is the study of people with reference to their exposure to air pollution in childhood. We studied the prevalence of respiratory infection in a sample of children all born in one week in March 1946. The original sample comprised 5 362 children. In 1961 when the children were 15 years old, 4 592 were living in Great Britain and on 3 866 of them there was information on the air pollution to which they may have been exposed. Four pollution categories were defined (these were based primarily on consumption of domestic coal in 1952) and data on respiratory infection were analysed by reference to these categories. The last paragraph of Douglas and Waller's important contribution to our understanding of air pollution and respiratory disease is best quoted in full.

'*The results are simple and consistent: upper respiratory tract infections were not related to the amount of air pollution, but lower respiratory infections were so related. The frequency and severity of lower respiratory tract infections increased with the amount of air pollution. Boys and girls were similarly affected, and no difference was found between children in middle class families. An association between lower respiratory tract infection and air pollution was found at each age examined, and the results of the school doctors' chest examinations at the age of 15 suggest that it persists at least until school leaving age.*'

We hope to select a sample from the original group and assess lung function, as well as question the subjects (now aged 24) about subsequent respiratory symptoms.

To investigate further the hypothesis that pollution in infancy might be of importance in the development of chest illness later in life we have been looking to see if there are differences in prevalence of respiratory symptoms and simple measures of lung function in young people who were born just before, and just after the 1952 'smog'. Thus the relevance of a massive exposure to pollution in the neo-netal period could be assessed. The results so far (on 2 000 people) fail to show any marked effect of this episode, but do indicate that a history of infection of the lower respiratory tract in childhood (and the present habit of cigarette smoking) are significantly related to respiratory 'status'.

There is a distinct possibility that we are beginning to discern a clearer picture of the development of chronic bronchitis: air pollution could play its part by facilitating in some way the infection of the lower respiratory tract in early life in such a way as to sow the seeds for later impairment of function.

Some recent laboratory work has yielded results which are compatible with those derived from the epidemiological studies mentioned above. The growth of haemophilus influenzae (a micro-organism normally found in the sputum of patients with chronic bronchitis) is affected by particulate matter collected from the air of London. This particulate matter is predominantly coal smoke and contains stimulating and inhibiting substances. The effect of smoke on the growth of this organism is complex; the results of this work may give further support for the case for the abolition of smoke.

5. OXIDES OF NITROGEN

Combustion processes can yield several oxides of nitrogen; NO_2, N_2O_4 and NO are the common ones, (nitric oxide (NO) being the predominant oxide emitted by motor vehicles). They have different chemical properties and different toxic effects and, therefore, the practice of lumping them together as NO_x is to be deplored and, for technical reasons concerned with analytical processes, can produce misleading figures for their concentrations. There was a reason for this slack habit: it used to be assumed that NO was rapidly oxidized to NO_2 and therefore barely had an independent existence. It is now realized that, in the low concentrations found in ambient air after it is dispersed from the source, it can persist for many hours and therefore must be considered as a potentially toxic pollutant.

Most of the attention given to oxides of nitrogen as pollutants of the ambient air has been in relation to the part they play in the Los Angeles type of photochemical 'smog' which, though annoying, has not been unequivocably demonstrated to have pathological effects of any magnitude. But nitrogen dioxide (NO_2) is well known as a toxic gas in industrial environments where the MAC is 5 ppm. It has a strange effect in that it can produce, if inhaled in high enough concentrations, pulmonary oedema which occurs only after a latent period.

The concentrations found in polluted air are rarely more than 0·1 to 0·2 ppm and could not produce pulmonary oedema; but there is evidence from good animal experiments that concentrations, not much higher than those occasionally found in outside air, can produce cellular alterations and structural changes resembling those seen in some human lung disease; and it would seem that exposure to this gas can also enhance mortality resulting from exposure of animals to infectious aerosols. There is a hint from some American work that lower respiratory tract infection in infants is more prevalent in areas where NO_2 concentrations are high. More thorough work is needed before this is accepted.

Very little work has been done on the toxic effects of nitric oxide (NO) mainly because it used to be thought that it was oxidized immediately to NO_2. Indeed, some of the effects seen following exposure of animals to high concentrations of NO were undoubtedly due to the NO_2 produced by oxidation; but NO is a very active molecule, capable of forming addition compounds with haemoglobin as does CO, and more work on its true effects is long overdue. It is present in concentrations of up to 1 000 ppm in main stream cigarette smoke.

6. OTHER POLLUTANTS DERIVED FROM FUEL

These are legion and in this essay their inclusion would hardly be justified. There is adequate material to be read, good and appallingly bad, on the significance of lead and other trace elements.

7. CONCLUSION

This informal review, written without detailed references, is intended rather as an *aide-mémoire* than as a paper. It sounds warnings and gives some good news. It is to be hoped that it encourages the preservation of a sense of perspective.

A D BRADSHAW*

10 The ecological effects of pollutants

SUMMARY

The ecological effects of coal mining and sulphur dioxide emission are considered. Both are problems which have been worked on for a long time, and about which a great deal is known; but it is concluded that, while their effects on the environment can be good as well as bad, there are still, in both cases, important and unanswered questions, and a poor application of those solutions that have been established.

1. THE CONCEPT OF POLLUTION

We tend to think that pollution refers to discharging some single chemical or other substance into the environment, which makes that environment less satisfactory for life; but the word has a wider meaning: to soil, to defile, to taint, to violate. A discussion of pollution caused by fuel has therefore a wider remit than the problems of the discharge of noxious substances; we must rather consider any violation of the environment, because from an ecological point of view every disturbance of the environment, from whatever cause, must be considered and its effects and importance assessed. This means that the ease with which its effects can be corrected must also be considered.

Pollutants must be considered at six separate levels:

    (1) What is happening or being discharged?
    (2) Where does it have its effects?
    (3) What effects does it have?
    (4) Do they matter?
    (5) Can they be corrected?
    (6) Are they being corrected?

Some of these questions have been considered in the previous papers. This paper will consider the ecological problems caused by pollutants (meaning the effects they have generally in natural, or semi-natural environments) and will select two problems: the one from the gaining of

*Professor, Department of Botany, The University of Liverpool.

fuel, and the other from using it. It will examine them in detail, to see what is known about them, what needs to be known, and what lessons can be already learned.

2. COAL MINING AND THE DESTRUCTION OF LAND

2.1. Effects

Coal lies underground in narrow seams buried under layers of other material. This covering material may have to be pushed aside as in open-cast mining, or thrown up from below as in deep mining. The act of removal of the coal may cause subsidence. All this affects the land surface which may become covered with large areas of sterile waste material, or submerged under water.

The land surface carries the soil, a fragile structure produced by the effects of climate and plant and animal life acting over many millennia, which may be cultivated for agriculture. We certainly cannot afford to lose any of it, particularly because of the threat of over-population and a world food shortage. Even if this does not transpire, none of us is happy any longer to see countryside transformed into an unattractive, biologically sterile, wasteland.

The destruction of the soil is highly localized, just within the mining areas, and perhaps this is why it has been permitted. Certainly it is why there has never been any real concern shown, until recently, by Members of Parliament who do not live in coal-mining areas. But the effects are not totally localized: a waste heap 1 000 yd distant violates the environment hardly any less than one that is only 50 yd distant. The effects are cumulative, so that total whole human environment of an area may be destroyed even if the waste heaps constitute only 50% of it. They not only dissuade people from living there, but also make farming impracticable, and reduce disproportionately the amount of wild life. They also attract industries that are environmentally undesirable, such as junk yards and scrap piles.

One further product of mining is acid mine drainage produced by the oxidation of pyrites to give, in effect, sulphuric acid. If this is released into streams it can pass many miles downstream, causing environmental disturbance over a wide area.

So the many by-products of coalmining that violate our environment, such as spoil heaps, subsidences, opencast workings, acid mine drainage, appear totally unacceptable; but mining must continue. How, then, can these problems be lived with, and contained in an acceptable and economically feasible manner? We are looking for solutions that are effective and economic. But before proceeding we should be quite certain that the effects of coalmining, in their relation to the environment, are all 'evil'. A pollutant must be able to be demonstrated to be a pollutant; otherwise it may be something else which we perhaps need not bother about, or something that we welcome. The opposite of 'pollution' is 'improvement' or 'betterment' as well as 'purification,' or 'repair'; but could pollution in any sense be termed 'improvement'?

2.2. The Norfolk Broads

In East Anglia there are a series of lakes, the Norfolk Broads, which are renowned both for their beauty and wild life. They total about 5 000 acres. So much are they valued that 2 500 acres of them are now nature reserves, a higher percentage of the total area than for probably almost any other type of environment. In 1960, Lambert, Jennings and Smith[1] showed that they were the first major open-cast fuel workings in Britain, and were in operation in the 12th and 13th centuries. The authors, in elaborate detail, have indicated the deep peat cuttings, the old trackways serving the workings, and the boundaries between different workings, as the trappings of a mediaeval industry. It must have been quite a substantial undertaking, as well as quite a mess, for there to have evolved such places as Hickling Broad, which has an overall area of 1 300 acres.

How many of those preserving our countryside have considered this when they have passed across these unique areas? There are other areas of the same peat deposits in Norfolk, and in Cambridge, which were not exploited in this fashion. Today they are dull, flat, areas of countryside, albeit productive.

So the disturbance from winning fuels is not necessarily bad, and could well contribute greatly to our environment.

2.3. Spoil heaps

However the Broads are a particularly favourable environment where recolonization by water, plants and animals could proceed rapidly; and they also have had 500 years for this process. By contrast, later coal-mining areas have produced landscapes of spoil much less favourable for colonization. The spoil is grossly deficient in plant nutrients especially nitrogen: phosphorus and calcium, and can be very acid.[2] This is sufficient to prevent much colonization. There is also the fact that the material has only been able to be colonized for a short time.

These areas, in all the major coal fields, are a major affront on our environment. There are one or two areas where colonization has proceeded some way, as in Somerset.[3] None of these has achieved the status of a nature reserve like the Norfolk Broads.

With the recent awakening of our interest in the environment much of this legacy of past mining is now being tackled, despite the fact that it covers many thousands of acres in the form of spoil heaps and degraded ground. If soil is fragile, and can only be formed under natural conditions over many millennia, how can restoration be tackled?

Top soil is an obvious answer, and where it can readily be found it should be used, and spread over the offending material. It is a ready-made restoration agent; its use a simple earthmoving exercise; but it requires skill and understanding, for top soil can deteriorate in storage and transport; and can be enormously expensive (up to £3 000 per hectare) if it has to be carried more than a short distance. It may also be impossible, if top soil is just not available in that region, or without destroying other already existing sites.

We now know a great deal more about soils, their structure, their properties, the way they contribute to plant growth, and the way they are formed. In essence, soil development from parent material is characterized by the development of a satisfactory structure that is to a large extent due to the presence of organic matter and micro-organisms; and by the development of a satisfactory store of nutrients available to plants (especially nitrogen, calcium, potassium and phosphorus) in a labile form in the organic matter, and attached to the soil particles themselves.[4] The development of a fertile productive soil takes a long time because there are initially shortages of nutrients and slow growth of plants. But because of a modern fertilizer industry (itself dependent on coal) we can now obtain the necessary plant nutrients cheaply, thus fertility of any material can be rapidly improved.[5] For colliery spoil, once an adequate structure has been created artificially by machinery, and the appropriate lime and fertilizers added, in most cases very satisfactory grass and clover growth can be obtained, with ensuing rapid development of a good soil—as has been shown by much careful and patient work.[6,7] In this new soil the added nitrogen, phosphorus and other nutrient elements will be readily available and a satisfactory structure will be built up by the growth of the plants and the activities of micro-organisms.

At Whalleys Basin in Wigan an excellent soil and grass sward has been achieved by the Lancashire County Council from raw spoil in 15 years. At Bickershaw, arable crops are being taken off land that was once colliery spoil. Thus the rapid restoration of degraded land and waste materials is possible by modern techniques without the use of top soil. The costs may be considerably less than by the use of top soil, and indeed may allow restoration to be carried out where otherwise it might be economically difficult.

But colliery wastes vary considerably in their composition. Some wastes, due to the presence of iron pyrites generating sulphuric acid by oxidation, are extremely acid with pH as low 2·5 to 3·5 and require a great deal of lime; others have high phosphate requirements. It may be difficult to deal with these: they must be recognized in time and treated accordingly, by high lime, and other dressings, as required. Appropriate analytical and experimental techniques for recognition of the problems are available and can be built into a proper assessment programme.[8] Some materials may be so extreme that they cannot be dealt with, but if they are recognized in time they can be buried under other more favourable materials.

Similarly we now know much more about the growth of trees on colliery spoil in this country. There are several areas where excellent woodlands are developing,[9] but again there have been problems due to the failure to recognize the poverty of the materials or the ecological requirements of different species. Tree species require reasonable conditions and adequate supplies of nutrients just like other species; lack of these leads to poor growth, and also therefore greater liability to damage by other factors, such as the competition from other plants, vandalism, fire. Tree species also have their own idiosyncrasies; there is scientific skill in choosing the right species for the right site.[10]

2.4. Subsidence

Subsidence is another important product of coal mining. Generalization about its effects is impossible. It can destroy villages, or only cause minor cracking of houses; it can flood whole farms or merely disturb a few drains. Minor subsidence can readily be dealt with. Major subsidence, leading to flooding, is a different order of problem. In many cases the areas have been turned into lakes, and have become valuable both for recreation, fishing and wild life, such as the Lancashire Flashes. Gradually, usually without help, nature has been creeping into these areas. Potentially, in time, they could become as beautiful as the Norfolk Broads.

2.5. Open-cast mining

There can be few other operations that disturb the environment on quite such a violent and grand scale as open-cast mining. 50 to 100 ft of overburden is stripped off and dumped in piles that may be 150 ft high. But five years later, the untutored person could pass by and not realize the violation that had occurred.

A systematic planning beforehand of the movement of materials, the careful storage of top soil and subsoil, its careful return after mining, the restoration of drainage, the rebuilding of fences and other permanent structures, and care for the soil in the growth of the first year's few crops, means that a total and almost perfect agricultural restoration is possible.[11,12] Indeed, the ground may be even more productive afterwards.

Here is a destruction of the environment whose effects are known, where there is ample skill available to know how to cope with those effects, and where that skill is being applied properly. It is an expensive operation, but the interesting fact is that it is economic.

In other parts of the world the situation is not quite so rosy, and although enormous strides have been made in America to ensure that strip-mined areas are now properly restored, the restoration is still not up to British standards on account of the excessive costs that would be entailed. At least that is what is claimed.

2.6. Acid mine drainage

Since coal deposits have formed in lagoons under anaerobic conditions, reduced sulphur-containing compounds are also accumulated not only in the coal but also in the surrounding bedrock. These compounds, mainly iron pyrites, oxidize on exposure to air and water, releasing ferrous and ferric sulphate, hydrated ferric oxide and sulphuric acid. We now know a great deal about the process.[13]

The products can be very toxic, especially since they can cause the pH of the water containing them to drop to 2, or below; if the water is kept alkaline the main problem is not acid toxicity, but large quantities of ferric hydroxides in suspension.

In the Appalachian coalfield of the US there is a high pyrite content to the coal and the bedrock and topographical conditions allow a flow of water and air through the workings. As a result in Appalachia there are immense acid mine drainage problems affecting 6 000 miles of streams and 15 000 acres of open water.[14]

The number of troublesome mine drainages in the UK are very few because of low pyrite content, and little drainage flow from the workings. However they do occur, although the treatment of such drainage waters is not difficult, provided the volume is not great. Acidity can be removed by treatment with alkalis (such as lime) and the iron hydroxides by oxidation and settlement in lagoons. Where in this country it is necessary this is being done, and we have no serious problems from this cause.[15] Elsewhere, however, the excessive amounts of drainage, some coming from open-cast overburden, make effective treatment impossible.

2.7. The present situation and control

From all this it almost appears that we can be content with the present situation. Certainly the problems are known and the solutions are available, and in many cases they are being applied. At the moment, thanks to governmental intervention, the backlog of violation of the environment from mining is being broken.

But can we really be complacent? There seems to me to be several reasons why we cannot.

Firstly, having apparently found a solution we seem often to apply it without question, thinking that it must be the only solution. The use of top soil regardless of expense, when other cheaper solutions may be possible, is an example. There has been little encouragement of research into other solutions except by research councils and a few local authorities.

Secondly, having applied a solution we may not stop to see if any subsequent aftercare is necessary; yet the upgrading of soil cannot be achieved overnight. There are many examples of reclamation work reverting because no one has continued the process of upgrading, and the new vegetation cover, instead of becoming a self-reliant system, collapses. Often this is because the wrong scientist, the engineer, rather than the biologist, is involved.

Thirdly, by dint of having a solution we may not notice that it is not the best solution. At the moment in the furore of reclamation there is a tendency to consider that if the land goes back to agriculture the job has been well done. But in many situations, especially after open-cast mining, or in the restoration of old mining areas, advantage should have been taken to create a more complex environment including hills, lakes and woodlands, to encourage wild life and provide interesting and biologically valuable diversity to the landscape. This is particularly important in some of the older mining areas where the present landscape is very drab. The omission is disappointing since excellent examples of what can be done are available, especially in the Stoke-on-Trent scheme; and the opportunity to take a creative view of landscape restoration will never come again.

Finally, while we may have a good solution its application may not be as widely applied as it should. The particular place where this is true is in the coal industry itself. In this country, as long as industry is active the steps required by law to deal with waste heaps are minimal. The reason that is usually given is that any such restoration measures could not possibly be undertaken without upsetting production. This is obvious nonsense in most situations, in the coal industry as much as anywhere else. The ultimate perimeter can be set and steps taken to begin full-scale and final restoration, not by the planting of a pathetic row of boundary trees, but by the implementation of a properly conceived landscape plan that recreates the total environment in a systematic manner. If a pit has a life of 20 years, and restoration is not begun until that life is finished, then the children who grow up near it will never have had a decent environment to enjoy. Since low-cost techniques of grassing are available[16] there is no reason why a temporary cover should not even be provided on material that might be shifted in a very few years. It is worthwhile noting that in the Westphalia province of Germany, which included the Ruhr coalfields, restoration from the very beginning is now required by law. In this country the active or recently active pits can be some of the worst violators of our environment.

3. SULPHUR DIOXIDE

Smoke has been recognized as a pollutant since the 13th century: the particular part played by sulphur was recognized by 1600. Today probably we have more papers on sulphur dioxide than on any other pollutant, and we have a national survey of its emission and distribution, and legislation and an inspectorate to provide control over it.

It would be easy to think that there was no more to be said on it. But it is becoming increasingly clear that the more we know about any pollutant the more we realize there is still more to know about its effects can be properly understood. The effects of sulphur dioxide are some of the most studied, and yet the least understood problems of the environment at the moment.

3.1. Emission

The emission of sulphur dioxide is well documented both locally and nationally. The national trends have shown a peak in the early 1960's and a fall since then;[17] but the fall is small, less than 10%. What happens in the future will depend on the sulphur content of the fuels that become available. We cannot automatically assume that the trend will continue downward.

3.2. Dispersion

There has been a remarkable reduction in the concentrations of sulphur dioxide near the ground in large cities in the United Kingdom. In the North West region of Britain this has been 40% in seven years. It has partly been due to changes in the use of fuel but more particularly to the 'high stack' policy, so as a result in the same region, the concentration in country sites has risen by 30%.[18]

Since high concentrations can have very marked effect on all living organisms, this change in dispersion pattern has obviously brought relief to the urban regions; but on the principle that what goes up has to come down in some form or other, and as total emissions are not changing greatly, it ought to be known where it is all going. Certainly it is going into country districts. It has been suggested that it is going a good deal further, and that one country's sulphur dioxide may well land in another country.[19] Just who receives whose sulphur dioxide, and what percentage gets exported, or imported, is still a matter for enquiry. but it is clear that there is substantial export from the evidence of Drs Holdgate and Reed.*

What happens to the sulphur dioxide is complex and is only now becoming understood.[20] It can become oxidized to sulphate and probably is mostly in this form in long-distance transport. It comes down to the ground sooner or later, either by dry deposition, or in rain. Sulphur dioxide is very reactive and is therefore rapidly removed by dry deposition; sulphate will mainly be removed by rain. The balance between sulphur dioxide and sulphate depends on the rate of oxidation. Oxidation does not occur directly by either oxygen or ozone. It depends on a complex ozone olefin reaction.[21] Olefins are produced by internal combustion engines, and by plants.

Thus the problem of sulphur dioxide becomes two problems: that of sulphur dioxide transported only short distances, and that of sulphate which is transported longer distances. Since sulphur dioxide and sulphate have very different effects, they must be considered separately.

3.3. Effects of sulphur dioxide

The direct effects of sulphur dioxide, whether on human beings, plants or animals, are well known. Because of their structure, plants are more susceptible than other organisms and there is a great deal of literature showing a variety of effects.[22,23,24] Some of these effects can be completely disastrous, particularly where very high levels have been produced by emissions from smelters, but serious effects can equally be produced by sulphur dioxide from the burning of fuel.

Major damage occurs when sulphur dioxide concentrations reach about 500 $\mu g/m^3$. The critical level varies for different species. Some, like privet, can tolerate ten times as much as others like clover.[24] Others, such as lichens, are far more susceptible than clover and will be killed at concentrations above 50 $\mu g/m^3$, and they have therefore been used as a clear indicator of sulphur dioxide distribution.[25,26]

Damage of this sort is obvious, and has usually created public concern, the loss of humbler plants, such as lichens, has not caused much outcry; although nonetheless, their loss to the environment is very conspicuous, and is to be deprecated.

There is increasing evidence that low levels of sulphur dioxide can cause reductions in growth without visual symptoms. Such growth reduction can be confused with effects due to other causes, such as exposure, soil poverty, or soot. But careful work has shown that rye grass growth can be reduced by half by the presence of about 200 $\mu g/m^3$ of SO_2.[27]

This experiment should shatter any complacency there may be about the effects of sulphur dioxide because the

*Paper 8: The fate of pollutants.

level of sulphur dioxide concentration is within the range of concentrations found in ordinary conditions. At present, average concentrations are about 110 μg/m³, are higher near cities, and may fluctuate up to 300 or even 500 μg/m³ quite often.[18]

There is great need for further work of this sort because of the serious implications. In the past such effects have been difficult to recognize because of confusion with other factors. It is only careful experimentation that can reveal them. There are a number of questions that must be asked, such as the effect of fluctuations in concentration, the nutritional state of the plant, and climatic factors; all these are known to change the appearance of acute effects. Present survey methods are not adequate for recording fluctuations. Equally, the response of different species should be ascertained: grass is not usually considered very sensitive. Forest tree growth, also, must be examined: in the past the direct effects of sulphur dioxide at normal concentrations have been difficult to disentangle[28,29] although conifers are more susceptible to high levels than other plants. Conifers still cannot be grown in many cities.

There is some evidence that two pollutants together can combine to have greater effects than each separately. This synergism may be important in some situations.[30,31] It needs further examination, notwithstanding that if there were serious effects we would have noticed them.

We need also to know far more about the effects of sulphur dioxide on natural vegetation. The effects on lichens have already been mentioned. It is remarkable that in the polluted areas of the Pennines, bog moss (Sphagnum spp.) is almost completely absent.

Analysis of the peat shows that it disappeared in the 18th century, at a time when atmospheric pollution increased violently. It is difficult to believe that this was not due to the effect of sulphur dioxide.[32] The rye grass from the same general region has been shown to be remarkably tolerant to sulphur dioxide.[27] The disappearance of the bog moss is not by any means a trivial matter: it was one of the main components of the vegetation and, since the disappearance of the bog moss there has been very severe erosion of the peat.

Sulphur dioxide is, then, more of a trouble maker than we might expect; and we need to know much more about the extent of its effects. We must be able to take a balanced view, realizing that it is not having disastrous effects, but we must be able to quantify *exactly* what it is doing.

3.4. Effects of sulphate

At distances from sources of emission, increasingly large proportions of the sulphur fall-out are as sulphate. The sulphate radical is not in itself a toxic compound at all. Nearly all soils contain large quantities of sulphur as sulphate; and since sulphur is an essential element for plant growth, where sulphur is found to be deficient (as in the Coast Range of California) it has to be added to get good crop growth.

For this reason sulphate fall-out is valuable as a source of sulphur. It is significant that sulphur deficiencies are very uncommon indeed in this country, and in West Europe in general. The value of sulphur dioxide, even as a source of sulphur for grass growth, has recently been demonstrated.[33]

But the sulphate radical cannot exist by itself. It usually falls to the ground either as sulphuric acid, or as ammonium sulphate. In either form it is a source of acidity. The ammonium radical is broken down in the soil by micro-organisms leaving sulphuric acid.

As a result of this property sulphur emissions have recently come under close scrutiny. The soil is an open system, although quite well buffered; but if acid is continually added cations (such as calcium) will be removed. Careful studies have shown that calcium, in particular, is in a sensitive balance in many soils.[34] As a result there is now considerable concern over the effect that sulphate input is having on the calcium status of woodland areas in Scandinavia.[19] There is no doubt that sulphate must be causing a loss of calcium and that, ultimately, since calcium is important to both soils and plants, there must be decrease in productivity of the forests. The problem, however, is to know how much loss is being caused, and what factors are operating to prevent it, and what is the point at which the loss will have serious effects on forest yield. The other problem is that a proportion of the sulphate falling is of natural origin, and cannot be attributed to fuel.

This is a serious effect which needs investigation in this country as well as elsewhere. It can affect the growth not only of trees, but of all sorts of vegetation. It may well be partially the cause of the erosion of Pennine peats. It is certainly a cause of the quite remarkable acidity and poverty of soils in parks in industrial cities such as Liverpool and Manchester. But it must be related to other factors, such as dissolved carbon dioxide which can have similar effects.

In areas where there is a large store of calcium in the soil the acidification and calcium loss can be disregarded. Equally it can be disregarded wherever liming is a regular part of agricultural practice. Nevertheless, sulphate fall-out will cause a significant and cumulative loss of calcium.

A final consequence of such sulphate deposition will be in rivers. If there is acidity this will find its way into rivers, with possibly serious repercussions.[19] But the evidence is not easy to produce in clear-cut manner, and often there will be no change in the acidity of the river, but an increase in calcium level which, although it could cause changes, would not be harmful.

3.5. The present situation and control

Sulphur emission is certainly harmful. Just how harmful it is can only be revealed by further work. At the moment control of emission is expensive, and only operated in one or two selected places where it has obvious effect on buildings.[35] There is, nonetheless, pressure to operate control of emission elsewhere. Decisions on extending control can only be made on proper assessment of the total damage in all its forms over the whole country. At the moment we are far from obtaining such an assessment.

4. CONCLUSIONS

With any subject that has been worked on for a long time it is easy to believe that all the answers are known. This is an attitude that should firmly be resisted. The two examples chosen have no novelty at all. Yet in both there are large unanswered problems. It is perfectly likely that the problems when they are answered will turn out not to be serious; but this will not be known until the necessary research has been completed. Exploitation of our environment will continue as long as man continues to exist. In the past, the fuel industry's record of care for the environment has not been particularly good. There is now a new sense of responsibility. This has to be backed up by sufficient and careful investigation, and a willingness to see that the findings are acted upon.

5. REFERENCES

1. LAMBERT, J. M., JENNINGS, J. N., and SMITH, C. T. 1960. The making of the Broads: a reconsideration of their origin in the light of new evidence. Royal Geographical Society and John Murray, London.
2. CHADWICK, M. J., CORNWELL, S. M., and PALMER, M. E. 1969. Exchangeable acidity in unburnt colliery spoil. *Nature*, **221**, 161.
3. DOWN, C. G. 1969. The problems posed by colliery waste tips in Somerset. Proc. Bristol Nat. Soc. **31**, 625-630.
4. RUSSELL, E. J. 1961. Soil conditions and plant growth, 9th Ed., Longmans, London.
5. COOKE, G. W. 1967. Control of soil fertility. Crosby Lockwood, London.
6. Civic Trust 1964. Derelict Land: a study of industrial dereliction and how it may be redeemed. Civic Trust, London.
7. Newcastle upon Tyne University 1972. Landscape Reclamation. Vols 1 & 2. I.P.C. Science and Technology Press, Guildford.
8. FITTER, A. H., HANDLEY, J. F., BRADSHAW, A. D., and GEMMELL, R. P. 1973. Site variability in reclamation work. *Journ. Inst. Landscape Architects* (in press).
9. CASSON, J., and KING, L. A. 1960. Afforestation of derelict land in Lancashire. *Surveyor*, **119**, 1080-1083.
10. WOOD, R. F., and THIRGOOD, J. V. 1955. Tree planting on colliery spoil heaps. Forestry Comm. Research Branch Paper 17.
11. HUNTER, F. 1953. Opencast coal sites reclaimed. *Agriculture Lond.*, **60**, 335-336.
12. DAVIES, W. 1963. Bringing back the acres. Opencast coal. *Agriculture Lond.*, **70**, 133-138.
13. Ohio State Univ. Res. Foundation. 1971. Acid mine drainage. United States Env. Prot. Agency Water Quality Off. Water Pollution Control Res. Series DAST 52.
14. KINNEY, E. C. 1964. Extent of acid mine pollution in the United States affecting fish and wildlife. United States Dept., Interior. Fish and Wildlife Circ. 191.
15. GLOVER, H. G. 1973. Acid and ferruginous mine drainages. NATO Adv. Study Inst on Waste Disposal and Renewal and Management of Degraded Environments, (in press).
16. BRADSHAW, A. D., and HANDLEY, J. F. 1972. Low cost grassing of sites awaiting redevelopment. *Journ. Inst. Landsc. Arch.* **99**, 17-19.
17. Royal Commission on Environmental Pollution 1971. First report. HMSO Cmd. 4585.
18. Warren Spring Laboratory 1972. National Survey of Air Pollution 1961-71, Vol. 2. HMSO.
19. Sweden, Royal Ministry of Agriculture. 1971. Air pollution across national boundaries. The impact on the environment of sulphur in air and precipitation. Sweden's case study for the UN conf. on the human environment. Kungl. Boktryckeriet, Stockholm.
20. PENKETT, S. A. 1972. The effects of wide scale dispersion of sulphur pollutants. Atomic Energy Res. Est Report 7306.
21. COX, R. A., and PENKETT, S. A. 1971. Oxidation of atmospheric SO_2 by products of the ozone-olefin reaction. *Nature*, **230**, 321-322.
22. BOVAY, E. 1969. Effets de l'anhydride sulfureux et des composés fluorés sur la végétation. Proc. First Europ. Cong. on Influence of Air Pollution on Plants and Animals. Wageningen 1968, 111-135.
23. BRANDT, C. S., and HECK, W. W. 1968. Effects of air pollutants on vegetation. In Air Pollution ed. A C Stern, 2nd ed., Academic Press, New York, 401-443
24. Agricultural Research Council 1967. The effects of air pollution on plants and soil. Agric. Res. Council London.
25. GILBERT, O. L. 1965. Lichens as indicators of air pollution in the Tyne Valley. Brit. Ecol. Soc. Symp. 5, 35-47.
GILBERT, O. L. 1970. A biological scale for the estimation of sulphur dioxide pollution. *New Phytol.*, **69**, 629-634.
26. BARKMAN, J. J. 1969. The influence of air pollution on bryophytes and lichens. Proc. First Europ. Cong. on Influence of Air Pollution on Plants and Animals. Wageningen 1968, 197-209.
27. BELL, J. N. B., and CLOUGH, W. S. 1973. Depression of yield in rye-grass exposed to sulphur dioxide. *Nature*, **241**, 47-49.
28. SCURFIELD, G. 1960. Air pollution and tree growth. *For. Abstr.*, **21**, 339-347, 517-528.
29. LINES, R., and PHILLIPS, D. 1963. Atmospheric pollution. *J. Forestry Commission*, **32**, 97-100.
30. MENSER, H. E., and HEGGESTAD, H. E. 1966. Ozone and sulfur synergism injury to tobacco plants. *Science*, **153**, 424-425.
31. WILLIAMS *et al.* 1971. *Environmental pollution*, **2**, 57-69.
32. TALLIS, J. H. 1964. Studies on southern Pennine peats. III The behaviour of Sphagnum. *J. Ecol.*, **52**, 345-354.
33 COWLING, D E., JONES, L. H. P., and LOCKYER, D. R. 1973. *Nature*, **243**, 479.
34. OVINGTON, J. D. 1962. Quantitative ecology and the woodland ecosystem concept. *Adv. Ecol. Res.*, **1**, 103-192.
35. Institute of Fuel, 1973. Report: Energy for the Future. Institute of Fuel, London.

GÖRAN A PERSSON PhD*

11 Problems, policy and practice in the international field

SUMMARY

The environmental pollution issue will continue to grow despite short-term fluctuations in intensity. Legislative control of environmental pollution in Western European nations is in a state of flux: new central control agencies have been created, and there is a trend toward uniform regulations throughout a given country. Environmental pollution from the combustion of fuel will also probably be controlled in the future by economic measures.

Most environmental pollution problems are connected with the extraction, transport, storage, processing and use of fuels. Air pollution from fuel combustion in stationary sources is a problem of great importance for the fuel and power industry. Desulphurization of residual fuel oil will be general in the Northern European countries by 1980.

With existing regulations for the control of motor vehicle emissions photochemical smog will soon present acute problems in many European urban areas. This will lead to a shift in public attitudes, and restrictions on automobile traffic. Lead-free petrol will not necessarily be a requirement in the future.

Increasing concern about the disposal of nuclear waste may cause a delay in the growth of nuclear power.

Conservation of fuel should be encouraged. Fostering a more efficient and rational use of energy will make available the economic resources to meet advanced environmental objectives.

1. INTRODUCTION

The state of the environment is a major topic for public debate in most countries, and environmental policy is becoming accepted as an important part of the welfare of society. Continuous and progressive industrialization, with the availability of cheap energy, has led to an increase in material standards of living; while, at the same time, there is increasing awareness that this process has resulted in harmful effects on the environment. The concept of a high standard of living should include a healthy environment. For this reason, society has to take more and more active measures.

*Head of Air Protection Division, Swedish National Environmental Protection Board.

There is still a lingering conviction among some industrial and government representatives, that public concern over pollution is a fad that will pass if it can successfully be ignored for a while longer. It seems clear that the environmental pollution issue will not disappear, but will, instead, continue to grow despite short-term fluctuations in intensity. The principal brake on the movement is the fear—justified or not—of economic dislocation and unemployment.

2. INTERNATIONAL TRENDS IN ENVIRONMENTAL POLICY

Legislative control of pollution in Western European nations is at present in a state of flux: much new legislation is being passed, and more is proposed. The legal or administrative measures coming into use in various countries are, in fact, strikingly similar in their essential features, despite national differences in legal and political systems.

In any one country, frequently there are seen to be wide variations in policies pursued by local governments, or other officials. This is as true of those countries with centralized government as it is of those having federal governments. Variable regional regulations, in conjunction with local governments or other official administration, have been one of the strongest factors inhibiting effective legislative control of environmental pollution. Local bodies are often guided by the fear of losing industry to other regions that have weaker regulations, or by the desire to attract new industrial enterprises through lenient regulations.

This is leading to the consolidation of pollution control activities in central government agencies. New central environmental control agencies have been created in several national governments, and others soon will be. Along with this trend, there is also a corresponding one toward establishing uniform regulations throughout a given country.

Several of these new agencies are independent ministries, whose principals are officers of cabinet rank; and this is a trend likely to be pursued more strongly in the future. It is probable that the new environmental ministries will not attain, for a considerable period, either the political power, or the effectiveness of the older, established ministries. However, environmental ministers can be expected to play increasingly influential roles in national political affairs, and effective political power in environmental affairs will gravitate increasingly toward central governments.

The existence of laws does not of itself provide an indication of the effectiveness of control. Enforcement policies may be more important than the laws themselves, and laxity in enforcement has been common even in countries where legal machinery for control has been available.

One characteristic of all regulations is that they are rather inflexible and, thus, may not be very well suited to reducing damage to an optimum level from the point of view of allocation of general resources. This is a serious disadvantage: resources are too scarce to satisfy fully all the needs of society. Thus, in some countries, there is a growing interest for complementing legislation with economic measures. Such practice is already current in Eastern Europe, and is, perhaps, seen to be most advanced in East Germany.

One apparent approach is that which involves charging a fee for environmental damages. The optimum fee on a particular activity is equal to the marginal social damage it generates. In practice, however, it is impossible to arrive at the optimum because a reasonable cost-estimate of marginal damage is not readily obtainable. An alternative possibility involves the pricing and standards technique. Fees can be used as a tool to reach the politically-based, environmental 'goals', e.g. ambient air quality standards, or a specific reduction of total emissions.

The pricing and standards technique possesses one important property: it is 'least-cost' in realizing its aims. Another merit appears to be its potentiality in giving commerce and industry (and, indirectly, technicians, product designers and researchers) incentives to find new techniques and new products whose effects are less damaging. One drawback is the difficulty in determining in advance the effect of these economic measures.

To sum up, there are reasons to believe that a policy compounded of regulations and economic measures will develop. Studies in the US and Sweden indicate that environmental pollution from the use of fuels is an interesting field for the use of economic measures. A tax on sulphur in oil and coal, and on lead in petrol has been proposed. A dynamic view of the matter is necessary if

there is to be success in working out rational environmental policies; yet although attention and measures vary widely from country to country, the development of a specific issue in one nation is not independent of developments in others. Action by one country tends to set a precedent for action elsewhere. In particular, events in the US often furnish a guideline for European countries, although the reverse is occasionally true. Within Central Europe, the precedents are often set by Germany, and by Sweden in Scandinavia.

3. FUEL AND THE ENVIRONMENT

Most environmental pollution problems can be traced back to the extraction, transport, storage, processing and use of fuels. Some of the more evident problems are listed in Table 1.

Environmental problems can also be grouped into local, regional and global problems. Pollution of the air and oceans are examples of problems involving all three aspects.

Again, there are the differences in problems that stem from continuous emissions, and those that are caused accidentally. While it is difficult to compare air pollution, with thermal pollution from a power station, it is almost impossible to weigh these problems against the risk of an accident in a nuclear power station.

3.1. Use of fossil fuels

A recent OECD study has confirmed that during the next decade energy demand in OECD countries will continue to be met overwhelmingly by the use of fossil fuels. However, the pattern of fuel usage (as between solid, liquid and other forms) is likely to be very different in North America, Western Europe and Japan. There will also be certain major differences between the Mediterranean basin and the rest of Western Europe. Furthermore, different countries expect widely differing trends in fuel consumption over the period considered, both as to overall consumption of each type of fuel available and as to fuel preference within different sectors of consumption. This means that the situation in respect of emissions would be variable between countries, and areas within countries, with emissions in some areas increasing proportionally far more than the overall average, whilst in other areas it would even decrease during this decade. Consequently pollution control strategies should also be different, offering different priorities in the development of abatement methods.

In OECD Europe, the 1968 consumption of liquid fuels in stationary sources, 279 million tons oil equivalent, is expected to increase to 671 million tons by 1980; in North America from 189 million tons in 1968 to 318 million tons in 1980; and in Japan from 69 million tons in 1968 to 272 million tons in 1980. In OECD Europe and Japan liquid fuels are expected to account for about 60% of the total energy demands for stationary sources in 1980, with gaseous and solid fuels accounting for about 20% each.

In North America coal and natural gas will account for 83% of the projected 1980 fuel requirements, with coal contributing 29% of the total and gaseous fuels 54%. Between 1968 and 1980 there is expected to be a 90% increase in the consumption of coal. On the other hand, for Europe as a whole during this period, there will be

TABLE 1   Fuel and environmental effects

Process	Environmental effect						
	Air pollution	Noise	Thermal pollution	Oil spill	Radiation	Aesthetic impairment	Accidents (disastrous)
Extraction							
Coal						x	
Oil				x			x
Uranium						x	
Transportation							
Pipelines				x		x	
Tankers				x			x
Trucks		x		x			
Processing							
Oil	x	x					
Coal	x						
Uranium					x		x
Conversion							
Power station	x		x				
Heating appliances	x						
Nuclear power plant			x		x		x
Mobile sources	x	x					
Storage							
Radioactive waste					x		x

a projected swing away from use of coal, by an average of approximately 30%. However, increasing oil prices, political problems in the oil-producing countries and delays in building of nuclear power stations may change the picture.

At present the world's fleet of some 200 million cars takes close to 12% of world oil production. Road vehicles as a whole take about 21% of the oil and 10.5 of gross energy. There is no substitute for the present propulsion fuels in sight.

3.2. Air pollution from stationary sources

The pollutants formed by combustion of fossil fuels are sulphur dioxide, nitrogen oxides and particulates. Oil is gradually becoming the dominant fuel. Emission data for typical oil fired stationary sources are given in Table 2.

There are three different problems connected with these emissions: A *local* problem to which all the pollutants mentioned contribute, a *regional* problem from the sulphur dioxide, and its acidification of the rainfall; and eventually a *global* problem, not only from carbon dioxide, but from sulphur dioxide and its conversion to fine sulphate particulates.

The solutions to these problems are quite different. The local problem—high concentrations of pollutants in the vicinity of the sources—can be solved with a limited impact on the fuel industry. Tall stacks are in many situations an effective solution to the local problem.

Regional and global problems require control of the total emissions of sulphur dioxide; and as the desulphurization of fuel oil has both practical and economical advantages to the desulphurization of flue gases capacity desulphurization at refineries will be required.

International pressure to solve the problems of pollutants air-borne across national borders will be a considerable impetus, but internal policies on air pollution will still be determined primarily by domestic political considerations affected by local problems. European nations, whether part of the European Economic Community or not, can be expected to follow their own internal demands. If, for example, the current OECD study of long-range transport of sulphur oxides confirms allegations that Norway and Sweden are injured by fall-out of these pollutants, those two nations are unlikely to obtain any substantial redress until sulphur oxide emissions are curtailed at national source as a matter of domestic policy.

The determination of threshold concentrations that avoid damage to health, vegetation, metals and building materials is the subject of widespread and often bitter controversy. Recently an expert committee of the World Health Organization has recommended long-term goals for maximum levels of sulphur oxides and suspended particulates. These figures, with some modifications, will probably be used in most countries as tools in the planning for cleaner air in the urban areas. If these figures can be reached the practical solution will be to use clean fuels in urban areas—especially in sources emitting near ground level.

In several countries, some measures have already been taken to reduce sulphur dioxide emissions. Usually these measures consist of restrictions on the sulphur content of fuels used domestically, and in commercial establishments that discharge their effluent gases at low elevations. The usual policy in respect of utility plants has been to discharge the flue gases from high stacks to permit dilution, and to continue burning fuels with relatively high sulphur content. However, the rapid growth in the use of electricity is leading to a corresponding increase in emissions of sulphur dioxide from thermal power plants, and to disputes as to whether dilution alone can ensure adequate local air quality.

The proposed policy in Germany is to require reduction of sulphur dioxide emissions from all sources, ultimately including power plants; the latter are expected

TABLE 2 Technical data and emission factors for oil-fired stationary sources

Type of installation	Individual heating appliance	Small district heating plant	Large district heating plant	Diesel machinery with waste heat boiler	Combined power and heating plant	Power station
Type of fuel oil	Distillate		Residual			
Size (fuel effect)	0·03 MW	1 MW	50 MW	60 MW	300 MW	800 MW
From the installation delivered energy/ton oil						
heat MWh	7·0	9·3	9·5	3·4	6·8	—
power MWh	—	—	—	4·6	3·4	4·8
Primary fuel efficiency	60%	80%	84%	70%	90%	42%
Sulphur content of fuel						
alt. 1	0·5% S		2·5% S			
alt. 2	0·2% S		1·0% S			
Emitted pollutants per ton consumed						
alt. 1	10 kg SO$_2$		50 kg SO$_2$			
alt. 2	4 kg SO$_2$		20 kg SO$_2$			
	0·3 kg particulates		1·5 kg partic.	1·5 kg partic.	0·5 kg partic.	0·5 kg partic.
	5 kg NO$_x$*		5 kg NO$_x$*	25 kg NO$_x$*	15 kg NO$_x$*	15 kg NO$_x$*
Emitted pollutants per MWh useful energy†	1·5 kg SO$_2$	1·2 kg SO$_2$	5·7 kg SO$_2$	6·9 kg SO$_2$	5·3 kg SO$_2$	12 kg SO$_2$
	0·6 kg SO$_2$	0·5 kg SO$_2$	2·3 kg SO$_2$	2·7 kg SO$_2$	2·1 kg SO$_2$	4·8 kg SO$_2$
alt. 1	0·04 kg partic.	0·03 kg partic.	0·2 kg partic.	0·2 kg partic.	0·05 kg partic.	0·1 kg partic.
alt. 2	0·7 kg NO$_x$	0·6 kg NO$_x$	0·6 kg NO$_x$	3·4 kg NO$_x$	1·6 kg NO$_x$	3·6 kg NO$_x$

*Calculated as NO$_2$.
†The losses of power at distribution from a power station have been assumed to be 13% and from in urban areas located combined power and heating plants 10%. The losses at distribution of hot water have been assumed to be 7%.

to be controlled partly by desulphurization of fuel oil, and partly by flue gas treatment. A new directive of the Land Northrhine-Westfalen concerning licensing of new power plants in congested areas demands either the use of very low sulphur fuel—0·5% in large power stations—or stack gas desulphurization with a similar effect on emissions.

In France, power plants are using high stacks for dispersion of the effluent. Emission limits have been imposed for the very large power plant being constructed at Fos sur Mer, where the presence of aircraft runways limits the permissible stack height.

In Italy, no clear policy on power plant emissions has emerged.

In the Netherlands, the growth of sulphur dioxide emissions has been limited by the expanding use of natural gas for power generation. However, the growing shortage of energy will probably soon enforce the reservation of gas for domestic and commercial usage, with a consequent reversion to oil in power plants.

In Sweden, the emission of sulphur dioxide has been increased by the growing use of fuel oil, but it is planned that the country will use only low-sulphur fuel oils by 1980. Current restrictions on fuel oil sulphur content are 1% for the Stockholm, Gothenburg and Malmö areas. In Norway and Denmark similar restrictions on the sulphur content of fuel oil have been introduced for the Oslo and Copenhagen areas.

To sum up, the greatest impact of air pollution control of fuel combustion in stationary sources will be on the petroleum industry's products. Desulphurization of residual fuel oil will probably become a requirement in most of the northern European countries within the next five years, becoming general by 1980. During most of the remainder of the decade, the allowable maximum sulphur content of residual oil will probably be 1%, although lower levels are likely to be demanded in some areas with particularly critical air pollution problems.

Within the next three to five years the maximum sulphur content of distillates will be 0·3% in most European countries.

3.3. Air pollution from motor vehicles

A number of metropolitan areas in Europe have combinations of automotive traffic density and meteorological conditions that are conducive to severe photochemical smog problems. The appearance in some localities of photochemical smog has already been reported. Its presence in some other areas is fairly obvious but is ignored or acknowledged very grudgingly, if at all, by local authorities. One reason for this attitude is that the effects may be partly concealed by the presence of other types of pollutants. Another is probably that, in some localities, air quality has actually improved with respect to such pollutants as dust, smoke, and sulphur dioxide, and the growth of photochemical smog has not yet offset the improvement. Other, more immediately obvious pollutants from motor vehicles—carbon monoxide, smoke, and odours—have received more attention.

Nevertheless, it can be expected that automotive air pollution in general, and photochemical smog in particular, will soon present acute problems matching the growth of the automobile population. Vehicles currently being produced for the European markets need only minimal air pollution controls to fulfil the present regulation* and the principal car-manufacturing nations are moving only very slowly towards more stringent standards. Sweden has urged adoption of the forthcoming US standards to avoid a further increase in the total emissions of hydrocarbons and nitrogen oxides in urban areas, but most other European nations are evidently resisting any such move. The appearance of severe automotive pollution problems will thus coincide with the existence of a high density of vehicles which, after they have left the factory, cannot readily be fitted with controls.

It is clear that the use of the car for personal transportation is at the centre of a rapidly converging group of social, political, and economic problems. The previously conspicuous environmental problems—air pollution, noise, traffic congestion, space for parking and roads—are being joined by the interest for more efficient use of fuel. As a means of transportation, the car is extremely inefficient in its utilization of energy; public transportation can carry passengers at a far lower unit consumption. This is especially evident for American cars, but is true also in a European context.

On environmental grounds alone, it will probably become necessary to restrict automotive traffic and to revitalize and extend public transportation within the next decade. Although limited measures of this type are currently being taken in a number of European metropolitan areas, more forceful efforts are probably not politically acceptable at present. The environmental effects—including air pollution—resulting from the rapid growth of automotive traffic in Europe will probably become critical within a decade, leading to a shift in public attitudes.

The type of controls for automobile exhausts will affect the fuel requirement. With a catalytic reactor lead-free petrol will most probably be required. It is questionable, however, whether the catalytic reactor will actually be applied on a large scale in Europe. Even in the US it looks as if it will be used for a relatively short period before alternative approaches are pursued to control automotive air pollution.

The alternative is a different type of power source. Stratified-charge engines of the type developed by Honda evidently offer substantial reductions in carbon monoxide and hydrocarbon emissions, possibly in the control of nitrogen oxides for future requirements.

External combustion engines, such as the Stirling-cycle, present a feasible approach to a propulsion system that can combine minimum pollutant emissions to the atmosphere with performance comparable with that of the conventional internal combustion engine. Well-designed external combustion devices should be capable of attaining lower pollutant emissions than any internal combustion engine or gas turbine combustors, and have generally favourable characteristics for vehicle propulsion. They also afford a reduction in engine noise, which is in itself a major environmental problem.

Currently, several nations are introducing regulations to require reduction of the lead content in petrol. Most countries are aiming at levels of 0·4 g lead per litre. A few years ago a complete ban on lead in petrol looked likely to be in force in at least some of the European nations by 1980. If the catalytic reactor is not chosen for

---

*EEC Regulation Nr. 15.

the control of vehicle exhaust, the health effects of lead will probably not be serious enough to motivate such a step. Restrictions on automobile traffic, and the introduction of external combustion engines, and possibly of gas turbines, will lead to reduction in the production of petrol, or of motor fuels in general. Restrictions on automobile traffic may come even earlier; but if so, the primary objective is likely to be conservation of fuel rather than air pollution control. It is clear that this development will have a profound effect on the nature and quantity of motor fuel produced.

3.4. Nuclear power problems

Nuclear power is in many ways, less disturbing for the environment than power from fossil fuels. The problems generally connected with nuclear power are radioactive emissions during normal operation; heated water discharges; accidental radioactive emissions; and containment, transport and disposal of radioactive wastes.

Even if the growth of nuclear power will (1) help abate air pollution, (2) help reduce traffic and noise in the area surrounding the power plant and (3) will reduce the risks for oil spills, the public in most countries is alarmed about the growth of nuclear power. One reason for this is the amount of articles designed not to inform but alarm the public. We are approaching a state of 'maximum feasible misunderstanding', and the development must be taken seriously.

The concern over reactor waste, accumulating with the projected growth of nuclear power, is understandable. But the progress made in concentrating and solidifying this waste gives hope for an acceptable solution. This policy is much more favourable from the environmental point of view than the dilution policy normally chosen for traditional pollutants. There is, however, all reason to agree with the conclusions made in the recent report from the working party authorized by the Council of The Institute of Fuel* that a foolproof nuclear waste disposal

system must be developed as a matter of extreme urgency.

The problems of nuclear waste disposal are further complicated by the linkage to a political problem: the spreading of the potential to produce nuclear weapons. The need for an international control system is quite evident, but the experience so far is not encouraging.

A nuclear power station requires significant amounts of cooling water. This raises the question of what has been called thermal pollution. The discharge of waste heat is not limited to nuclear power but the size and low efficiency of these stations have stressed the problem. Siting power stations at the shore is helpful but this is not possible in all countries. Looking at the international situation it is unlikely that thermal pollution will be the limiting factor for nuclear power.

4. FUEL CONSERVATION

There is increasing concern for a more efficient use of energy in most industrialized countries. Some believe in an energy crisis in the near future, others in an environmental crisis if the use of fuel continues to grow. None of these are well-grounded, even if there is a risk of temporary crisis, through shortage of oil and gas.

The main reason for fostering a more efficient use of energy is the existing waste of energy. Without sacrificing we can make important savings at least on a long-term basis and allocate investments to other sectors of our society. On the other hand, some additional investment costs may be involved in the savings of energy. The savings shown in Table 3 are feasible within a high-energy society. If we want to change our way of life drastically there are several other possibilities. It is unlikely that there is a majority for such a change in any country at the moment.

The most striking example of the waste of fuel is the typical American car with its large body and multi-horsepower performance. Fuel consumption in Europe is only half the figure for the US; but equally, European cars are extravagant in consumption. A specific consumption of 1 litre per 10 miles would be a more reasonable figure than 1 litre per 10 km.

If we are prepared to contain the demand for energy at the present rate of growth a more efficient and rational use of energy would make it possible to allocate the economic resources to meet the environmental objectives ahead. Higher fuel prices would also support a fuel conservation policy.

5. CONCLUSIONS

In the context of increasing demand for energy, and the present fuel supply situation, there can be no conclusion other than that abatement of environmental pollution from the extraction, transport, processing and use of fuels will be needed to avoid deterioration of environmental conditions. Abatement of air pollution from combustion of fossil fuels in stationary and mobile sources will be a matter of great concern in most countries.

The Scandinavian nations, the Netherlands, Switzerland, and the German Federal Republic can be expected to follow an increasingly energetic policy in controlling air pollution, probably attaining full force shortly after the end of the decade.

TABLE 3 Examples of possible energy savings. Sweden

Action	Saving of energy			
	% of energy in specific area		% of energy (total)	
In specific area	Short term	Long term	Short term	Long term
Industry				
Rational operation	1	2	0·4	1
Changed manufacturing methods	—	2–5	—	1–2
Changed structure of industry	—	5–20	—	2–9
Other changes	—	2–5	—	1–2
Transportation				
Increased vehicle occupancy	2	5	0·2	0·5
Encouragement of public transport	—	10	—	1
Cars with lower petrol consumption (30%)	—	30	—	3
Rational freight transport	2	5	0·1	0·2
Encouragement of rail transport	5	10	0·2	0·5
Domestic				
Rational firing and ventilation	1	2	0·5	1
Decreased indoor temperature (1–3°C)	5	15	2	7
Better insulation, heat recovery etc.	—	10	—	5
Improved temperature control	—	2	—	1
Structural changes	—	5	—	2

Energy for the Future, published July 1973, and available from The Institute of Fuel.

In Belgium, Italy and Spain respective governments are not subjected to the same degree of pressure from popular feeling as is experienced by governments of Northern European countries.

France appears to be in an intermediate category. Popular feeling is growing. However, the attitude of the public has evidently not yet been translated fully into effective governmental policy.

The general limits for sulphur in residual fuel oil and distillates will be 1% and $0\cdot3\%$ respectively. In European countries control will start in urban areas, and will gradually be extended to cover the whole country. Economic measures could be a substitute for regulations to attain a specific national emission level.

Due to more stringent exhaust emission standards fuel quality specifications will probably not require lead-free petrol within this decade. Legislation governing the use of private cars in urban areas, and the specific fuel consumption of the internal combustion engine, will lead to a slower growth of petroleum consumption. How fast such legislation will develop is difficult to forecast.

The limited supply of clean fuels, questions of security of supply, and the rational use of natural resources, make fuels management an important aspect in pollution-control policies. Ideally this calls for some means of allocation of clean fuels at local, national and international levels. At local and national levels, usage of different types of fuels in various sectors can be encouraged through pricing mechanisms, fiscal policies, standards and regulations. On the international scale, the situation is more difficult. Each country will doubtless wish to insist on having a wide range of options as regards access to clean fuels to implement its air pollution control policies; but it seems doubtful whether a satisfactory distribution of the limited world supply of clean fuels can be attained without some measure of international agreement. A better planning of investments in desulphurization plant would also be possible in a European policy in fuel management.

The experience of international co-operation, when it comes to discussions of control actions is, however, not encouraging.

12   Energy trends

SUMMARY

The paper reviews past trends and possible future developments in energy consumption, and discusses the inter-relation between these and environmental effects arising from energy activities.

Section 1 comments on the role of government in the energy sector and, particularly, in relation to the environment. Section 2 examines post-war trends of total energy consumption as a basis for projecting future demand. Some of the problems of energy forecasting and the principal factors involved in making projections are discussed. The paper then goes on in section 3 to look at the distribution of energy consumption between individual fuels, dealing first with past behaviour, and then trying to assess likely future levels. The discussion concentrates on the UK, but the figures are briefly related to their world context.

Future developments in the supply of various types of energy, and their relation to consumption, are dealt with in section 4, and section 5 outlines some of the principal environmental consequences of energy activity in the past, and likely changes in the future.

1. INTRODUCTION

1.1. This paper discusses the role of government in the UK in relation to energy, and the consequences of energy activities for the environment. In the balance of the energy system the Government represents the consumers of energy and those who are affected by the operations of the energy economy. In the context of the theme of this Conference, it particularly represents those who are affected by activities associated with supplying energy. At the same time, the Government's responsibilities for the nationalized industries involve it in matters affecting the supply of energy, but again it discharges this function with regard to the interests of consumers, and those affected by the operations of the industries.

1.2. In representing the energy consumer the Government seeks to ensure plentiful supplies at the lowest overall cost to the nation. In doing so it has to have regard to the future as well as to the present, and to the effect of energy costs on those in other sectors. Thus its tax policies and its attitude to the plans of the nationalized energy industries have to take account of these, as well as narrower fiscal and public expenditure considerations.

1.3. The Government acts specifically on environmental problems associated with energy in a number of ways. It operates smoke control legislation and maintains nuclear and alkali inspectorates concerned with control of pollution from nuclear installations and other sources. It exercises control over power station-siting and positioning of overhead lines; and government provisions have encouraged, for example, the re-landscaping of exhausted open-cast coal production sites and other disused areas associated with coal production. It is also worth recalling in the present climate of concern about availability of adequate energy resources for the future that, in the immediate post-war years under conditions of energy shortage, the government took steps to promote conservation policies, including the setting up of the National Industrial Fuel Efficiency Service (NIFES) and a reviewing of standards for housing construction with regard to conservation.

1.4. Trends in the use of individual fuels have been influenced to some extent by taxation policies. Social and regional policies towards areas associated with coal mining have influenced the trend of coal production, and the pattern of power station fuel usage. The Government also commissions research and assists the electricity and nuclear construction industries in promoting the development of nuclear power. Similar Governmental actions are likely to continue to influence patterns of fuel use in this way in the future, and have to be seen as part of the general climate within which energy policy evolves.

1.5. The Government's future policy will contain new elements as the international situation changes. The UK is working with its partners in the EEC towards a common energy policy, and the UK may also be affected by developments in agreements on ownership of continental shelf and deep-sea resources. Other international pressures for agreements on supply and demand, such as the OECD agreement on pooling oil resources in crises, will

---

*This paper was prepared principally in Economics and Statistics 2 Division, Department of Trade and Industry.

tend to influence both longer-term policy, and governmental action.

1.6. The starting point of our examination of energy trends is the demand for energy expressed by the final consumers who actually use it. This demand is met by the industries supplying primary fuels, such as coal, oil and natural gas, and secondary fuels, such as electricity, which use primary fuels as inputs. Thus the demand by final consumers for energy becomes a demand for a mixture of primary fuels. The conditions under which these fuels are available to meet the requirements influence the choice between them and operate on demands to determine their actual consumption. In particular, the price mechanism enables conditions of availability to influence the combination of the various fuels used by the secondary energy industries, and also consumer's choice between them. Prices of fuels may reflect the cost of making them available or, of course, they may be influenced by policies of producers or governments through fiscal, and other devices.

1.7. Section 2 of the paper begins by looking at the demand for total energy, and section 3 at the demand for individual fuels, and at certain important features of their conditions of supply. Section 4 is concerned with reserves and the likely effects of levels of reserves on the supply situation; while section 5 outlines the consequences of developments in the UK energy situation for the environment.

1.8. The paper attempts to make clear the difficulties of preparing energy forecasts, and underlines the generally slow adjustment of trends in the energy sector and the complex inter-relationships between demand, supply and price. Subject to these difficulties, several broad features emerge. First, it seems likely that the demand trends of the post-war period will tend to persist, but in the longer term will probably respond to rising real prices of energy. This will lead to reduced rates of increase in demand, and ultimately to the development of alternative sources of energy and new technologies. The timing of these developments depends critically on the pattern of new discoveries of reserves, and their costs, and is therefore extremely uncertain.

1.9. Population growth is expected to continue.[1] This growth is likely to be associated with increasing demand for energy. However, the mere growth of population by itself is not regarded as a major factor in determining the level of energy demand.

1.10. A second general theme which is likely to emerge is that pollution and other environmental effects will increase, but at a slower rate than energy consumption, because the movement towards cleaner fuels will continue and pollution control will be maintained. Particular environmental problems will be associated with the development of large refineries, power stations and oil and gas terminals. One important development, as the scale of operations grows, will be the increasing risk of extreme environmental effects from accidents at large installations: nuclear power stations, drilling rigs, terminals and tankers.

1.11. The pollution effects of energy are likely to be curbed to some extent by the rise in real energy prices, but this will be slow and for the achievement of particular and acceptable levels of pollution, some more direct forms of government intervention are likely to be needed. The capacity of such intervention to achieve its objectives will depend upon the relative importance which is

attached to other (often competing) policy objectives.

2. DEMAND FOR ENERGY

2.1. The demand for energy by final consumers is the energy actually needed by them for heating, for power to carry out work or for any other purpose. Energy actually consumed for these purposes is measured in useful therms and is described in this paper as useful heat. The demand for useful heat generates a requirement for energy to be fed in to users' equipment. This is described as heat supplied and, if the efficiency of an item of equipment is known, the required amount of heat supplied to meet the useful heat demand on the equipment can be determined. In the same way as useful heat requirement is a demand on the consumers' equipment, so the requirement of heat supplied to provide this useful heat is a demand on the energy supply system.

2.2. The total consumption of energy in the UK depends on the aggregate demand for useful heat, and the consumer's choice of fuels and equipment to meet this. It depends, also, on the efficiency with which the primary fuels are converted into secondary forms of energy by the industries concerned. Table 1 shows estimates of total useful heat consumption and the corresponding total consumption of primary energy for the UK in 1950, 1961 and 1971, together with the aggregate heat supplied to final consumers in those years. It is important to note that, although we are dealing with demand, the observed consumption figures are not necessarily measures of demand but may represent either demand or supply, depending upon the circumstances. However, it is judged that supply constraints have not persisted over the greater part of the period from 1950 to the present, and that consumption in this period might reasonably be interpreted as measuring demand.

2.3. The table shows that the consumption of useful heat has risen steadily, and that the total inland consumption of primary energy has kept in step. The overall average efficiency of the system, the proportion of primary consumption which is actually used by the consumer, has remained more or less the same throughout the period. Useful heat as a proportion of heat supplied to final consumers has risen, reflecting the increasing switch to fuels, (particularly to electricity) which are used at higher efficiencies than those they replace. The use of progressively more efficient equipment, although certainly an important trend, is not represented in the figures where the efficiency with which each fuel is used has been assumed constant throughout the period.

2.4. In forecasting future energy trends two major independent variables should be taken into account:

TABLE 1 Inland consumption of primary energy, heat supplied to final consumers and useful heat consumed by final consumers in 1950, 1961 and 1971 in thousand million therms. UK (Figures in brackets indicate percentages of consumption of primary energy)

	1950	1961	1971
Inland consumption of primary energy	60·0 (100·0%)	68·5 (100·0%)	82·6 (100·0%)
of which heat supplied to final consumers	46·1 (76·8%)	50·2 (73·3%)	56·9 (68·9%)
Useful heat consumed by final consumers	20·6 (34·3%)	23·5 (34·3%)	29·0 (35·1%)

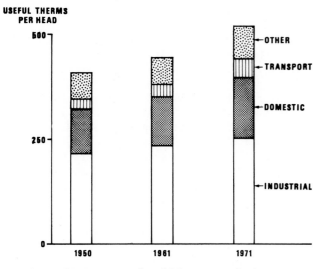

FIG. 1   Final consumption of useful heat per capita *by sectors in 1950, 1961 and 1971*

population, and the level of economic activity. It is also useful to examine demand in the principal energy consuming sectors, because there are substantial differences between them in the nature of energy use. Fig. 1 shows estimates of total energy demand divided between the principal sectors: industrial, domestic and transport, for 1950, 1961 and 1971. Fig. 2 shows trends in total consumption by final users both *per capita* and per unit of gross domestic product over the period 1954 to 1971.

2.5. From Fig. 1 it is clear that useful heat consumption per head has risen steadily, and that there has been growth in all sectors. Fig. 2 confirms that *per capita*

consumption of useful heat shows a clear rising trend, but that it is subject to considerable short-term fluctuation, suggesting that there is no simple relationship between population and energy consumption and that other factors are involved. Useful heat consumed per unit of output has fallen, again with substantial short-term variation, but there is some suggestion that the rate of decrease may be slackening. Energy consumption is thus growing more slowly than total output, although the decreasing steepness suggests that the rates are tending to converge. Again, the fluctuation around the trend suggests that the relationship is complex.

2.6. The use of *per capita* figures removes the effect of population growth which, in this context, embraces a number of demographic and social factors. In practice, the effect of these factors may be more complex and influential than the simple use of population numbers would suggest. Domestic demand, for example, is likely to be more directly affected by the number of households, their composition and the number and average size of rooms occupied, while transport demand will be more directly affected by vehicle numbers, and by the characteristics of motive power. Thus, because population has shown a steadily rising trend it could appear to explain increases in energy consumption, even though much more is involved than population numbers alone.

2.7. Average incomes, and various social improvements, have shown a fairly steady rising trend and have also generally led to greater use of energy; but problems also arise in the use of a measure of economic acitvity as an explanatory factor. Gross domestic product, as with population, embraces a wide range of effects such as changes in the pattern of consumers' expenditure on

FIG. 2   Trends in useful heat consumption by final users per head of population (solid line) and per £ of gross domestic product at constant (1963) prices (broken line). UK, 1954-1971

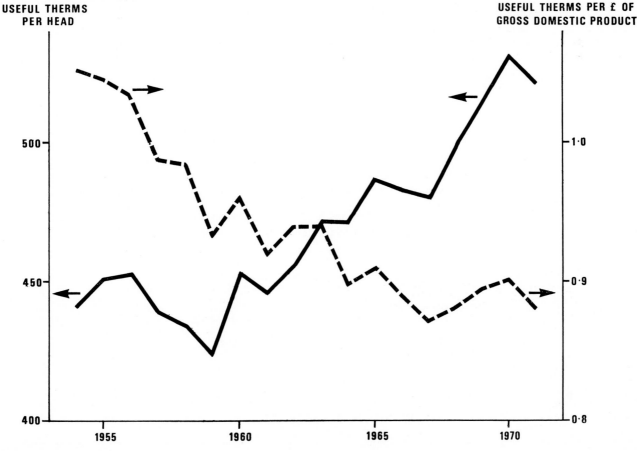

domestic energy demand, or in industrial production on industrial demand. As with population economic activity will tend to explain general trend movements which, however, cannot be causally attributed to it. Also, like population, its use in forecasting requires considerable care.

2.8. The relationship between the growth rate of total primary energy consumption and that of gross domestic product is known as the energy coefficient.[2,3] Historical evidence for the UK suggests that the coefficient as defined has remained more or less constant with energy demand growing at around 0·7 of the rate of GDP growth. As suggested earlier the corresponding relationship for useful heat to GDP is a rather more relevant variable and the evidence suggests that this has recently been rising towards 1:1 ratio. The difference between the two indicators reflects the increasing efficiency of energy use by final consumers.

2.9. A further factor, likely to be of considerable importance in the future, is the real price of energy. Past experience provides little guide to the overall price-elasticity of demand for energy—although it seems reasonable to assume that any substantial increase in the real price will result in a direct reduction of the demand for useful therms, and for their more efficient use. This will affect the demand for heat supplied and hence that for primary energy. This point is considered again in section 4.

2.10 Forecasts also need to make allowance for factors likely to change the nature and direction of past and current trends. The most important of these are probably 'saturation' effects. For example, a crude assumption that fuel use in transport might continue to grow at its past average rate (3·5% p.a. from 1950 to 1971) would soon become constrained by limitations of road space. Similarly, a constant rate of domestic energy growth must eventually reach a level at which the implied energy consumption per household becomes unrealistic. A limited amount of information on likely saturation effects can be obtained by examining the experience of other countries, particularly the USA, where economic activity and energy consumption levels *per capita* are very much higher than in the UK. For example, in 1970 world primary energy consumption per head was around 500 therms compared with about 1 500 therms in the UK and over 3 000 therms in the USA, thus suggesting that world and UK levels are still far from saturation. The approach can be extended to individual sectors and, for example, in the case of domestic consumption use can be made of information on the energy-using behaviour of high-income groups obtained from the Family Expenditure Survey, or similar special enquiries. For the industrial

sector, consideration of the most modern plant-in-use can suggest a possible trend for future developments.

2.11. It is clear that forecasting future levels of total and sectional energy demands is highly complex. However, one set of plausible estimates suggests that the demand for useful heat in the UK in 1980 might be between 35 to 40 000 million useful therms compared with 29 000 million in 1971. Assuming that the relationship between useful heat consumption and primary energy requirement remains much as in Table 1, this would imply primary energy consumption between 100 and 114 000 million therms or between 390 and 450 MTCE compared with 323 in 1971. A substantial rise in primary energy consumption therefore of between 20% and 40% seems likely between 1971 and the end of the decade.

2.12. Looking beyond this country it is clear that the relative importance of the UK in world energy consumption will continue to decline in the future. World primary energy consumption grew at about 3·6% p.a. between 1950 and 1960, and at about 4·4% p.a. in the following decade. The corresponding figure for the UK in both periods was 1·6%. After correcting for population growth, however, the difference is somewhat less; world *per capita* energy consumption growing at 1·7% p.a. between 1950 and 1960 and at 2·3% p.a. between 1960 and 1970, while the UK rate was 1·1% p.a. in each period. The effect of these trends was to reduce the UK share of primary consumption from 7·2% of the world total in 1950 to 4·5% in 1970. Between 1971 and 1980 world energy consumption is likely to grow by between 40 and 60% compared with the range of 20 to 40% suggested for the UK.

3. DEMAND FOR FUELS

3.1. From dealing with the general demand for energy, this section turns to the demand for individual fuels. Table 2 gives figures for the consumption of primary energy by fuels in 1950, 1961 and 1971; Fig. 3 shows total final demand in terms of heat supplied by fuels for the same years, with the corresponding figures for the industrial and domestic sectors in 1950, and 1971.

3.2. Table 2 shows that the dominant trend in primary fuel consumption over the last twenty years has been the replacement of solid fuel by oil. Natural gas has made a substantial impact since 1967 when supplies from the North Sea first become available. Nuclear power still accounts for only a very small part of total primary energy. Final demand in Fig. 3 shows much the same pattern but with the secondary fuels, gas and electricity, playing a prominent part in the replacement of coal. As far as final consumers are concerned, diversification of

TABLE 2   Inland consumption of primary energy by fuels in 1950, 1961 and 1971. UK

	1950 Consumption (thousand million therms)	Share of total (%)	1961 Consumption (thousand million therms)	Share of total (%)	1971 Consumption (thousand million therms)	Share of total (%)
Coal	56·7	89·7	53·7	72·1	38·8	42·9
Petroleum	6·2	9·8	19·9	26·7	41·2	45·5
Natural gas and colliery methane	—	—	—	—	7·2	8·0
Nuclear electricity	—	—	0·3	0·4	2·7	3·0
Hydro electricity	0·3	0·5	0·6	0·8	0·5	0·6
Total	63·2	100	74·5	100	90·5	100

fuel use has also been an important feature of the recent past: the proportions of the different fuels used becoming more similar over time. The separate figures for the domestic and industrial sectors show that the patterns of replacement vary substantially between these sectors. Oil, now the principal fuel in most sectors, is least important in the domestic sector. In the transport sector (not separately shown) oil replaced coal almost completely between 1950 and 1971.

3.3. Price factors have obviously played a part in these switches in demand for fuels, but it is difficult to determine their effect explicitly. The growing use of electricity, for example, reflects the large premium that consumers are prepared to pay for its convenient characteristics. In the choice of fuels for domestic central heating, capital cost, and running rate, are probably assessed jointly and the outcome of the comparison is influenced by particular aspects of the installation under consideration. Moreover, expectations about future prices may be more important than current levels, and this probably contributes to the stability of consumption trends under changing prices. The preparedness of consumers to pay a premium for particular characteristics, other than a basic supply, covers much of the demand for energy outside the bulk heat use of power stations, and very large industrial consumers. The size of the premium varies with the particular use to which the fuel is put. Domestic users, and industrial consumers, requiring clean and very controllable heat for process purposes will be prepared to

pay a large premium, while other users would be willing to pay only a little above the bulk heat price to secure a slightly cleaner, or more easily handled fuel.

3.4. Although substitution between fuels is partly related to relative prices, the uses to which energy is put limit the scope for substitution, which is also affected by supply considerations. Off-peak electricity is an example of the latter situation with its supply conditions precisely determined. Demand is also limited by technology: for example, the replacement of petroleum fuels for road vehicles would require the technical and commercial development of alternative technology so that any influence on fuel consumption could only be expected in the long term.

3.5. Changes in the pattern of consumption also take place relatively slowly because both demand and supply are subject to considerable time lags. The changes that do occur reflect the choices which new consumers make between fuels, and the extent to which existing consumers change the fuels they use. This latter effect is largely determined by the average life of fuel-using equipment, which tends to be long, with only a certain small proportion of the total stock being renewed each year. At the same time, changes in demand can only be met slowly, limited by the maximum rate of extension of production and distribution facilities. The energy sector, generally, is one in which long lead times play a dominant part.

3.6. The largest single consumer of primary energy is the electricity generating industry. The choice it makes

TOTAL FINAL CONSUMPTION

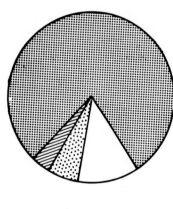

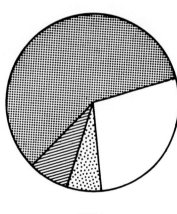

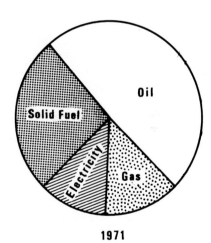

1950   **1961**   **1971**

INDUSTRIAL SECTOR CONSUMPTION

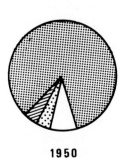

DOMESTIC SECTOR CONSUMPTION

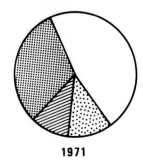

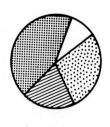

1950   **1971**   **1950**   **1971**

FIG. 3   The split by fuels of total final energy consumption in 1950, 1960 and 1971 and those of industrial and domestic sector consumption in 1950 and 1971. Figures are on a heat supplied basis and are for the United Kingdom

between the fuels it uses reflects the overall movements of primary consumption, although its switch from coal to oil has been considerably slower. Its future policy will clearly have a very considerable impact on the pattern of primary consumption so that any forecast of future primary energy consumption must take into account the criteria used by the electricity industry in decision-making.

3.7. As in the case of estimating overall demand for energy, forecasting the pattern of fuel use requires consideration of a large number of factors. One plausible set of assumptions for 1980 would suggest that, of the total primary energy demand of 100 to 114 000 million therms, nuclear and hydro-electricity might account for between 4 to 7 000 million therms, with the actual outcome depending essentially upon the performance of existing nuclear stations and those currently under construction. The share of natural gas in primary fuel consumption will depend upon the availability of supplies from new sources and is likely to lie between 17 to 25 000 million therms. It is currently estimated that oil from the UK sector of the North Sea will supply about 30 to 43 000 million therms by 1980. The balance of the total primary requirement estimated in para. 2.11 would therefore lie between 25 to 60 000 million therms approximately, and would be shared between coal and imported oil.*

3.8. The shares of the fuels in world primary energy consumption in 1970, the latest year for which figures are available, were similar to those shown for the UK in Table 2, with coal accounting for about 35% of total primary consumption, petroleum for about 43%, natural gas for about 14% and primary electricity (nuclear and hydropower) for the balance of about 7%. As in the UK, petroleum consumption was the fastest growing component but, unlike the situation in the UK, world coal consumption has increased throughout the period since 1950 although the world-share of coal decreased between 1960 and 1970. Over the next 10 to 20 years, the world pattern seems likely to follow a similar pattern to that suggested for the UK, except in the case of coal where the availability of low-cost resources will almost certainly keep world coal production rising. In the longer term the pattern must be less certain with the possibility of a substantial increase in the share of nuclear energy if the technology is developed successfully and with trends in the use of other fuels reflecting their longer term availability.

4. SUPPLY OF FUELS

4.1. The demands for energy which we have been considering can only be met if reserves are available, and this has to be looked at both in terms of physical quantities and cost of supply. This cost covers the cost of physical extraction and transportation, and any other charge which the supplier may make. In assessing the availability of supplies to the UK, the total demand from other countries on world reserves must be taken into account.

4.2. There has been much discussion of levels of energy reserves which have produced widely divergent estimates[4,5] chiefly depending upon differing assumptions

about the rate at which new discoveries will be made. However, all of them accept that resource supplies are finite so that, at some point in the future, reserves will be substantially depleted. Prices are therefore expected to rise in real terms. In section 2 it was suggested that increasing real energy prices would be expected to affect the level of demand by leading consumers to satisfy their requirements more efficiently by using more efficient equipment, and making better use of waste heat; and secondly to reduce their demands in total: in other words through conservation.

4.3. Increasing real prices will also produce a greater incentive to explore for further supplies of fuels used at present, and to develop alternative sources of energy: sources that at present are unacceptable on cost grounds. Substantial work is in hand now on utilization of solar, tidal and geothermal energy and on nuclear fusion, but past experience of developing new technology suggests that commercial use could be extensive only in the long term.

4.4. The success of developments in conservation, and new technology induced by rising costs will themselves tend to influence the future path of costs. It is worth recalling that, often, rising pollution standards are equivalent to reducing the effective level of reserves, and therefore add to rising costs.

4.5. Whatever may be the level of reserves there is little doubt that, as we have seen in section 2, the pressure of world demand on them will continue to grow in the next decade. The path of energy costs over time, however, remains very uncertain because of its dependence upon the very unpredictable rate of discovery of new reserves. We have become accustomed to a high rate of technical innovation in the past half century. It may well be that this will continue, and that many of the anxieties now being expressed will prove to have been groundless. However, in the light of current knowledge it would be imprudent to place too much reliance on the inevitable uncertainties of scientific discovery especially in regard to their timing.

5. CONSEQUENCES FOR THE ENVIRONMENT

5.1. The impact of energy activities on the environment has reflected and will continue to be affected by both the growth in total energy use discussed in section 2 and the changing pattern of individual fuel use covered by section 3. The rising consumption of primary energy tends to produce more pollution. This can be offset to some extent by control which, in turn, will be affected by changes in fuels used, and by developments in technology.

5.2. The replacement of coal by oil in industry, and in power stations, has probably reduced pollution at the point of consumption; but one important consequence has been the relocation of the pollution associated with production of the primary fuel from coal-mining areas to refinery sites. This affects different interests but generally reduces the effect on amenity. Increasing oil use has, however, introduced a substantial and increasing risk of marine pollution. Growing use of oil and production from the UK continental shelf will both continue to increase this risk. Rising natural gas consumption has led to benefits in amenity both by replacing the less 'clean' fuels, coal and oil, and by displacement of town gas and its associated plant. Again, there is relocation of environmental effects to terminals and storage facilities

---

*Examples of more detailed forecasts based on two of the many possible scenarios are given in reference 6.

but, as in the case of oil, the overall impact is much reduced. The persistence of this advantage into the future will clearly depend upon the continuing availability of natural gas. A growing share for nuclear electricity reduces some kinds of pollution, although it adds to the energy system a risk of extreme environmental effects and its use requires the maintenance of very stringent pollution control. The risk increases with the extent of production activity.

5.3. Increased consumption of electricity requires more generation capacity, though in terms of the environment, this is partly offset by the increasing size of successive generations of stations. The number of public supply power stations in the UK fell from 332 in 1950 to 295 in 1970, but the larger size of the later stations and the decision to locate nuclear power stations in remote places has had marked effects on visual amenity in certain localities. The need to build gridlines associated with these stations has also added to environmental effects; although, at the other end of the distribution system, progress has been made with underground local lines in areas with high standards of visual amenity. There is likely to be less insistence in the future that nuclear power stations be isolated, but there will be a continuing tendency for all types of power station to occupy and affect sites in areas of high amenity value— simply because of the shortage of suitable sites outside such areas.

5.4. The increasing preference of domestic, industrial and commercial consumers for cleaner secondary fuels is a marked feature of the last 10 years, transferring pollution away from domestic and factory chimneys to power stations, to refineries and to smokeless fuel and gas production plants where it has been controlled, or has affected different localities. Enormous improvements in air quality have followed this transfer, but relocation of this type raises questions of comparing pollution in the narrower sense with broader environmental considera-

tions, and of comparing environmental effects on different sections of the community.

5.5. In summary, the forecast growth in energy consumption must be expected to add to the risks of environmental damage, but this can be expected to increase at a slower rate than energy consumption because of the steady movement towards cleaner fuels, and the increasing efficacy of pollution control. This control itself, by insisting on more rigorous standards for fuels, may contribute to supply difficulties, and will tend to push up the real price of energy with the effects on consumption and production that have been discussed in previous sections. The principal issues for the environment at present, and in the foreseeable future, are likely to lie in the design and location of very large facilities which will tend to be in coastal areas. Potential pollution will, of course, continue to increase requiring increasing control activity to contain it. Growing populations with rising incomes are imposing increasing demands on scarce energy resources at a time when new standards of social ethics are setting up counter forces of conservation and pollution control. It is the delicate and difficult task of governments to reconcile these conflicting and competing tendencies having regard to the diversity of the interests they represent.

6. REFERENCES

1. Office of Population Censuses and Surveys: Population Projections 1970–2010, London HMSO 1971.
2. DARMSTADTER, J., TEITLEBAUM, D., and POLACH, G. Energy in the World Economy, published for Resources for the Future Inc. by the Johns Hopkins Press, Baltimore and London, 1971.
3. BROOKES, L. G. Energy and Economic Growth. *Atom*, **183**, Jan. 1972
4. KING HUBBERT, M. Energy Resources, in Resources and Man. Committee on Resources and Man, National Academy of Sciences —National Research Council. FREEMAN, W. H. for National Academy of Sciences, San Franscico, 1969.
5. World Power Conference Survey of Energy Resources, 1968.
6. RAY, G. F. Energy in Medium-term Forecasts Reassessed, *National Institute Economic Review*, **62**, Nov. 1972.

F E IRELAND, BSc, CEng, FRIC, FIChemE, FInstF*

13   Legislation: its impact and future trends

SUMMARY
The paper begins by explaining the application of Common Law and Statute Law. It then goes on to discuss in turn planning, discharges to air, radioactivity and discharges of liquid wastes. Under each of these headings the various environmental laws applicable to fuel and its use are mentioned, their impacts are described and as far as possible future trends are predicted. At the time of writing the paper the Water Bill was being presented to Parliament and it has been assumed that the important parts have been passed. Any deviations can be corrected at the Conference.

Environmental control of fuel in its winning, processing and use, begins with planning. There is today a greater awareness of the functions of planning, and there is a good chance of avoiding the social mistakes of the past.

It is probably in its air pollution effects that fuel has its greatest national impact on the environment and local and central authority control are described together with progress made. An attempt has been made to estimate the cost of controlling air pollution. There are small sections on radioactivity and its control at nuclear power stations, and disposal of solid waste.

The paper concludes with the section on discharges of liquid wastes and its estimated cost in 1970.

1. INTRODUCTION
The impact of fuel, its winning, processing, use and effluent disposal, on the environment forms a complicated pattern, covering as it does such things as aesthetics, noise, air and water pollution, residue disposal, transport, land use, public health and amenity, effect on animals and vegetation, etc.

The law concerning pollution in its wider aspects has two forms—common law and statute law. Common law has evolved through actions brought in the courts and the consequent rulings made by judges. Statute law is the written law as defined by Acts of Parliament and

*HM Chief Alkali and Clean Air Inspector.

administered by officers of central or local government. In this paper it is not intended to discuss common law in detail. Suffice to say that damage or nuisance must be proved. Generally, the court will grant an injunction preventing new discharges or giving time to carry out remedies.

2. STATUTE LAW
British laws on control of industrial activities date back to the Industrial Revolution of the 19th century. These laws have been revised from time to time and it is these latest which are described in the paper.

3. PLANNING LAWS
Like any other activity, the production and consumption of fuel involves, directly or indirectly, the use of land— the mining of coal, processing of oil at a refinery, storage of gas, generation of electricity, or the ultimate use of any of these fuels in factory or home.

The use and development of land is controlled under the Town and Country Planning Act 1971 which places day-to-day responsibility for planning administration on local planning authorities, i.e. county councils and county borough councils, although functions of the county councils can be delegated to county district councils.

Section 22 of the Act defines development. It means the carrying out of building, engineering, mining or other operations in, on, over or under land or the making of any material change in the use of any buildings or other land.

Section 23 of the Act provides that planning permission is required for development and this is normally obtained by applying to the local planning authority (although certain development is permitted generally by the Town and Country Planning (General Development) Order 1973). The application must be for a well-defined purpose which is classified under the Use Classes Order and, if permission is given, the owner cannot convert the use from one class to another without obtaining fresh planning permission. Nor can he extend operations within a class indiscriminately. An appeal may be made to the Secretary of State for the Environment if permission is refused or granted subject to conditions.

In considering a planning application, the authority are required to have regard to the provisions of the development plan, which sets out the local planning authority's policy and proposals for the development of their area for up to 20 years ahead, and to any other material considerations, which means material planning considerations. Pollution is capable of being a material consideration and while, as a matter of policy, planning control is not used to duplicate specific statutory controls (e.g. under the Alkali and Clean Air Acts), it is used where specific legislation does not adequately cover amenity considerations.

Planning permission will normally be required for development as defined in connection with the production, processing, distribution and consumption of fuel (the latter will often be ancillary, of course, to the main use, e.g. as in factories and homes). Where a Government Department authorizes a local authority or statutory undertaker to carry out development under a statutory provision, that Department may direct, under section 40 of the Act, that planning permission be deemed to be granted. The power is used sparingly but it is sometimes used in the case of generating stations, overhead power lines, pipe lines and underground storage of gas.

If there is great local controversy over a planning application, or well-known public opposition, the Secretary of State may arrange for a Public Inquiry to be held. The Inspector's report then helps the Secretary of State to make his decision.

It is unrewarding to criticise decisions taken in the past, when priorities were different from today, but we have inherited some unfortunate environmental situations. There is now much better control, with a proper awareness of the functions of planning in relation to the environment. Much more expert advice is available to help take good planning decisions.

When planning permission has been granted for development, the owner must then satisfy the details of other controlling laws governing the environment.

4. DISCHARGES TO AIR

It is probably in its air pollution effects that fuel has its greatest national impact on the environment. Control is divided between local and central authorities, with central government holding an overall watching brief. Local authorities, through their Public Health Departments, are the guardians of their own local environments. Out of several hundred thousand industrial enterprises in this country it is estimated that at least 30 000, possibly up to 50 000, have emissions to air which require the special attention of control authorities. Slightly more than 2 000 are registered under the Alkali &c Works Regulation Act 1906, the remainder being controlled by local authorities under the Public Health and Clean Air Acts.

5. PUBLIC HEALTH ACT 1936

This Act allows local authorities to take action when there is a public nuisance. It is not the best way of controlling emissions to air because no action can be taken until nuisances have occurred, and then only if nuisance can be shown. The Act is useful as a form of 'clearing house' for miscellaneous emissions that are not specifically controlled under other Acts. The Public Health (Recurring Nuisances) Act 1969 strengthens local authority powers by permitting them to anticipate nuisances from industry and requiring ameliorative action to be taken to avoid nuisances. The Public Health Act was the backbone of control for many years and has proved to be extremely useful in its limited way.

6. CLEAN AIR ACTS 1956 AND 1968

Following the London 'smog' of 1952 the Government set up a Royal Commission under the Chairmanship of Sir Hugh Beaver (the Beaver Committee). This Committee made a very thorough investigation of air pollution knowledge at the time, including an estimated cost of the effects. It appreciated that the two main problems were particulate and gaseous pollution, with special emphasis on sulphur dioxide. Recommendations were made that smoke, grit and dust should be attacked immediately and that research and development should be carried out on sulphur dioxide with a view to assessing its importance as a pollutant and preventing or ameliorating its emission at a later date. The Beaver Committee also recommended that certain troublesome industrial processes, with special technical difficulty in overcoming problems of prevention, should be scheduled under the Alkali Act.

The Clean Air Acts 1956 and 1968 regulate emissions to air from the combustion of solid, liquid and gaseous fuels and regulate the heights of new industrial chimneys. The important features about these Acts are that they operate a system of standard setting, prior approval of equipment, continuing inspection and penalties for wrongdoers, which seem to be the essentials for good control. The Acts set a limit of Ringelmann 2 on smoke emissions, whether from chimneys or not.

The Secretary of State is given powers to set limits on grit and dust emissions from chimneys. He has already issued regulations setting limits for emissions from the burning of solid or liquid fuel, where the material being heated does not contribute to the emission. A Working Party is studying miscellaneous processes where the material being heated does contribute to the emission and its recommendations should allow the Secretary of State to set further regulations for these classes of industrial processes.

Regulations have been issued to allow local authorities to require works in their areas to measure their emissions of grit and dust by methods specified in the Regulations.

A booklet has been issued to local authorities and been published by the Department of the Environment recommending the procedure for determination of chimney heights. It has received international recognition.

One of the major functions of the Clean Air Acts is to permit local authorities to set up Smoke Control Areas and these have been an important reason for the improvement in Britain's environment.

Control of burning colliery spoilbanks and their offensive emissions to air is firmly placed as the statutory duty of colliery managers. Whereas in the late 1950s the alkali inspectorate was being asked to advise local authorities and colliery managers on 20 to 30 burning spoilbanks each year, this consultancy service has fallen to nil, and is a measure of the improved control.

The impact of the Clean Air Acts has been striking, as

can be judged by the recently published *National Survey of Air Pollution 1961 to 1971* by Warren Spring Laboratory. The evidence of improvement in the environment is here for all to experience.

Nearly 5½ million domestic premises are now covered by Smoke Control Orders. In 1956 it was estimated that approximately half of the smoke emitted to air in Britain came from domestic fires and half from industry. Both sources have made good progress with their respective problems and, today, it is estimated that about 80% of that remaining comes from the domestic fire and 20% from industry. Nationally, smoke emissions have been reduced by 65% despite a 9% increase in population and a 29% increase in gross energy consumption.

In London, there is 50% more winter sunshine since 1958, average visibility has increased from 1½ miles to 4 miles in the same period and the persistence of early morning mist has decreased remarkably. There are far more bird and plant species thriving than a decade ago.

As a bonus, ground level concentrations of sulphur dioxide have been reduced by 30 to 40%, mostly by the change in heating habits of householders, but partly by the more advanced policy of dispersion from suitably tall chimneys.

This country still burns 14 million tons of coal domestically, most of it in open grates. We cannot rest content until the bulk of this is burnt in smoke-reducing appliances or is replaced by cleaner fuels.

7. ALKALI ETC. WORKS REGULATION ACT 1906

The first Alkali Act was passed in 1863 to control emissions of hydrochloric acid gas from the first stage of the Leblanc process for making alkali. Gradually, other processes were added and the 1906 Act is a consolidating Act. Today, 60 different types of process are scheduled under the Alkali Act and over 2 000 works are registered for the operation of over 3 000 processes. Many of the scheduled works are concerned with fuel, its processing, combustion and by-products.

So far as fuel was concerned up to 1958, the alkali inspectorate had been occupied mainly with the by-products of coal carbonization in relation to their offensiveness and toxicity, the only substantial fuel processing industry scheduled being oil-refining. The 1956 Clean Air Act added smoke, grit and dust from registered works to the list of noxious or offensive gases controlled under the Alkali Act. The 1958 Alkali etc. Works Order added to the schedule those fuel processing and fuel using works where there was technical difficulty in controlling emissions to air. These include electricity, gas and coke, producer gas, iron and steel, ceramic and lime works as primary users of fuel and several others where fuel usage and its emissions are secondary to the main process. Typical by-products works scheduled under the Alkali Act are gas liquor, sulphate of ammonia, tar, benzene, pyridine and acid sludge works.

Works scheduled under the Alkali Act have to be registered annually.

A condition of first registration of a scheduled works is that, at the time of registration, it is fitted with the proper appliances, to the satisfaction of the chief inspector, for meeting the provisions of the Act. An important exception to this condition is any works which was in existence at the time when that class of works became scheduled.

All scheduled works must use the best practicable means (a) to prevent the emission of noxious or offensive gases, and (b) to render harmless and inoffensive those gases necessarily discharged.

The expression 'best practicable means' has reference not only to the correct use and effective maintenance of equipment installed for the purpose of preventing emissions to air, but also to the proper control, by the owner, of the process giving rise to the emission.

As with the Clean Air Acts, the Alkali Act uses the setting of standards, prior approval, continuing inspection and penalties in order to obtain proper control.

The Alkali Act is in its 110th year and it is difficult to describe briefly the impact it has had on industrial emissions, because it has passed through many phases. We really ought to consider its impact since the Second World War and more so since 1958, when its activities were so greatly expanded. The Act gives to the inspectorate wide powers of control over emissions from registered works and the impact seems to be in the policy adopted by the inspectorate to administer the Act, rather than in the provisions themselves. The inspectorate has always sought to gain improvements by seeking the co-operation of managements and in educating people working in industry as to their responsibilities to the community. Great success is claimed for this method, although critics may not agree; but there are imperfections in everything and more forceful measures have to be taken against the recalcitrants, who are definitely in a small minority.

Estimates of emissions to air from registered works are attempted periodically, so as to obtain factual information on which to assess progress. No great accuracy is claimed, but they do put matters in perspective. Some examples are given for the fuel industries.

Electricity works were scheduled in 1958 and there was not a lot of factual information about emissions of grit and dust available at the time, so there is controversy. The inspectorate has estimated emissions of the order of 1 million tons per year. A survey in 1972, based on emission tests, fuel consumption and gas volumes, gave a result of just below 200 000 tons per year, despite a considerable increase in fuel consumption. This figure includes emissions of particulate matter from oil-fired stations. Whilst dealing with electricity works we must not forget gaseous pollution, of which sulphur dioxide is probably the most important, although we must not overlook others such as oxides of nitrogen. The policy adopted in Britain of combining waste gases from separate furnaces to a common chimney and using a properly designed chimney of adequate height, efflux velocity and gas temperature, has ensured that ground level concentrations are negligible. This policy has contributed to the 30 to 40% fall in ground level concentrations of sulphur dioxide throughout the country since 1960.

Cement works are an example of the use of fuel in a process where the material being heated contributes to the emission. About 1960 it was estimated that the loss of cement works dust to air from chimneys was of the order of 1 to 2% of the clinker made. Based on production at the time, this meant an emission of about 100 000/200 000 tons per year. A more precise survey in 1972, based on emission tests, showed an escape of about 40 000 tons per year. With improvements in hand, there should be an appreciable further reduction in 1973.

The saga of the iron and steel industry is too large to feature in detail here, but substantial reductions in emissions to air are claimed, e.g. sinter emissions from 1 to 2% of sinter made reduced to below 0·2% and the target for the future, when all strands are fully equipped with modern arresters, less than one-fifth of this; blast furnace emissions virtually eliminated with daily 'slips' at 15 grains per cubic foot (35 g/m³) reduced to about once every six months.

Revolutions in industry are taking place continually and the alkali inspectorate's policy allows these to take place in a rational manner and at the same time keeps control over emissions to air in as practicable a manner as possible. Since vesting day in 1949, the number of gas works in Britain has fallen from 1 050 to 96 at the end of March 1972, 463 being registered in 1958 when the process was scheduled. The industry changed from one based on coal carbonization to a petrochemical industry and then to natural gas.

The coal-fired bottle oven for pottery manufacture, a heavy emitter of smoke, has virtually disappeared. There were 2 000 bottle ovens in Stoke-on-Trent in 1939, 295 in 1958 and none today, the last shutting down in 1967.

In the heavy clay industry, there has been a rapid change to smokeless firing. At the end of 1961 there were 2 224 coal hand-fired, downdraught, intermittent kilns making smoke, whereas at the end of 1972 there were only 80, excluding salt glazing and blue bricks.

8. COST OF AIR POLLUTION CONTROL

The Beaver Committee estimated in 1954 that the annual cost of the effect of air pollution in Britain, mainly from the combustion of fuel, was about £250 million. Beaver later said he thought it was probably nearer £350 million.

Many authorities all over the world have attempted to put a cost on the effects of air pollution, but the estimates are unreliable because much depends on the financial assessments placed on amenity and aesthetics. In 1972, the Programmes Analysis Unit, a joint unit of the Department of Trade and Industry, and the United Kingdom Atomic Energy Authority, published their report entitled *An Economic and Technical Appraisal of Air Pollution in the United Kingdom.** It represents the views of the authors in 1970, when their main conclusions were as follows:

(1) The mean excess economic costs in conurbations over those that exist in non-conurbation pollution levels is £130 million per annum.
(2) The mean total national economic cost of air pollution could be £400 million per annum, of which about two-thirds is probably due to domestic fire emissions.
(3) Pollution from industrial sources and electric power generation could be currently costing a mean of £120 million per annum, 60% of which arises from industry.
(4) Motor vehicle pollution appears to be costing a mean of £35 million per annum.
(5) The largest mean damage costs are in the areas of agriculture and health and are estimated to constitute 80% of the current economic costs.

---

*HMSO (Code SBN 7058-0182-9) price £5·00.

There are many uncertainties in the figures and some industries disagree with the results.

In 1968 the alkali inspectorate made a survey of the amounts industry had spent on air pollution control at works scheduled under the Alkali Act over the previous 10 years. There was a steady rise in expenditure during the period and in 1968 it was estimated that these industries were spending at the rate of about £40 million per annum. It can be guessed that the remainder of industry under local authorities was spending a like sum, giving a total of about £80 million per annum. In 1972 another survey of the cement industry showed that air pollution control costs had at least doubled since 1968 and, if we assume the same for other industries, we find that today the cost of industrial air pollution control is about £160 million per annum. These costs will include non-combustion processes which are difficult to separate from the combustion, but the latter will predominate. They exclude domestic fires and transport.

Let us look at some of the costs of air pollution control and see where our priorities lie.

Air pollution control measures at a large modern power station involve a capital cost of £4 to 5 million. and a large integrated steel works is at least as much. The operating and maintenance costs are additional.

To desulphurize residual fuel oil from an average of 2·6 to below 1·0 per cent sulphur would cost between £100 and £200 million in capital, probably nearer the latter, plus about £100 million per year operating costs. This would reduce sulphur dioxide emissions from about six million to five million tons per year. Would it be worthwhile? Or could we use this large sum of money more effectively in some other way?

This great technological age demands bigger furnaces, larger industrial process units, greater power stations and huge oil refineries, and the watchdogs demand greater and greater standards of prevention. How far can we go, and how much can industry and our economy stand? There are a lot of gaps in our knowledge which only research and investigation can fill. The Department of the Environment is working to that end.

9. FUTURE TRENDS

Because the British system of control has evolved over a long time it is firmly based. Whether it can stand up to modern demands is a subject for controversy. With its flexibility for change as technology advances, I believe it can cope, provided adjustments are made from time to time, and I hope it will survive international pressures. Our entry into the Common Market is bound to have influences which cannot yet be predicted and our representatives are engaged in negotiations. We have before us suggestions for much tougher standards of emission, greater penalties, the use of charges for pollution, extensive monitoring and uniformity of laws, standards and implementation.

At home, local authority reorganization must have an influence. The larger and stronger local authorities are likely to have their own specialist teams of scientists and sampling experts to control emissions and ground level concentrations. Looking into the future is difficult, but we have to plan ahead to satisfy future generations. Preparations are being made for the proposed Environmental Protection Bill.

10. RADIOACTIVITY

The present legislation dealing with solid, liquid and airborne effluents from the nuclear power industry is the Radioactive Substances Act 1960.

The United Kingdom Atomic Energy Authority was set up by the Atomic Energy Authority Act 1954. The Act is largely administrative, but it does stipulate that it is the duty of the Authority to ensure that nothing they do can cause a radioactive hazard to anyone, anywhere. The main points of control of radioactive wastes were:

(a) All waste disposals must be authorized by the Secretary of State for the Environment and the Minister of Agriculture, Fisheries and Food.
(b) Ministers may attach such conditions to the authorizations as they think fit.
(c) Persons appointed by the Minister may enter and inspect authority premises and take samples of wastes.

This Act has been mentioned because, before legislation controlling their operations actually existed, the Central Electricity Generating Board accepted an obligation whereby nuclear power stations were subjected to controls similar to those imposed by law on the UKAEA. This informal arrangement continued until the coming into force of the Nuclear Installations (Licensing and Insurance) Act 1959.

The Radioactive Substances Act 1960 came into effect on 1 December 1963 and it gave permanent effect, in substantially the same form, to the temporary controls to which discharge of waste by the UKAEA and nuclear site licences were subject by virtue of the 1954 and 1959 Acts.

It is not possible to lay down hard and fast rules for the disposal of wastes, because nearly all authorizations are unique in having some special problem which needs to be taken into consideration. Some of the major points common to all authorizations are as follows:

Broadly, the recommendations of the International Commission on Radiological Protection and the Medical Research Council are used to assess the hazards of a discharge. When disposals are made to a local authority tip for controlled burial, there must be adequate cover and there must be no possibility of contaminating local drinking water supplies. Liquid waste disposals to a sewer necessitate an assessment of the dilution and a knowledge of any process to which the sewage is subjected. Discharges of gaseous wastes require a knowledge of the concentration of activity in the plume, and an assessment as to whether this constitutes a hazard to anyone.

It is still true to say that the majority of radioactive discharges are many orders of magnitude below levels that would give rise to doses not regarded as acceptable by the ICRP or MRC.

There are some wastes for which local disposal is not appropriate. The Act gives the Secretary of State power to provide other facilities, and this has given rise to the National Disposal Service operated by the UKAEA.

11. DISPOSAL OF SOLID WASTES

There is not much to say about the legislation on disposal of solid wastes from the fuel industries, because very little special legislation exists to control it.

Planning permission has to be obtained to deposit spoil from mining operations and thereafter control of com-

bustion is regulated by the Clean Air Acts. The Stevens Committee is examining the planning control over mineral operations, including restoration, e.g. should the equivalent of the Ironstone Restoration Fund be set up to ensure restoration when operations cease.

The Deposit of Poisonous Wastes Act 1972 excluded wastes from power stations and mining, although it does include objectionable wastes derived from by-product plants.

12. DISCHARGES OF LIQUID WASTES

Although this section is written before the passing of the Water Act 1973, the legislation has been anticipated and the section is written as though the Act has been passed by Parliament, as is likely to be the case by the time the Conference is held.

Prior to the passing of the Water Act 1973 there were nearly 1 400 authorities in England and Wales bearing responsibility for, or having an interest in some aspect of pollution control of rivers and streams. This figure included 29 river authorities and 1 200 sewerage authorities who, between them, exercised control over sewage and industrial effluents discharged to rivers and public sewers.

Discharges of industrial effluents to public sewers were controlled by local sewerage authorities, using a consent or agreement procedure authorized by the Public Health (Drainage of Trade Premises) Act 1937, and the Public Health Act 1961. The local sewerage authorities exercised control over the volumes and chemical composition of industrial effluents discharged to public sewers and, in this way, two main objectives could be achieved; firstly, the control of the quantity and quality of the polluting material contained in the effluents so that the treatment works could operate effectively and so produce final treated effluent discharges to rivers to meet the statutory quality standards of river authorities, and secondly, to enable the cost of treatment and disposal of the wastes to be recovered by the authorities.

Discharges of industrial effluents to rivers were similarly controlled by river authorities by a consent procedure whereby the quantity and quality of polluting material (and heat was classified as polluting material) permitted were limited to levels appropriate to the circumstances. Generally, in practice this meant that the effluents had to comply with the so-called 'Royal Commission Standard' (30 mg/l of suspended solids and 20 mg/l biochemical oxygen demand—BOD). It was generally accepted that this 'standard' was required as a 'norm', even if the dilution available in the river was more than the 8:1 envisaged by the Royal Commission. In some circumstances, in particular where the amount of dilution was low, standards more stringent than those mentioned above were applied.

From this it will be seen that the effluent control procedures as they affected discharges both to sewers and to rivers were similar. A right of appeal to the Secretary of State for the Environment in the case of dispute concerning the standards was available, and the system permitted great flexibility of approach to solutions to the co-related problems of river pollution prevention, and the technical and financial problems of industrial effluent treatment. One important consideration for the coal industry should be mentioned at this stage, namely, that mine waters were exempt from control providing they were discharged from the mine workings in their natural state. Powers to exercise control over these discharges of mine water could be obtained by river authorities by application for orders, but few such orders had been granted in the past.

Prior to 1973 the pollution control powers of the river authorities were mainly restricted to non-tidal waters. They exercised no control over discharges of effluents to the sea by pipeline, and in estuaries and tidal waters of rivers their powers were limited, unaltered 'pre-1960' discharges being in most cases exempt.

The Water Act 1973 replaced the authorities mentioned above with 10 large regional water authorities exercising all of the powers of the bodies they supplanted.

It is important to note, however, that the new authorities (with the exception concerning mine waters to be mentioned later) will still exercise control under the existing legislation already described which, as has been shown, is capable of very flexible application. The possible impact on the fuel industry of the new structure of the water industry stems not from new legislation, but from the extension of the application of existing statutes to a greater area of control by fewer authorities than heretofore. Government policy will determine the magnitude of the impact.

At the present time the disclosure of any information concerning river authority effluent discharge consent conditions is a criminal offence. River authorities are required to maintain a register of particulars of conditions of consents but inspection of the register is restricted. Under common law a riparian owner is entitled to the water of his stream without sensible alteration in its quality, and if it is polluted he can maintain an action against the polluter without proving actual damage and may obtain an injunction to prevent continuation of the injury.

The Royal Commission on Environmental Pollution considered the only value of confidentiality about trade effluents was to protect industry against the risk of common law actions, and that some protection was needed in appropriate circumstances. The Government accepts this view and the fact that rivers must continue to be used to receive effluents after the necessary treatment. The cost of removing all impurities, even if practicable, from the effluents would be unacceptably high. The Government therefore do not consider that riparian owners should continue to be able under common law to obtain an injunction to stop a discharge which conforms with the conditions imposed by a regional water authority, or that a discharger should be liable for compensation for damages caused by such a discharge which complies with those conditions. However, they consider an owner whose property is harmed by a discharge should be entitled to compensation which should be the responsibility of the regional water authority.

As has been stated, full control has been exercised for many years over discharges to inland non-tidal rivers, and it is likely that the greatest impact of the changes will be felt by dischargers to tidal waters, estuaries and the sea. It is intended that the regional water authorities will exercise the main provisions of the Rivers (Prevention of Pollution) Acts to tidal and coastal water within the three-mile limit. This means that all effluent discharges to these waters will become subject to the same sort of controls as discharges to inland waters. However, the flexible approach permitted by legislation, as with inland non-tidal waters, permits a programme of implementation of controls, the severity of which and the cost of compliance with can be determined to suit local and national economic circumstances. As would be expected, the pressure to bring about the improvement of inland non-tidal watercourses, many of which may be needed to supply drinking water, is greater than the pressure to improve the conditions of estuaries, since the latter may be considered less important on grounds of amenity, and will not be used for water supply. It follows from this that if any economies are to be made in the programme of water pollution prevention, it is likely they will be made in these areas.

As previously mentioned, mine waters have not been penalized under the Rivers (Prevention of Pollution) Act 1951 if they were discharged from the mine in the same conditions as they were raised or drained from it. It is intended that legislation will be introduced to control these discharges by extending the consent procedure of the Act to them. River pollution from these sources generally arises on two accounts: firstly, discolouration by oxidation of the iron salts contained in them and, secondly, adverse effects on the ecology of the rivers due to acidity and oxygen depletion.

Considerable expense will be incurred in the treatment of these waters, both in capital installations and continuing costs of chemicals for treatment, together with costs of disposal of large volumes of intractable sludge. The National Coal Board estimate that it will cost the Board an extra £7 million a year in operating costs, and

TABLE 1 Points of discharge, volume and cost for industry concerned in fuel production

Discharges taking place to:	Industry	Volume × 1 000 gpd Total industrial eff't	Cost £1 000s Estimated as required in 1970
Non-tidal river	Petroleum ref'g	15 675	176
	Gas & coke prod.	32 024	467
	Elec. gener'n	4 656 615	341
	Mine—process	35 494	610
	Mine—water	197 047	—
Tidal river	Petroleum ref'g	549 665	2 080
	Gas & coke prod.	33 237	4
	Elec. gener'n	8 545 447	99
	Mine—process	57	Nil
	Mine—water	5 665	—
Canal	Petroleum ref'g	1 134	Nil
	Gas & coke prod.	1 428	Nil
	Elec. gener'n	114 557	60
	Mine—process	292	2
	Mine—water	4 083	—
Sea	Petroleum ref'g	Nil	
	Gas & coke prod.	5 903	
	Elec. gener'n	4 238 809	
	Mine—process	2 848	
	Mine—water	14 565	
Total	Petroleum ref'g		2 256
	Gas & coke prod.		471
	Elec. gener'n		500
	Mine—process		612
			3 839

£10 million in capital expenditure to deal with pollution arising from this source, It is a matter for debate as to how the cost should be carried by the community, and in this connection discharges from disused mines pose particular problems. On the credit side however, treated mine waters will make a valuable contribution to much needed future water resources.

The River Pollution Survey of England and Wales (2, 1970), suggested the likely cost to industry of their part in restoring rivers to what river authorities considered would be a satisfactory condition. Table 1 on page 154 gives the details for those industries that may be concerned in fuel production. Additionally, it gives volumes discharged, including those taking place to the sea (details taken from the DOE Report on Discharges to Coastal Waters) for which no cost figures could be included. As mentioned earlier, the proposals for dealing with environmental pollution would include extending the authority of regional water authorities to the sea, and some additional remedial costs could be incurred on account of this.

It should be noted that the costs do not include those incurred in providing pre-treatment plants before discharge to sewers, for treatment at a sewage disposal works. These costs are, however, relatively low as reported in the 1970 Survey.

Although river authorities had no control over pre-1960 discharges to controlled tidal waters, they were asked to indicate in the Survey the extent of any remedial work that they would wish to see carried out. In many cases river authorities considered that although the discharges could not be described as satisfactory, they were not unduly embarrassed with them at that time. They were of the opinion, however, that considerable treatment may be advisable in the future and any change in legislation to bring pre-1960 discharges under control could result in additional remedial costs.

In the Survey, forward planning for industrial effluent treatment was not enquired into, as this was not considered feasible. The Survey was to be updated in respect of costs (it was already updated yearly on river quality), but this was not likely to be before 1975.

Professor J K PAGE, BA, FIES*

14   Energy, environment and planning

SUMMARY

The paper stresses the need to think about the problems of the fuel industries on an activity-environmental systems analysis basis. Environmental planning must be a strategic objective of corporate planning, and not a tactical response to stresses arising from engineering activities planned without proper consideration of environmental factors that are likely to arise eventually.

The problems of the planning control of various facets of the energy chain are then considered, starting with mining the resource. The importance of environmental models for planners is stressed, together with the need for establishing planning constraints to protect the environment.

Controls must balance environmental costs against economic benefits. The problems of the fuel-processing industries are discussed, and it is pointed out that national environmental benefits are often achieved at the expense of severe local environmental costs.

The paper considers the problems of the impact of transportation of fuels on the environment. It concludes on the role of planning in urban atmospheric pollution control. Particular emphasis is placed on the significance of meteorological factors in determining interactions between pollution-producing activities and pollution-sensitive activities.

The need for planners to seek a deeper meteorological understanding is stressed.

1. THE PLANNING PROCESS

The physical planning process is concerned with the satisfactory control of all activities (existing or future) involving the use of land; to secure a socially desirable, and socially acceptable overall pattern of physical development. This can only be done by adopting appropriate techniques of control, backed by legislation, to ensure the following aims are met:

(1)   The preservation of the long-term biological viability of the international, national and local

*Professor, Building Science, University of Sheffield; Chairman Environmental Group, Yorkshire and Humberside Economic Planning Council.

environments, e.g. the sea must not be allowed to become poisoned.

(2)   The safeguarding of the social needs of existing and future communities. In particular, society must be adequately protected from adverse environmental developments. The planning process must also preserve a certain open-endedness against the future to allow for future social and technological changes. The process is therefore cybernetic rather than mechanistic.

(3)   The development of appropriate employment opportunities for the local community. However, the environmental impacts of such industrial and commercial developments must be properly regulated to meet the demands imposed by Aims 1 and 2.

(4)   The safeguarding of the development potential of valuable natural resources, e.g. minerals, but subject to the safeguards set out under Aims 1 and 2. Where possible valuable minerals must be protected against alternative developments that would prevent their future exploitation—though compromise is frequently necessary, e.g. the presence of existing communities.

The ecological factors concerned with the long-term future of the biosphere are so fundamental, that they take priority over the social issues. Technologists can no longer (at the present scale of industrial activity) afford to overlook the long-range biological implications of their actions. Biologically-informed protest will increasingly add its weight to the present protest movements based on social arguments. The solution to the ecological problems set by industry will require knowledge of the interactions with biological systems, and an understanding of the detailed workings of the global atmospheric and water resource systems that transport pollutants and effluents from one place to another. The planning assessment must become multi-dimensional and multi-professional.

Much technological development in the past has failed to respect the basic environmental needs of the individual. The actual planning process invariably involves an attempt to 'strike a fair balance' between a 'wide spectrum of conflicting interests.' There are considerable planning tensions between industry and the populace—which are the results of poorly-considered industrial developments.

These tensions will only be alleviated by technological design procedures, and planning processes, that give far higher priority to environmental design factors in the engineering design process.

In a competitive world, the industrial problem essentially lies in finding a right balance between environmental control costs, and consequent social benefits. Therefore, all environmental control policies must be based on a sound economic foundation. Consider an energy process that may be controlled by different levels of investment to give different outputs. Fig. 1 shows a typical case where the higher the investment, the higher the standard of pollution control achieved. Very high levels of pollution control need disproportionally high investments. A second cost function may be established showing the social costs of different levels of environmental pollution. Below a certain threshold there may be no significant social costs. The sum of the two costs gives total costs. In economic terms, the optimum control policy for the country as a whole is to aim for minimum total costs. Exemption from effective planning controls reduces the damage charges placed on any industry that is causing pollution. If the developer is not forced to carry the true environmental charges, his short-range financial policy is to save money on pollution control equipment. As a result, the community has then to carry the pollution costs. The policy that the polluter must pay is very important; but it must be tempered by an understanding of what is reasonable on the part of the planners.

Unfortunately, the actual environmental penalties often fall on those who benefit least from some specific activity. Arguments of national interest appear to be of low importance to those who are locally affected by the winning of a particular mineral resource.

A balance has to be struck. This balance changes with shifts in social attitudes, and with rising prosperity. Clearly current social priorities (particularly among the more affluent) are moving towards a much lower tolerance of environmental pollution. This change, in turn, must force each industry to re-examine its historical attitudes towards its pollutants, to meet this growing protest.

Some technologists like to persuade themselves that present environmental protest is simply a wave in fashion that will soon decline.

The author believes the contrary situation will occur: that environmental protest will increase with the rise in educational levels, and the increase in prosperity. It will, furthermore, become more professionally-based, and better informed scientifically. This informed protest will make it essential to treat the problematical environmental design as a major technological problem common to all industrial organizations. Environmental design must, therefore, be right at the initiation of any new project. Thus the demand for improved environmental planning implies giving far greater weight in the engineering design process to:

(1) The reduction of adverse social impacts of engineering decisions;

(2) The reduction of adverse ecological impacts of engineering decisions.

This implies an important shift in the education of technologists, as well as in the environmental attitudes of management. It also implies full recognition of the social aims of the planning process. The new situation demands both a team approach to design, and a balanced set of objectives.

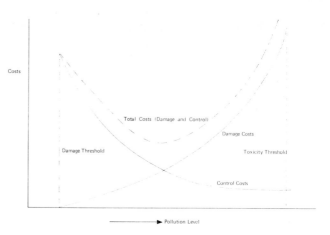

FIG. 1   Economic relationship between damage and control costs— A modified version of chart by K. Taylor. Note there may be a threshold below which there are no damage costs. The curve is likely to rise sharply once the toxicity threshold is reached

2. THE SYSTEMS APPROACH IN PLANNING

This essential relationship between industrial activities and environment implies that technologists must think in terms of optimizing the overall design of the 'Activity Environment System.' The environment of a system has been defined by Hall as the set of factors outside the system:

(1) A change in whose attributes affects the system;

(2) Whose attributes are changed by the behaviour of the system.

The scale of today's industrial activity, and the impact on the environment in consequence, has become so large that industrial systems designers cannot deal with the environmental problems simply at technical level, implying minor alterations to details at completed design stage. A strategic solution that would lessen, or mitigate the impact on the environment has, too seldom, been sought within the integral design process. For example, the aircraft industry has, over recent years, destroyed a substantial part of its development potential by failing to give sufficient attention to control of aircraft noise. This has made it almost impossible for the public now to accept new operating bases close to centres of population. Belatedly, the problem has been recognized, but only at the stage when it has become virtually impossible to find politically acceptable airport sites due to justly-based social protests. The outcome of the Roskill Commission illustrates the processes only too well. The need for comprehensive design processes is obvious. Systems analysis can help both developer, and planner, towards a complementary, and a more powerful analysis of their problems.

3. THE PLANNING PROCESS

Let us examine in more detail the various planning problems associated with the fuel industries, by breaking down the discussion into the following areas:

(1) Planning control of operations for actually mining the fuel resource.

(2) Planning control of operations for processing the basic resource chemically to produce processed fuels.

(3) Planning control of operations for the distribution of fuels to various markets in relation to pollution,

including planning policies for electricity generation, and distribution industry.

(4) Planning policies for the distribution and combustion of specific fuels in relation to clean air policies in particular urban complexes.

4. MINING THE RESOURCE

The location of fuel-winning activities are determined by geological factors. The fundamental planning issue is whether the mining of the resource should, or should not proceed at a particular place. This must involve balancing the economic benefits secured from the mining of the fuel, including local employment benefits, with the environmental penalties that will result from the fuel extraction process. In the past, economic benefits were given a higher priority than local environmental costs. As a result, many areas having association with extractive fuel industries now have an extremely run-down physical environment. This makes the development of alternative employment opportunities very difficult in such areas on the termination of mining. In the past the final land use was not considered important, and workings were simply abandoned as derelict when they had reached the end of their economic life, with all the associated destruction of the landscape. It is an historically determined situation which can only be dealt with by the injection of capital from central and local government. Some progress is being made towards clearing up this historic legacy through the relatively generous grant arrangements that exist for financing derelict areas.

The basic problem in dealing with such dereliction is to identify a viable long-term alternative use for the land, either for agriculture, recreation or urban/industrial development. It is no use simply to grass over a tip, and hope that the vegetation, if it survives the chemical environment of the soil, will provide an alternative use of the land. An awareness of the need to develop acceptable new uses now exists. For example, in some areas of Yorkshire coal tips are being developed to provide much needed local recreational facilities, such as golf courses. This ensures effective long-term landscape management after initial clearance of dereliction.

One aim of planning control should be to see that land is returned to acceptable uses as soon as extractive work is completed. The aim should also be to secure adequate control of the adverse impacts of the fuel-winning activities on the environment during the actual process of extraction. This implies planning constraints on the process itself. Such constraints are not always welcome to the extractive industries. Nonetheless, the UK is one of the most densely-populated countries in the world: the impact on the environment from the winning of minerals is particularly acute.

Unfortunately, many existing activities of the coal industry still do not come under effective planning control. This is because, as they were in existence prior to nationalization, they were exempted by General Development Orders from many of the physical planning requirements that would be imposed today. The effectiveness of current planning control techniques cannot be judged therefore, in those areas where exemptions exist from

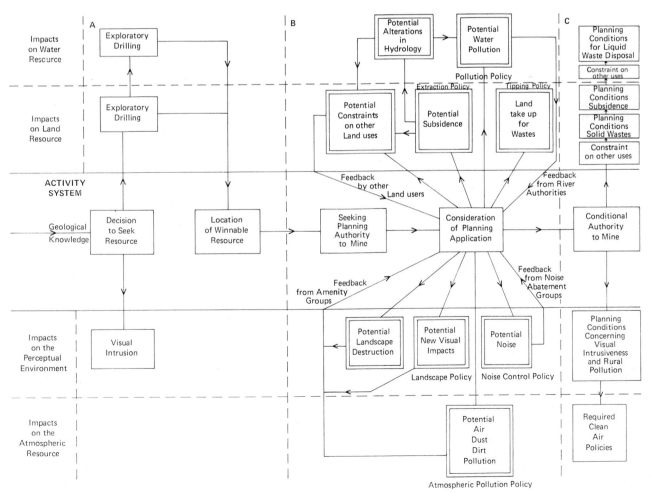

FIG. 2   Exploration and seeking of planning permission

General Development Orders. Whether this situation should remain inviolate in perpetuity is highly questionable, as no effective long-term environmental planning process can be established in the face of such exemptions, which perpetuate a situation where the economic benefits of production are not fairly weighted against the environmental costs to the community. A nationalized industry can be as ruthless as any other—especially in times of financial stringency.

New extractive activities, fortunately, do come under effective planning control; but resentment has been noted in many industrialists who have been used to previous exemptions under General Development Orders. For example, in Yorkshire, available land for coal tipping (exempted from planning control by the General Development Orders) is becoming exhausted. This has forced the NCB to find new land for tipping, and to obtain the associated planning permissions. Such planning permissions enforce conditions that include substantial requirements for the eventual restoration of the site.

The planning problem in the future will be to achieve the right balance between costs in securing a better environment, and the consequent selling price of fuel. Locally it will be a question not only of environmental costs, but one also of employment benefits. Nationally, it will involve the question of energy competitiveness. How much local inhabitants should have to suffer the violation of the environment for the sake of national benefits is an important policy issue.

The basic stages of the planning process are shown in Fig. 2. Stage A is exploratory. Stage B involves the developer in seeking planning permission for active working. In order to assess the environmental impacts of the proposed activities properly, the planner has to construct a model of the environmental outcome of the proposed activities.

Outputs to the three basic sectors of the physical environment, the land resource, the water resource and the atmospheric resource, have to be identified. In addition, planners have to consider the environment perceived through our eyes and ears. Unsightliness and noise are important aspects of all mining activity. Forecasting these outputs requires a model of the process. An idealized conceptual model of a mining process is set out in Fig. 3, from which it may be seen that energy impacts on the environment are enclosed in circles. Subsidence is included in this category because it is an issue involving choice between release of potential energy by lowering the surface, or the successful build-up of strain in the rock structure by partial extraction. Hydrological factors are included in the energy category as hydraulic flows are related to potential energy gradients. It is not just the actual ground movements that count, but also the potential energy differences created between different points between which water is likely to flow. Mass flows are shown in triangles. These include solid wastes, liquid wastes and outputs to the atmosphere, including dust and dirt as well as smoke and sulphur dioxide.

On the basis of the study of the probable environmental impacts, considered in relation to all other aspects of land use in the area, and taking account of feed back from other parties with an interest in the environmental outcome, a decision has to be made, sometimes only after an extensive public enquiry whether to give authority to mine, and if so, what constraints to impose. These constraints are set down as Stage C of the planning process in Fig. 2. It will be noted that the constraint column contains constraints both on other land uses, as well as on the operations of the fuel industry. Industry itself should be properly protected by the planning process from the wrong types of adjacent developments, e.g. housing creeping towards noxious industry.

The effectiveness of planning conditions imposed will depend on the quality of the analysis made initially by the planning authorities. Accurate prediction of the environmental outcome will depend on the availability of an accurate description of all the mining activities within a time framework. It requires that the planning profession is able to predict the probable environmental outcome of the proposed activity. The developer, of course, may have an interest in concealing the true nature of his processes, and their consequent effect on the environment. Thus the personal experience and skill of the planner are highly critical for a successful resolution of the conflicts involved in setting soundly based planning controls. Lack of previous experience creates a special difficulty in areas unused to the huge problems thrown up by the fuel industries; and this will be an important factor in deciding the social effectiveness of the planning controls to be imposed on the activities of the oil industry in Scotland. Generalizing, the planning controls that are imposed may be ineffective, unbalanced excessively onerous on industry, partially successful, and so on. It remains the planner's task, however, to draft them as fairly as he can, with due attention to economic considerations, as well as to social needs.

Industrialists sometimes overlook the fact that planning is a public process. There will always be a number of people who will be adversely affected by any major change in land usage. These people have the right to make their views known as part of the planning procedure. Important differences may arise between the planners' viewpoint and those of the public. Such differences of opinion can be categorized as follows:

(1) Disputes over precise criteria to be used for acceptance of a planning proposal; e.g. the public may feel that permissible levels of noise are set too high; or that tipping allowed is too close to houses, etc. This raises issues of how close is acceptable, i.e. the definition of planning standards.

(2) Disputes over the accuracy of the planners' prediction of the probable outcome. Alternatively, accusations of failure of the planner to predict outcomes obvious to others; or imputations of concealment of the consequences of activities by the developer; collusion between planner and developer, etc.

(3) Disputes over the potential effectiveness of the controls to be imposed to meet established planning standards/criteria accepted as reasonable by both sides.

(4) Disputes arising from misconceptions about the planning process, and the nature of the legal powers of planners.

(5) Disputes concerned with the value to be put on existing land uses, e.g. conservation of existing assets, such as outstanding landscapes; i.e., the proposed change might be acceptable in itself if it were not going to destroy something particularly valuable that already exists.

The developer may of course try to influence the public, attempting perhaps to persuade local inhabitants that the economic benefits are great and the environmental consequences insignificant. Conservationists, however, often see through this approach. The penalties for concealment are often expressed in excessive time delays.

Fig. 3 indicates an idealized environmental systems model for mining an underground mineral. The planner's function at the next stage of the development is to monitor continuously the actual environmental outcomes against the planning conditions laid down as part of the planning consent. A control feedback loop exists through the planning machine, but it is only effective if the feedback channel is actively used. Experience shows that many local authorities have devoted too little resource to enforcing their planning requirements. Action must be taken as soon as the actual outcome begins to deviate from the agreed outcome. It is to be hoped that the strengthening of local government by reorganization into larger units will ensure more effective use of such control loops. Indeed, such controls should be extended to environmental restoration of the land. This would ensure that effective terminal land uses are open on the completion of mining. Society will otherwise be left with substantial dereliction that will have to be remedied at public cost.

More attention should be paid to the potential usage of waste from mining. There are serious prospects of sand and gravel shortages in the UK; and waste material might find greater use in road-building, and similar. This would ease the economic problems of reclamation of tips. Some transfer of waste material between open-cast and deep-mining activities is now being achieved in Yorkshire, and deep-mining spoil can be used to restore the level of the land after open-cast mining. In some countries waste is returned underground, but the NCB claims this is excessively expensive.

The planning research areas that appear to deserve greater attention are:

(1) The development of improved models for predicting environmental impacts of fuel-winning activities to facilitate more effective planning control techniques;

(2) The development of better economic procedures (especially cost benefit techniques) for the assessment of environmental impacts arising from mining, applied at micro-economic, and macro-economic levels;

(3) The development of alternative, economic uses of waste products of mining to avoid excessive volume of tipping.

At the administrative level one might challenge:

(1) The capricious powers of the coal mining industry to cause serious subsidence affecting surface land uses above, without any legislative planning control by local government of the situation (contact being of a consultative nature, and associated with partial financial compensation for damage done). The surface oncosts to control damage from subsidence are not chargeable to the NCB;

(2) The exemption of large sections of NCB activity

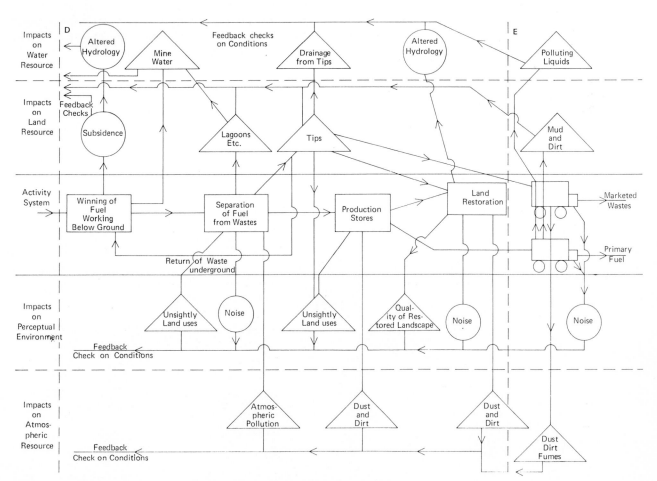

FIG. 3   The fuel-winning process with feedback checks against planning permission

from planning controls under General Development Orders issued over 20 years ago;

(3) The long delay between cessation of active tipping and land restoration, which may leave communities with serious dereliction for 30 years or more under present procedures.

At the educational level there needs to be developed:

(1) Planning competence to control complex industrial processes better;

(2) Competence of people in those fuel industries responsible for making major decisions involving land usage to handle economical issues better.

5. ENVIRONMENTAL ASPECTS OF PLANNING CONTROL

The fuel-processing industries are clearly of great significance in the energy chain. Enormous social benefits have accrued from smokeless fuel, which has made it possible for the UK to adopt a vigorous domestic Clean Air Policy. It must be recognized, however, that this success has not been achieved without some serious problems at local level. There are three planning problems associated with smokeless fuel plant:

(1) atmospheric pollution at a local level;

(2) obnoxious smells;

(3) pollutant discharges to river courses.

The smokeless fuel plants have been concentrated close to the primary points of fuel production (around Doncaster and Chesterfield, for example). The plants have become very extensive, and recent developments have had adverse effects on the local environment. Very strong objections were made at the planning enquiry for the Rossington plant about the likely pollution. The project, however, went forward because of the employment opportunities presented. The outcome has not been very satisfactory, as was predicted in much of the evidence given to the planning enquiry. Protest eventually became so strong that the case was taken up at Parliamentary level. Under the guidance of the Alkali and Clean Air Inspectorate some significant improvements have since been put in hand. Nonetheless, the local inhabitants are still very unhappy at the outcome.

Another example may be quoted—that of rivers which are grossly polluted from offensive discharges from certain firms manufacturing smokeless fuels.

The Yorkshire River Authority is now beginning to enforce very much higher standards under the vigorous leadership of their Chief Pollution Control Officer. In some senses there has been an exchange between adverse air pollution, and adverse water pollution. Areas that process fuels for nationwide use to help improve the nation's atmospheric environment have suffered some local environmental costs that have added to the environmental impacts of the mining industry itself. The rest of the nation, benefiting substantially, seldom recognizes the social costs that the mining areas are expected to bear in the national interest. Much higher priority clearly needs to be given to the overall environmental control of smokeless fuel plants.

The oil industry presents, in its refining activities, similar environmental problems of air pollution, offensive smells, water pollution, visual unsightliness, noise from flares, and so on. Again there is a need for improved standards and better planning controls. The 3rd Report of the Royal Commission on the Environment[2] has drawn attention to the serious problems of the pollution of estuaries, and the Government has indicated that it intends to end the present position concerning uncontrolled discharge of polluted effluents to estuaries. This

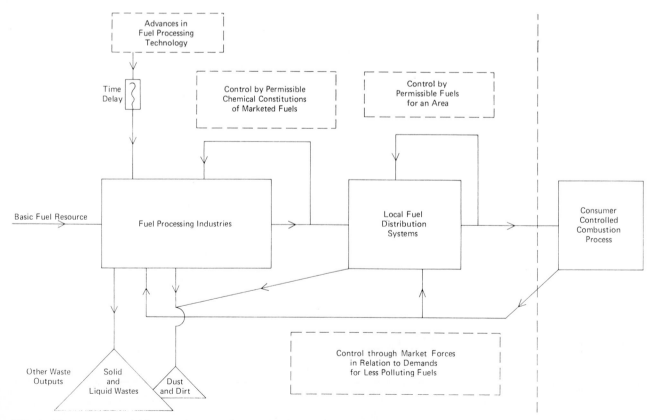

FIG. 4   The range of available control systems for control of atmospheric pollution from energy combustion systems

will have an important bearing on oil refining policy. Certainly, in Europe, stricter controls against river pollution has caused a switch in refining policy towards air-cooling rather than water-cooling. There has also been a recent tendency for refineries to move towards the centres of consumption. This is likely to raise important issues of plant noise, and planning vigilance must be directed towards this new aspect of refinery operation. Again the central issue is how much nuisance the local community should suffer for the benefit of the nation as a whole.

6. PLANNING POLICY AND DISTRIBUTION

Fuel has to be moved from point of production to markets. The environmental problems arising from the distribution of energy are:

(1)   transportation noise;
(2)   dust and dirt;
(3)   excessive vibration;
(4)   road congestion;
(5)   unsightly land usage; electricity pylons; overhead cables, etc.;
(6)   water pollution, especially movement of oil by water.

Obviously pipelines, once they are constructed, cause the least environmental disruption, and the domestic growth in the use of fuels of convenience (gas and electricity) has reduced the need to distribute fuels locally by road.

Rail transportation produces a relatively low environmental impact per unit weight carried, as the coupling of large numbers of wagons to a single traction unit reduces both energy requirements per unit load handled, and also noise frequency per hour. The abolition of the steam locomotive has vastly reduced atmospheric pollution from railways. A great deal of coal and oil is moved by rail, which confers important environmental advantages. The merry-go-round train system shifts huge amounts of coal from the mines to the electricity generating units, and with very little environmental disturbance.

However, increasingly large amounts of fuel are being moved by road with all the disadvantages of heavy road vehicles, excessive noise, very heavy axle loadings, and subsequent serious vibration. As an aid to productivity, the size of mineral-handling vehicles is increasing; and this aggravates further problems: carriage of finely divided particulate matter in open vehicles at high speeds on motorways leads to excessive aerodynamic removal of fuel dust from vehicles. Wet fuels may also be transported by road, causing offensive liquid discharges. Smokeless fuels are sometimes handled hot and still emitting significant quantities of sulphur dioxide. Many contractors' vehicles moving fuel are poorly maintained and emit substantial amounts of black smoke; are excessively noisy, and overloaded. Road transportation of heavy minerals always produces the most serious problems close to the primary points of distribution. There is a need to improve this current situation; but the intent should be to have long-range fuel carried by rail, or pipeline.

The electricity industry distributes its product mainly by overhead wires. From past policy of concentrating generation in the fuel-producing areas, like the Trent Valley and Yorkshire, a concentration of overhead grid lines from the power stations has accumulated that contributes to general unsightliness. Much of the landscape is very sensitive to visual intrusion, and while there is some underground transmission there is still concern at unsightly overhead grid lines.

Fuel distribution policy should not be seen simply as an economic matter, demanding merely the right infrastructure of roads, canals, railways, etc. The fuel industries, liaising with the Government, need to establish better control policies for the distribution of energy. The problems of distribution of oil from Scottish oil fields will need especial attention. Unfortunately the greatest impact in transportation will be most closely felt at the primary production points.

7. POLLUTION CONTROL IN URBAN AREAS

A final area that may be considered is the rôle of planning in atmospheric pollution control in urban areas. Many advances here have been, clearly, technologically based; but positive responses have often been the result of legal requirements enforced through local authorities, central government, or the Alkali and Clean Air Inspectorate.

There are four approaches fundamental to controlling urban atmospheric pollution:

(1)   control through permissible chemical contents of fuels that can be used in particular areas;
(2)   control through permissible techniques of combustion of these permitted fuels;
(3)   control through emission standards in conjunction with chimney height, and location policies;
(4)   control of environmental spaces around combustion sources.

The UK has tended to use these techniques outlined above at (1) to (3). Certain European countries, notably Germany, have used approaches based as at (4).

8. CONTROL BASED ON FUEL CONSTITUENTS

Fig. 4 indicates techniques for regulating atmospheric pollution by controlling permitted properties of fuels for combustion processes in a particular area.

Such controls arise as a result of national energy policies. Fuller[3] has recently reviewed the different policies of Western European countries. In Sweden for example the maximum sulphur content of fuel oils is limited to 2·5%. Sweden also has local controls for specific towns like Stockholm and Gothenburg where there is a limit of 1% sulphur. The City of London, too, has obtained 1% local limitation, and Manchester as a policy rather than a legal requirement) imposes a 1% value for fuels for municipal use. There is of course a danger that any country that does not establish national legal maximum permissible limits in sulphur may become the international 'dumping ground' for oil products of high sulphur content.

This raises the issue as to whether there should be a national policy for sulphur content of fuels. In view of the fact that oil fuels of low sulphur content are in limited international supply, an unrealistically low national maximum permitted level may not be the answer. A regional approach could be evolved, however, that established a high priority for low-sulphur fuels for the areas where the present sulphur dioxide contents of the atmosphere are highest. The properties of the local atmosphere and, in particular, its capacity to dispose of pollution effectively would be one critical factor needing assessment. Areas of low wind velocity and areas with frequent low level inversions are particularly sensitive to such pollution outputs.

The density of energy utilization per unit area is another factor. It is logical, for example, to suggest that South Yorkshire, with a high level of SO_2 emissions from the electricity generation industries and smokeless fuel industries deserves a policy demanding the use of low sulphur oil fuels in other industries to compensate. Natural gas has displaced a lot of high sulphur oil consumption in the Sheffield region, and has already brought a marked lowering in the relatively high atmospheric sulphur dioxide levels.[4] The process of SO_2 reduction could be taken further with more dynamic support from the oil industry, based on restrictions of sulphur content in adverse situations.

9. CONTROLS AND LOCAL PLANNING POLICIES

Finally, what can be done by improved local planning policies to reduce atmospheric pollution from fuels? Fig. 5 shows the control procedures open at a local level.[6] The controls include chimney height policy, activity location policy, real-time control during episodes of high atmospheric pollution. In addition, there are the control procedures that arise after improved combustion techniques. For listed industries, the technological control systems are normally under the control of the Alkali and Clean Air Inspectorate. For other activities, the control is vested in the local authority under the Clean Air Acts, and other public health legislation.

Clearly the Memorandum on Chimney Heights[7] should be a key document in considering the 'pay-offs' between alternative fuels and chimney heights, taking into account sulphur contents of fuels, the influence of neighbouring buildings, local topography, and existing background pollution levels. While applications for approval of new installations are examined both by the local authorities' technical officers (Public Health Inspector or Smoke Inspector) and by the planning committees, there is evidence that appearance often carries more weight among planners than effective pollution control. Aesthetic factors limit the acceptable chimney height; and so, indirectly, they influence the permissible fuels. This aesthetic consideration has been an important factor in giving preferential growth to natural gas and electrical installations in urban situations, especially in commercial and office buildings.

An alternative policy for meeting urban planning requirements, would be a much greater use of district heating. The effluent would be discharged through a single tall chimney of adequate height, as in Scandinavia.

A vital factor in the urban pollution situation is the quality of the spatial planning, which determines the relationships between pollution-producing activities and pollution-sensitive activities. Such a planning approach

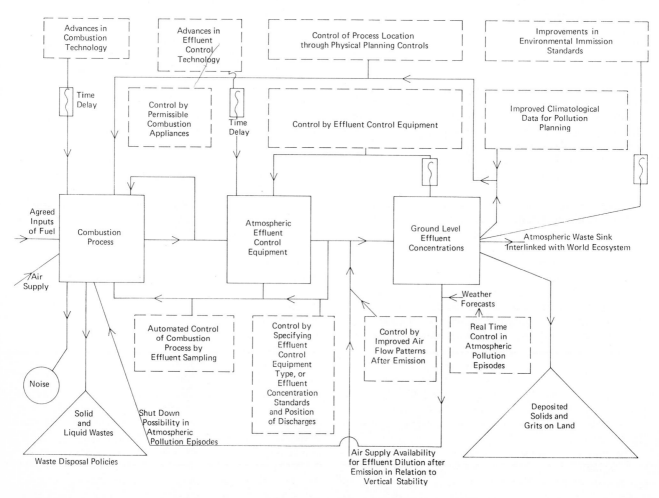

FIG. 5   Techniques for control of pollution by operational planning techniques, and legislative controls

demands an accurate analysis of the pollution outputs of different activities, as well as their pollution sensitivity. Only a proportion of industry is pollution-producing; and large sections of industry today, for example electronics, are pollution sensitive. The traditional zoning of towns into industrial, commercial and domestic zones is an over-simplification.

Furthermore, the present quality of the meteorological analyses performed for planning purposes is unsatisfactory and needs radical improvement if the pollution relationships between one area of a town and another are to be properly evaluated. This must involve much more than an analysis of prevailing winds. Information is needed about the vertical temperature, and wind structure above cities; the frequency of inversions, knowledge of the effects of terrain on air flow, including the air drainage patterns in still weather. It is a common experience in the UK, that the highest levels of pollution are encountered under anticyclonic conditions in winter, with subsiding air aloft when light easterly winds often prevail. Thus there is urgent need to introduce a more sophisticated meteorological analysis into the planning processes concerned with the prediction of pollution interactions between different urban zones. There is also an urgent need for a better understanding of building aerodynamics; and the environmental wind field around buildings. Current planning technology on atmospheric pollution control operates at a very primitive level. Fuel technologists and meteorologists have produced much basic information, but fundamental operational research, and meteorological studies, need to be put in hand to develop better operational procedures for planning for effective pollution control to complement recent successful developments in fuel technology.

10. CONCLUSION

This paper has attempted to approach the problems of the fuel industry from the point of view of controlling the adverse environmental effects through planning control procedures. This approach will demand a higher standard of environmental competence from planners, and a greater sympathy for the aims of environmental planning from industrialists. Industry needs to demonstrate its creative technological talents in environmental design just as much as in the basic process design. The philosophical attitudes of higher management will be most critical to the outcome, for the public now expects more of industry in environmental terms than has been hitherto achieved. The problem is one of planning the environment of the future. It is a task that demands vision.

11. REFERENCES

1. MEADOWS, D. K., MEADOWS, D. L., RANDERS, J. A., and BEHRENS, W. W. The limits to growth, Potomac Associates, Earth Island, London 1972.
2. Royal Commission on Environmental Pollution, 3rd Report. Pollution in some British estuaries and coastal waters. HMSO London (Cmnd. 5054), 1972.
3. FULLER, H. I. Review of European air pollution control measures. Symposium on Air Pollution, Dublin, March 1972. Ed. Dr. T. McManus. (1973) Publ. Technical Inf. Div., Institute for Industrial Research and Standards, Dublin, p. 7-38.
4. GARNETT, A., and READ, P. Natural gas as a factor in air pollution control. Institution of Gas Engineers, Communication No. 868, 109th Annual Meeting, 1972 (May). Inst. Gas E.
5. GILBERT, O. G. A biological scale for the estimation of sulphur dioxide pollution. New Phytol. (1970), 69, 629-634.
6. PAGE, J. K. Selecting an air pollution control policy. Symposium on Air Pollution, Dublin, March 1972. Ed. Dr. T. McManus. (1973) Publ. Technical Inf. Div., Institute for Industrial Research and Standards, Dublin, p. 39-65.
7. Department of the Environment (former Ministry of Housing and Local Government), Clean Air Act 1956, Memorandum on Chimney Heights, (2nd ed.), HMSO, 1967.

Professor J M BEÉR, DiplIng, PhD, DScTech, FInstF,* and A B HEDLEY, PhD, FRIC, MInstF, CEng*

15 Air pollution research: reduction of combustion-generated pollution

SUMMARY

The objective of this paper is to show how research can contribute to an air pollution control programme. Of the broad spectrum of pollution research the area of combustion-generated pollution is chosen because of combustion's significant contribution to the emission of pollutants. The nature of the major pollutants is discussed together with control methods available in areas of stationary combusiton sources, gas turbine combustors and internal combustion engines. It is shown that considerable reduction can be obtained by process modification, and it is argued that process modification is generally preferable to other methods, such as the use of 'add-on' devices. It is recommended that for the planning of research and development in this field a detailed long-term plan for research on air pollution control by combustion modification be commissioned.

1. THE OBJECTIVES OF AIR POLLUTION RESEARCH

The general purpose of any large-scale pollution research programme must be that of providing scientific information which, in turn, will facilitate development and application of improved control methods. The application of these methods must be economical and be with regard to particular (local) circumstances, such as population density, geographical and meteorological features of the region.

More specifically research objectives can be stated as:

(a) to develop diagnostic tools, i.e. methods and instruments for sampling and analysis of pollutants so that the distribution of their concentrations spatially and also in time can be determined;

(b) to obtain information on the physical-chemical nature of pollutants and on the processes of their formation and emission into the atmosphere.

(c) to obtain information on the effects of pollutants on both the living and the physical environment. This includes the physiological effects on man and animals, visibility reduction, agricultural damage and material damage;

(d) to develop new and improved methods for the control of emission of pollutants;

(e) to determine relationships between atmosphere (or ground level) concentration and that at the source of emission.

An air pollution control programme works effectively through the setting of legally enforceable emission standards. Fig. 1 illustrates the role that research plays in the

*Department of Fuel Technology and Chemical Engineering, University of Sheffield.

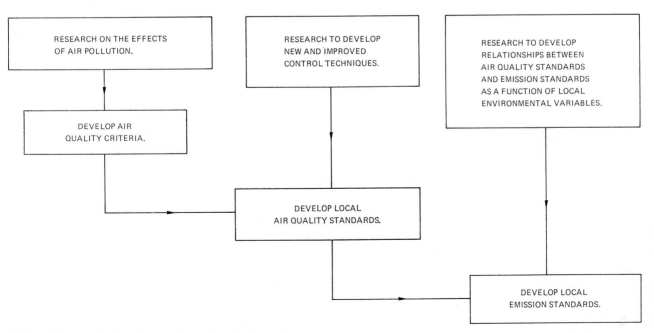

FIG. 1   The contribution of research to air pollution control programmes.

development of such a programme. Information obtained by research on the effects of air pollution is the basis of the first administrative step: that of developing and publishing *air quality criteria*. These give maximum ground level concentrations of pollutants for various periods of time (hourly, daily, yearly average) for different regions. Air quality criteria are not enforceable by law.

The next stage in the programme is the development and publication of *air quality standards* that will become legal standards. For these standards to be realistic the state of the art of control technology has to be taken into account together with the cost of implementing control. As these costs would have to be found from resources in competition with other expenditure socially well supported it is important that research pays sufficient attention to the economic features of various control methods, apart from looking at the technical feasibility of their use. On the basis of air quality criteria, and the information on control methods, air quality standards are developed and set. Air quality standards give maximum permissible local concentrations of pollutants for prevailing various periods of time.

As a next step *emission standards* compatible with the local air quality standards are developed. While ground level concentrations of pollutants, as specified by air quality standards, are important to protect the environment from harmful effects of pollution this cannot be done effectively without specifying the permissible concentrations at the emission sources.

The relationship between ground level concentration and emission source concentration of a pollutant is highly complex, and depends upon geographical and meteorological factors. It is the task of research to provide this information and thereby enable the setting of sensible emission standards for a region. In some countries the authorities have emergency powers to close down a sector of industry in a region if the ground level pollutant concentration exceeds that specified by air quality standards. If emission standards are compatible with air quality standards such action will only be required on the exceptional occasion of some rare meteorological condition, such as a prolonged inversion in a metropolitan area.

2. MAJOR POLLUTANTS

All industrial processes contribute to the pollution of our environment. By far the largest contributors to atmospheric pollution are, however, processes of combustion in power generation and industrial plants, in automobiles and aircraft engines and in domestic heating.

The pollutants receiving most attention are:

(*a*) Non combustible particulates
(*b*) Unburned gaseous and particulate matter
(*c*) Oxides of sulphur
(*d*) Oxides of nitrogen
(*e*) Combustion noise
(*f*) Secondary pollutants.

Fig. 2 represents the proportions of each of the major pollutants emitted by combustion and non combustion sources in the USA for the year 1966.[1] Table 1 shows the respective contribution of various energy conversion

TABLE 1   Pollutant emissions from combustion for energy conversion (after Battelle Memorial Institute[2])

Pollutant	Emissions, weight percent (1966)				
	Central station power plants	Industrial processing	Space heating and industrial steam generation	Gas turbines	Reciprocating-IC engines
Products of incomplete combustion					
Combustible particulate	18	6	45	1	30
CO	<1	<1	1	1	98
Gaseous hydrocarbons	<1	<1	1	1	98
PNA	<1	4	90	nil	6
Nitrogen oxides	21	3	17	1	58
Sulphur oxides	63	5	31	nil	1
Non-combustible particulate (ash)	72	3	25	<1	<1

TABLE 2   Forecast of US pollutant emissions from combustion of prime fuels for energy conversion 1975, 10^6 tons/year (after Battelle Memorial Institute[2])

Pollutants	Continuous combustion for energy conversion													Cyclic combustion for energy conversion				
	Central station power plants			Industrial processing	Industrial steam generation			Commercial and residential space heating			Gas turbines		Mobile IC engines		Stationary IC engines		
	Coal	Oil	Gas		Coal	Oil	Gas	Coal	Oil	Gas	Stationary	Aircraft	Gasoline	Diesel	Gasoline	Diesel	Gas
1. Products of incomplete combustion																	
Combustible Particulate	0.60	0.02	0.02	0.29	0.53	0.05	0.09	0.07	0.12	0.08	0.03	0.03	0.45	0.35	0.02	0.15	n
CO	0.11	n	n	0.02	0.11	0.01	<.01	0.33	0.03	<.01	0.06	0.39	41.6	0.20	1.7	0.10	0.55
HC	0.04	0.01	n	—	0.06	0.01	0.02	0.07	0.04	n	0.02	0.17	8.3	0.50	0.14	0.14	0.27
PNA(a)	n	n	n	0.30	n	n	n	4.9	0.10	0.10	n	n	0.58	n	n	n	n
2. NO_x	4.8	0.55	0.85	0.46	0.88	0.39	0.75	0.05	0.68	0.46	0.37	0.25	9.7	0.80	0.13	0.27	2.3
3. Combustion-improving additives																	
Lead	n	n	n	n	n	n	n	n	n	n	n	n	0.27	n	<0.1	n	n
4. Fuel contaminats																	
SO_x	28.5	2.05	n	1.39	3.49	1.04	<.01	0.65	1.15	n	0.15	0.09	0.33	0.13	n	n	n
Ash	5.4	0.01	n	0.31	0.62	0.03	n	0.09	0.08	n	n	n	n	n	n	n	n

(a) PNA shown in 10^3 tons/year
—, not available
n. emission considered negligible.

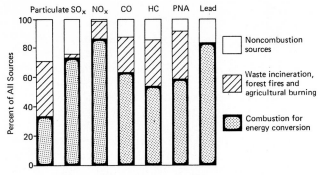

FIG. 2 Percentage contribution of combustion-source emissions to the total nationwide inventory of pollutant emissions in the USA 1966 (after Barratt, R. E.[1])

TABLE 3   Estimates of air pollution by smoke and sulphur dioxide from the main uses of fuels in great Britain in millions of tons (after Broadbent, D. H.[3])

	1938	1956	1967
Smoke			
Coal			
Domestic	1·71	1·26	0·75
Railways	0·26	0·24	0·02
Industrial, etc.	0·74	0·75	0·10
Total smoke	2·71	2·25	0·87
Sulphur Dioxide			
Coal			
Domestic	1·21	0·90	0·59
Electricity works	0·40	1·23	1·88
Railways	0·36	0·33	0·02
Industrial, etc.	1·65	1·68	0·97
Coke ovens	0·07	0·11	0·08
Gas supply	0·14	0·20	0·10
Total	3·83	4·45	3·64
Oil			
Domestic	—	—	0·01
Industrial	0·05	0·47	1·96
Road & rail transport	0·01	0·03	0·08
Marine craft	—	0·03	0·04
Total	0·06	0·53	2·09
Coke Total	0·24	0·35	0·28
Total sulphur dioxide	4·13	5·35	6·01

groups to the emission of pollutants for the same year.[2] While there will be differences in detail between these data and the corresponding figures for the UK, the trends shown in the table will be similar: stationary combustion sources are responsible for the major proportions of combustible particulates, sulphur oxides, nitrogen oxides and non-combustible particulate emissions. Reciprocating internal combustion engines generate a large proportion of the unburned carbon monoxide and hydrocarbons, together with a considerable amount of nitrogen oxides. The fact that IC engines emission occur at a very low level, and more usually in densely populated centres, accentuates the urban pollution problem. A more detailed breakdown for the distribution of pollutants in the USA predicted for 1975 in terms of million tons of pollutant emitted is shown in Table 2.[2]

Table 3 represents smoke and sulphur dioxide emissions in the UK for the years 1938, 1956 and 1967.[3]

It clearly illustrates results of the successful implementation of the Clean Air Act for smoke, but sulphur oxide emission trends are not so encouraging.

Nitrogen oxides emission data in the UK for the period

TABLE 4   Overall emissions of the oxides of nitrogen in the UK (after Derwent, R. G. and Stewart, H. N. M.[48])

Source	Notes	1965 Emissions (10³t)	1966 Emissions (10³t)	1967 Emissions (10³t)	1968 Emissions (10³t)	1969 Emissions (10³t)	1970 Emissions (10³t)
Transport							
Petrol engines	a	155	164	176	186	191	203
Diesel engines	b	48	51	54	57	60	62
Railways	b	2	2	3	4	4	5
		205	217	233	247	255	270
Domestic sources							
Coal fires	c	60	56	51	48	44	39
Smokeless-fuel fires	d	11	11	11	11	11	10
Gas appliances	e	8	9	10	12	13	15
Oil-fired central heating	f	3	3	3	3	3	4
		82	79	75	74	71	68
Commercial, Public service, industrial Fuel-and gas-oil central heating	g	28	35	40	44	48	53
Coal and coke	h	282	264	240	228	180	156
Gas	l	9	10	9	11	12	14
Industrial, fuel and gas oil	i	231	244	259	273	285	299
Incinerators	j	1	1	1	1	2	2
		551	554	549	557	527	524
Total low-level emissions		838	850	857	878	853	862
Power stations							
Oil-fired	k	46	52	48	46	56	86
Coal-fired	k	464	452	474	508	514	514
		510	504	522	554	570	600
Overall total emissions		1 348	1 354	1 379	1 432	1 423	1 462

1965 to 1970 are given in Table 4.[4] Both low level and total emissions of NO_x have steadily increased during the period 1965 to 1970. Low level emissions contribute well in excess of half of the total emissions and the largest single source of the total NO_x emission is the coal-fired power generation process.

There is no information on PNA (Polynuclear aromatics) emission in the UK but, judging from the data for the USA (Table 2), the most significant source of PNA is the coal-fired domestic stove. With the increase in recent years of gas and oil-fired domestic appliances at the expense of coal fires it can be assumed that PNA emission is gradually being reduced.

2.1. Non-combustible particulates

The non-volatile oxidation products of the inorganic constituents of the fuel are left as ash at the end of the combustion process. With grate-firing and slagging-combustion of solid fuel a large proportion of the ash stays in the boiler; nevertheless, small fly ash particles are carried by the moving flue gases and will eventually be transported out into the atmosphere unless their deposition is brought about by some means. With atomized oil and pulverized coal combustion a large proportion of the fuel ash is thus transported by the moving gases. Therefore the use of very large and reliable equipment, usually electrostatic precipitators, is essential to clean the gases. Modern precipitators can be highly efficient, but nevertheless a small proportion of the finer fly ash particles still end up in the atmosphere. Inefficient handling of once-precipitated material is frequently a source of pollution and, in general, there is a need to give more attention to the handling and disposal of its particulate material. In the atmosphere, the smaller the particles are, the longer they remain suspended, and hence they can act as condensation nuclei, catalytic surfaces for the oxidation of SO_2 to SO_3 or other chemical reactions, or simply reduce visibility. If present in high concentration at, or near the ground, they can impair health and reduce amenity values.

Although the particulate material leaving precipitators can be very fine, conditions can arise where fine particles agglomerate to form large particles that rapidly fall out of the atmosphere in the immediate vicinity of the plant.[5] Not enough is known about the chemical and physical properties of particle systems that result in such agglomeration.

2.2. Unburned gaseous and particulate material

The emission of combustibles is the result of incomplete oxidation of carbon and hydrogen in the fuel. Possible reasons for incomplete burning of the fuel are poor mixing of fuel air and combustion products in the combustion chamber; inadequate residence times at high temperature; and insufficient excess air. Most industrial combustion processes are mixing-controlled, i.e. the rate of the chemical reaction in the flames is high, and the overall rate of the process depends, therefore, mainly upon the progress of mixing of fuel and air in the combustion chamber. When mixing is inadequate, large amounts of excess air are required for completing the combustion which, in turn, leads to large heat losses by increasing the volume of the waste gases. Apart from these high exhaust gas losses, there are other reasons also (e.g. reduction of SO_3 and NO_x formation) for limiting the excess air levels in combustion processes. Hence the technical problem is to achieve complete combustion with a minimum of excess air. The designer has to achieve this by good aerodynamic design at a reasonable expense in terms of pressure energy.

Adequate residence time at sufficiently high temperature is another condition to be satisfied additional to good mixing in the combustor. This requires that the combustor should not be overloaded, and that the heat extraction during the combustion process be controlled, so that exhaust gases are not quenched before the combustion reactions can go to completion. This applies in all combustion devices such as boilers, gas turbines, or IC engines

The above-mentioned variables—residence time, excess air and temperature, however, are not independent, but are inter-related. Fig. 3 illustrates how gas temperatures and residence time change as a function of excess air for an oil fired combustion chamber.[1]

The physical chemistry of the formation and subsequent oxidation of soot particles is still not fully understood. There are, however, qualitative trends which, without the full knowledge of the detailed mechanisms, can be made use of in engineering design. The tendency to form soot increases with the carbon to hydrogen ratio in the fuel, with the rate of heating of the hydrocarbons during pyrolysis, and is strongly favoured by fuel-rich conditions in the flame. The overall process of formation of soot is fast, and its rate is believed to be controlled by the nucleation step. Once nuclei have been formed, the deposition of carbon on

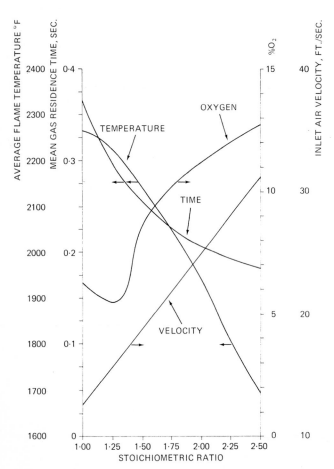

FIG. 3 Critical parameters affecting pollutant formation (after Barratt, R. E.[1])

these nuclei is very fast. The rate of oxidation of soot, in contrast to its formation, is a slow process with chemisorption as the rate-determining step, rather than boundary layer-type diffusion through a stagnant layer surrounding the particles. These are of the order of 200 to 1 000Å in diameter. Experimental data show good correlation with semi-empirical rate equations, assuming first order reaction with respect to the partial pressure of oxygen in the gas streams surrounding the soot particles. The rate equation is of the Arrhenius form, with exponential temperature dependence for the rate of oxidation of soot.[6] This points to the high sensitivity of the rate of burning of soot particles upon the gas temperature. Thus, when soot has been allowed to form, long residence time at high temperatures is required for its complete combustion.

The semi-empirical relationships between the rate of oxidation of soot, oxygen concentration and temperature can be used with good confidence in the range of the variables in which they were tested; and, as this range for temperatures and for oxygen concentrations well overlaps the practical cases in combustors, the relationships are directly applicable to design. A comprehensive survey of research carried out on the rate of burning of soot under conditions of elevated pressure was recently published.[7]

2.3. Oxides of sulphur

The sulphur oxides SO_2 and SO_3 are formed from the sulphur containing compounds in the fuel which react with the oxygen of the combustion air. These oxides have occupied the attention of combustion research workers for over 40 years; not initially because of the air pollution problem, but because of the severe corrosion that can result in combustion equipment due to the conversion of a relatively small amount of SO_2 to SO_3 which can condense as a liquid on cooled surfaces in the path of combustion gases. In boilers, only up to about 5% of the SO_2 is oxidized to SO_3. In other combustion appliances using very low sulphur content fuels, such as domestic gas water heaters and IC engines, over 40% conversion is possible. This is because SO_3 formation in flames is more dependent on oxygen atom content than SO_2 content.[8] It has been shown that SO_3 concentration in flames can greatly exceed thermodynamic equilibrium concentration based on the molecular reaction between SO_2 and O_2.[9]

It has also been shown that high flame concentrations of SO_3 may be frozen at that high level by rapid cooling of the combustion gases.[10]

Fig. 4 illustrates this point. It shows the relationships between typically-measured SO_3 concentrations and thermodynamically-stable SO_3 concentrations as a function of the temperature drop experienced by combustion gases as they pass through a boiler. It can also be seen that due to the short residence time the amount of SO_3 coming out of a boiler is very much less than it could be. These thermodynamic relationships, however, still apply outside the boiler chimney, and therefore the potential for the complete conversion of combustion generated SO_2 into SO_3, and hence to H_2SO_4 and other sulphates, is very great. The rate at which this occurs, however, is influenced by many factors. Recent research has shown that this process may be a great deal faster than previously assumed.

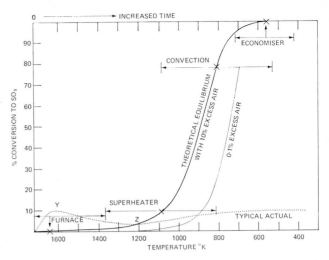

FIG. 4 *The variation of the theoretical equilibrium yield and possible actual yield of SO_3 with time in a boiler (after Hedley, A. B.[10])*

2.4. Oxides of nitrogen

Man-made NO_x represents only about 10% of the global, but because it is usually concentrated in heavily populated urban areas, this can result in average levels of up to 0.1 parts per million, as compared with natural background levels of less than one-hundredth of this figure. While the concentration of NO_x in polluted atmospheres is usually well below accepted levels considered to be a health hazard, the reasons for not allowing their emission into the atmosphere unchecked are mainly that they form secondary pollutants by further atmospheric reactions.

Nitrogen oxides are emitted in flue gases because of the oxidation of nitrogen to nitric oxide at the high temperatures generated by the combustion of fossil fuels with air. The major source of nitrogen is the air itself and it is the fixation of this nitrogen which is the main source of NO formed in turbulent diffusion flames, characteristic of industrial furnaces and combustion chambers.

Although it is thermodynamically unstable at low temperatures, the NO formed at high furnace temperatures remains virtually 'frozen' as the combustion gases are cooled in the exhaust system. The rate of oxidation of NO to NO_2 which is favoured by equilibrium at lower temperatures, is too slow to give significant conversion during the short time of residence in the combustion equipment. This reaction, however, has ample time to take place once the gases are released into the atmosphere. The rate of the chemical reaction between nitrogen and oxygen in the flame is highly temperature-sensitive because of the high energy of activation. The classical chain reaction mechanism postulated by Zeldovich[11] for NO formation between atomic oxygen and molecular nitrogen and atomic nitrogen and molecular oxygen respectively assumes atomic oxygen to be in thermodynamic equilibrium with molecular oxygen. This assumption has been shown to be invalid for flame regions where atomic oxygen levels are known to be higher than those predicted from thermodynamics. Therefore the initial rate of formation of NO is higher than that predicted by the equilibrium assumption.[12] These mechanisms involve only the elements nitrogen, oxygen, and hydrogen. However the existence of carbon-nitrogen bonded intermediate compounds has also been suggested as being responsible for the high initial rates of nitric oxide formation, and such considerations further complicate the problem of chemical kinetics of NO formation in fossil fuel-air flames. A special difficulty in predicting NO emission from turbulent flames is that because of the strongly non-linear dependence of NO formation upon temperature it is necessary to describe the flame pattern for such calculations not only in terms of time average but also in terms of fluctuating temperatures, velocities and concentrations.

A second significant source of NO_x emissions from fossil fuel combustion processes is the oxidation of chemically-combined nitrogen in coal and oil fuels. The relative importance of the reaction between oxygen and fuel nitrogen, will vary with the nature of the combustion process, and with the nitrogen content of the fuel.

2.5. Combustion noise

Most industrial combustion processes for the rapid transfer of heat by flames are inherently noisy due to the use of turbulent flames. Noise emission can become acute, especially with high intensity oxy-fuel burners used mainly by the steel and metal industries.

There is a typical noise spectrum generated by the combustion process that can be altered by conditions upstream (air-handling system) or downstream (furnace or room) of the burner. These interactions among the combustion system elements are not fully understood, but some general observations can be made.[13] It is assumed[13] that in industrial combustion processes which generate most of the noise in the 250 to 500 Hz frequency range the position of the peak frequency can be related to the parameter δ/S_L where δ is the flame thickness and S_L is the laminar burning velocity of the flame. Thus the expression $f_p \left(\dfrac{\delta}{S_L} \right) = a$ is constant,

where f_p is the peak frequency and the term δ/S_L is a function of air-fuel ratio. Because of the highly complex nature of turbulent diffusion flames, the use of flame thickness and of laminar flame speed for correlating combustion noise measurements must, however, be only a first approximation.

As in many areas of technology, in the field of combustion noise designers had to rely on semi-empirical data due to lack of more fundamental information. With improved facilities for measurements, a large amount of useful experimental data became available,[14] which were analysed to give guide lines for the design of burners and combustors. In another recent experimental study it was shown that by simply changing the mode of fuel introduction to the burner, the noise emitted could be reduced considerably.[15] This points to the importance, and necessity of further research in this area.

2.6. Secondary pollutants

Primary pollutants can undergo further chemical or physical reactions in the atmosphere to form secondary pollutants whose properties and behaviour may be entirely different from the former. The simplest example already mentioned is the further oxidation of nitric oxide to nitrogen dioxide. Another simple example is the formation of sulphuric acid aerosols due to condensation of the acid formed from the reaction of sulphur trioxide and water vapour. Sulphur dioxide can be absorbed by fog droplets and subsequently oxidized to sulphuric acid or sulphates the process being catalysed by dissolved salts in the droplets. Alkaline gases can be present in the atmosphere; and these can react with acidic oxide gases to form solid aerosols of substances (such as ammonium sulphate) often without the presence of liquid water, (in the form of droplets) being required at all.

The formation of photochemical smog is perhaps the most publicized example of secondary pollutant-formation. The processes are complex and involve many reactions whose mechanisms are, as yet, still not completely understood; nevertheless a simplified picture can explain the basic processes involved (Fig. 5).[16] The starting substances are nitrogen oxides and hydrocarbons. The critical first step leading to the formation of a group of oxidizing substances is the photodissociation of nitrogen dioxide; a reaction which is enhanced by the presence of sunlight. This produces atomic oxygen which combines with molecular oxygen to give ozone. The oxidants attack hydrocarbons producing reactive free radicals of several kinds that are also capable of attacking hydrocarbons and participating in other reactions as indicated. The significance of aldehydes in the participating reactions is also seen.

Pollution, especially in the form of aerosols, can increase or modify the activity of condensation nuclei and can therefore presumably influence the microphysics of rain cloud, or fog formation. This subject of weather modification resulting from air pollution has been one of some concern and controversy for several years and aspects of the problem have been reviewed recently.[17] There can be little doubt that combustion exhaust gases contribute significantly to the number of condensation nuclei present in the atmosphere above our cities. This contribution being greatest from smoke is shown by the nuclei numbers in Table 5.[18]

High concentrations of condensation nuclei should

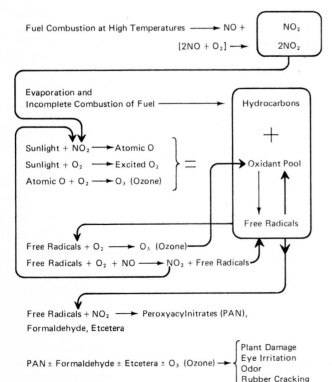

FIG. 5 *A simplified reaction scheme for photochemical SMOG Formation (after Tilson, S.[16])*

TABLE 5 Concentration of Aitken nuclei in different locations (after Stern[18])

Location	No. of places	No. of observations	Nuclei number/cm³ Average		
			Average	Max.	Min.
Large city	28	2,500	147 000	379 000	49 100
Town	15	4 700	34 300	114 000	5 900
Country (inland)	25	3 500	9 500	66 500	1 050
Country (sea shore)	21	7 700	9 500	33 400	1 560
Mountain 500–1 000 m	13	870	6 000	36 000	1 390
1 000–2 000 m	16	1 000	2 130	9 830	450
2 000 m	25	190	950	5 300	160
Ocean	21	600	940	4 680	840

produce a narrowing of the condensed droplet size distribution, thus inhibiting the growth of the larger droplets which would rapidly fall out of the atmosphere. The end result would therefore be a fine stable fog that could also contain other pollutants in its droplets such as sulphuric acid, ammonium sulphate, and so on, which would contribute to the stability. Because such a fog reduces the transmission of sunlight to ground level, we have conditions favouring temperature inversion, especially in valleys and geographical basins, and therefore a reduced opportunity for the break-up of the familiar smog condition by normal atmospheric circulation processes.

It can thus be seen that any decisions concerning air quality criteria, which are based on effects of primary pollutants only, will be strongly limited as they will not take account of the important secondary effects of these pollutants which arise from their interaction with other pollutants or naturally occurring aerosols present in the atmosphere.

3. CONTROL METHODS AT SOURCE
3.1. Emission from power stations and industrial processing plants
The principal pollutants emitted by fossil fuel-fired boilers in large power station are products of incomplete combustion, i.e., combustible particulate matter; carbon monoxide; nitrogen oxides; sulphur oxides and fly ash, and particulates formed due to the use of fuel additives.

With the doubling of electrical energy requirement approximately every decade, and because fossil fuel power generation may lose to nuclear energy its presently overwhelming relative contribution only gradually, it is a significant source of emissions worth special considerations.

There are two distinguishing features of power station boilers which have consequences for pollutant emission:

(a) Unit capacities have steadily increased to reach presently the range 660 to 1300 MW, mainly because of the need for reducing installation costs of boiler plants,
(b) Thermal efficiencies of boiler plants must be high because of the significant contribution of fuel costs to the total cost of electricity.

The consequence of increasing size is that the surface to volume ratio of combustors is steadily reduced with the result that high temperature zones arise in the central regions of the combustion chambers. The practice of low excess air combustion can be the cause of additional non-uniformity and demands high standards of air-fuel ratio control, and especially of aerodynamic design of the combustor to satisfy requirements for low emission of pollutants.

Combustible emission when burning pulverized coal depends upon:

(a) the characteristics of coal such as volatile matter, fineness of pulverized coal;
(b) the characteristics of the combustion process such as excess air, burner and combustion chamber aerodynamics and also the volumetric heat release rate and the rate of heat absorption in the boiler.

Table 6 represents examples of unburned carbon losses as a function of some of the above variables.[2]

When burning oil the same general pattern applies except that modern practice demands very low ($< 2\%$) excess air operation which makes it more difficult to maintain combustible emissions at a low value (max. 0·07 gr/SCF corresponding to about 0·3 lb/10^6Btu). For a given excess air level the solids emission depends strongly upon the quality of atomization (mean drop size and proportion of mass of drops with diameter in excess of 200 μm). Promising results of suppression of smoke emission by the use of water-oil emulsions were reported recently.[19] This technique, however, requires relatively large amounts of water (20% water; 80% oil) and it is questionable whether its use in large power station boilers would be economical.

Good aerodynamic design of burners is the other important condition to satisfy low solids emission. It is generally accepted that at least 8 in WG pressure drop across the burner register at full load is necessary to achieve efficient mixing in the combustion chamber. It is advisable also to maintain a sufficiently high pressure drop at part load. This can be achieved by variable burner geometry. The application of this principle to burner design requires further research and development.

The level of CO emissions from large boiler furnaces is low. It is normally kept below 0·01% at all loads in the flue gas. With low excess air operation CO concentrations of the exit gas, monitored by an infra red analyser, can be used for overall air-fuel ratio control. Efficient control, however, requires correct air-fuel ratios to be maintained in individual burners. Research

Table 6 Unburned carbon loss from pulverized-coal-fired boiler furnaces (after Battelle Memorial Institute[2])

Heat release rate, 1 000 Btu/ft³-h	Volatile (a) matter in coal, %	Coal smaller than 200 mesh %	Excess air, %	Unburned carbon lb/10^6 Btu input (b) Dry-bottom furnaces	Slag-tap furnaces
20	20	65	10	3	3
20	48	65	10	0·3	0·2
20	48	80	10	0·06	0·03
20	48	80	40	0·03	0·02
40	20	65	10	5	3
40	48	65	10	0·6	0·5
40	48	80	10	0·2	0·2
40	48	80	40	0·06	0·05

(a) Dry, ash-free basis.
(b) Based on 13 100 Btu/lb for coal.

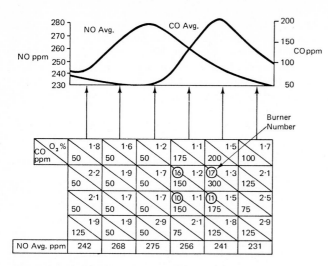

FIG. 6 Lack of O at one burner influences O_2 at all other burners and ultimately determines the final NO concentration (after Breen, B. P.[21])

studies indicate a good prospect of this being achieved by a feed-back control system based on an optical signal due to infra red flame emission.[20]

The importance of burner uniformity is illustrated by Fig. 6 which represents measured CO, NO and O_2 distribution in combustion products produced by a matrix of burners. A group of burners in the right hand corner of the matric (Nos. 10, 11, 16 and 17) produced a peak CO due to poor air distribution. To avoid unacceptable CO emission the total air supply was increased with the result of producing a peak emission of NO from the burners on the left hand side. It is clear that low excess air combustion can be operated efficiently only when accurate and reliable air-fuel ratio is available for individual burners.

There are no emission standards of NO_x published

for combustion sources in the UK. Proposed legislation in the USA for emission from new installations is shown in Table 7.[22,23]

Typical levels of emission from conventional combustion plant shown in Table 8 indicate that considerable research and development is required for the proposed standards to be achieved in boiler plants.[24]

High NO emission from boilers is partly due to the relatively long residence time at high flame temperatures which the combustion products spend in the large combustion chambers (thermal NO formation), and partly due to the preferential formation of NO from the nitrogen chemically bound in the fuel under typical lean mixture conditions in the furnace.

Systematic experimental studies carried out with a large number of boiler plants,[25] and those at the International Flame Research Foundation at IJmuiden,[20] show that the level of NO emission from flames can be reduced by combustion modifications.

Table 9 shows estimated NO_x reduction and control costs for a 1 000 MW plant using oil and gas fired system.[27]

It is worth noting that the combination of control techniques is particularly promising (the negative costs in Table 9 refer to savings). Two-stage combustion allows a fuel-rich primary zone to form that is known to be beneficial for reducing not only thermal but also fuel NO formation. While combustion modification as a method shows distinct promise for NO_x emission control in power plant, further research is necessary to investigate possible side effects on boiler availability (slagging of heating surfaces), and on combustion stability over wide load ranges.

In industrial processing plant NO_x formation is generally more difficult to control because of the lower rate of heat absorption in the combustion chamber. In regenerative furnaces, such as open hearths or glass melting tanks, it is not practical to consider combustion modifications in the furnace. An alternative approach would be to use a fuel-rich after-burner to destroy the NO_x in the flame gas; and, after the recovery of the heat produced by combustion, to introduce a second afterburner operating at a lower temperature and hence remove the combustibles left from the fuel-rich first stage.[28] A diagram of such a scheme is shown in Fig. 7. This example illustrates the possibilities of process modifications for cases where the particular technology

TABLE 7   USA federal legislation limits for NO_x emission (after Gills, B. G.[23])

	NO_x maximum emission (2 h average) expressed as NO_2		
	lb/10⁶ Btu	g/1 000 k cal	ppm @ stoicheiometric fuel:air ratio
Gaseous fuel	0·20	0·36	245
Liquid fuel	0·30	0·54	370
Solid fuel	0·70	1·26	860

TABLE 8 NO_x emissions from a high-intensity combustor and from various types of conventional combustion plant (after Shaw, J. T.[24])

Approx. temperature of flue gas at hottest °C	Fuel	N chemically combined in fuel, w/w (dmmf basis)	Type of plant	Composition of combustion air	NO_x reported in flue gass ppm v/v, dry basis
2 300	Coal	1·4%	High intensity combustor	Oxygen-enriched air	10 000 to 13 000
1 500 to 1 700	Coal	Probably 1-2%	A range of coal-fired power station boilers	Air	200 to 1 400
1 500 to 1 700	Oil	Not disclosed	A range of oil-fired power station boilers	Air	110 to 800
1 500 to 1 700	Natural gas	Negligible	A range of natural gas-fired power station boilers	Air	50 to 1 500
1 500 to 1 700	Cracked, residual fuel oil	1·0%	An oil-fired power station boiler	Air	425
	Paraffinic fuel oil	0·2%		Air	215

TABLE 9 Estimated NO$_x$ reduction and control costs for a 1 000 MW boiler by selected combustion modifications (after Salooja, K. C.[27])

Control method	NO$_x$ reduction % Fuel used— Gas	Oil	Control cost Annual $1 000 Gas	Oil	$ per ton NO$_x$ reduced Gas	Oil
Low excess air	33	33	—95	—297	—5	—30
Two-stage combustion	50	40	0	0	0	0
Low excess air plus two-stage combustion	90	73	—95	—297	—2	—14
Flue gas recirculation	33	33	202	202	12	20
Low excess air plus flue gas recirculation	80	70	107	—95	3	—5
Water injection	10	10	144	179	27	60

(steel, glass, cement) does not permit combustion modifications in the furnace itself. In many countries air quality standards regarding SO$_2$ concentration are legally enforced. Typical figures are the American standards of 0·1 ppm SO$_2$ on a 24 hour average and/or 0·02 ppm SO$_2$ as an annual average of ground level concentrations. Normally the sulphur in the fuel will appear as oxides of sulphur in the combustion products. With coals commonly burned today the SO$_2$ concentration in the flue gases range from 800 to 4 000 ppm. Methods of reduction of sulphur oxide emission include desulphurization of fuel, desulphurization of waste gas, use of sulphur-absorbing additives, and new combustion systems with sulphur retention. Fuel and waste gas desulphurization

are technically, and in certain cases economically feasible, but are not discussed here because they are outside the scope of this survey. The use of additives may involve the introduction of finely dispersed alkaline-earth compounds in the form of dolomite, or limestone, which react with SO$_2$ and can then be precipitated from the waste gas by usual gas-solid separation methods. Results of field and laboratory tests indicate total removal of SO$_3$ and up to 98·6% removal of SO$_2$. The estimated cost of removal of sulphur oxides and fly ash from a new 500 MW unit burning coal with 10% ash and 3% sulphur is about 0·7 p/10^6 Btu of coal.[29]

The most important new combustion system capable of removing sulphur in the combustion process is fluidized-bed combustion. By the use of limestone or dolomite as additives in a quantity 1·8 times stoicheiometric, sulphur oxides could be reduced to meet rigorous emission limits (100 ppm v/v SO$_2$). Cost calculations based on a 660 MW unit showed that compared with the wet scrubbing system the retention of sulphur in the fluidized bed was cheaper, particularly for coal sulphur contents of less than 4%. In Table 10 the wet scrubbing process, and the fluid bed process, are compared with the Monsanto CATOX flue gas desulphurization process.[30] The latter was chosen because it has been tested on a fairly large scale for a sufficiently long period. Table 10 clearly shows the potentials of the fluid bed combustion process for purposes of desulphurization.

Esso (UK) has carried out experiments for the US Environmental Protection Agency using a fluid bed of lime particles for the gasification and desulphurization of heavy fuel oil.[31] Their study demonstrated that continuous gasification, regeneration, solids recirculation and sulphur removal was feasible. Both the gasification and the regeneration of solids took place in

FIG. 7 Schematic diagram of fuel-rich after burner system for regenerative furnaces (after Battelle Memorial Inst.[2])

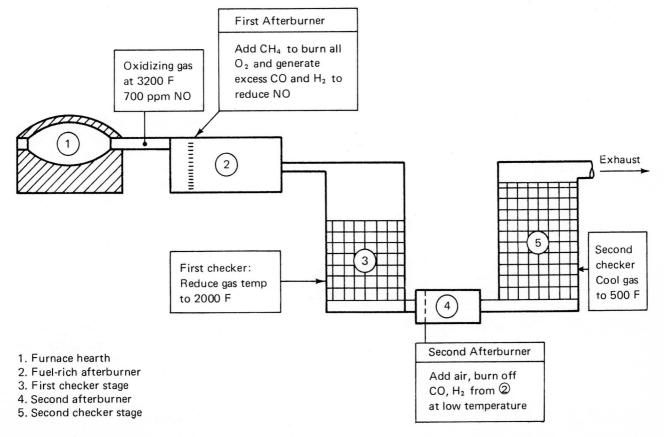

First Afterburner

Add CH$_4$ to burn all O$_2$ and generate excess CO and H$_2$ to reduce NO

Oxidizing gas at 3200 F 700 ppm NO

Exhaust

First checker: Reduce gas temp to 2000 F

Second checker Cool gas to 500 F

Second Afterburner

Add air, burn off CO, H$_2$ from ② at low temperature

1. Furnace hearth
2. Fuel-rich afterburner
3. First checker stage
4. Second afterburner
5. Second checker stage

TABLE 10   Cost comparison of desulphurization process (after Thurlow, G. G.[30])

Sulphur oxide control process	% increase in station cost Station load factor %	Capital	Generation
1. Monsanto CAT-OX	50	13-19	29 (24) †
	85		17 1(3) †
2. Limestone * injection/ wet scrubbing	50	3	9 ‡
	85		7 ‡
3. Fluidized bed boiler/ limestone*	50	<1	3
	85		2·7

*Limestone addition at 150% stoicheiometric requirements for 100% sulphur removal.
†Allowing a credit of £6/ton for the H_2SO_4 produced.
‡Includes reheating the stack gases.

fluidized beds. While results of this investigation show promise for large-scale application, they point also to several areas that require additional study to improve long-term process operability.

Mechanical (cyclone) and electrostatic precipitators have been developed to a high standard. Further improvements could be obtained if agglomeration of fine particles could be promoted. This affects primarily mechanical precipitators. In electrostatic precipitators the possible lowering of the sulphur content of the exhaust gas presents a problem. Efficiency of precipitation falls when the electrical resistivity of the fly ash exceeds values of about 2×10^{10} ohm cm. SO_3 present in the flue gases and adsorbed on the surfaces of particles lowers the electrical resistivity and hence makes them easier to capture in electrostatic precipitators. It is a research problem to find some additive, or to modify the combustion process so as to compensate for this action of SO_3 in low sulphur-emission systems.

There are very few dependable data available on the effects of fuel additives. Most commercial additives claim to perform functions in the fuel as listed below:

Combustion catalysts.
Corrosion inhibitor.
Demulsifying agent.
Dispersant.
Gum inhibitor (anti-fouling agent).
Pour point depressant.

Recently the Environmental Protection Agency has carried out a systematic study with 206 different additives, and burning a distillate fuel oil.[32] The conclusions of their investigations were as follows:
(1) No additive reduced NO_x or SO_x emission and only 17 out of 206 reduced particulate emission.

TABLE 11   Proposed exhaust emission standards for new aircraft gas turbine engines

Thrust lb force	6 000	6 000 to 29 000	29 000
CO	2·2	2·1	1·7
Hydrocarbons	1·0	0·4	0·4
smoke number	35	25	20
NO_x	3·7	3·2	3·0

Emissions of CO, hydrocarbons and NO_x in lb mass 10^3 lb force-h /cycle.
Smoke number refers to a filter smoke spot reflectance test.
Cycle refers to simulated landing and take-off operation.

(2) Only metallic additives containing cobalt, iron and manganese appreciably reduced particulate emission. However, the unknown toxicity of new emission they create makes their use questionable.

As a result of their investigation, the EPA have concluded that when burning distillate fuel oil, burner modifications are a more suitable route to air pollution control than the use of fuel additives.

3.2. Emission from gas turbine engines

Gas turbine engines are used mainly for aircraft propulsion, but are also increasingly employed for power generation in stationary plant and as automotive engines, particularly in heavy lorries. The pollutants of interest for aircraft in airports are CO, hydrocarbons and aldehydes (odour) which are emitted at idle and taxiing operations. During approach, and take-off, visible smoke is a primary pollutant. During high altitude cruise the principal pollutants are combustible particulates (invisible smoke) and oxides of nitrogen. Industrial stationary gas turbines do not operate for extended periods at low load, and thus do not emit significant quantities of CO, hydrocarbons and aldehydes. At high power levels the emission of combustible particulates and of NO_x is similar to those from aircraft gas turbines. This is, however, often due to the fact that stationary gas turbines are of the types originally designed for aircraft application. The less stringent operational conditions of stationary turbines (no necessity to satisfy high altitude relight requirements) gives the designer additional freedom to improve conditions by combustion modification.

At present automotive gas turbines can meet US emission standards except for NO_x: the emission level of which is four times higher than the proposed standards.

The proposed 1979 emission standards for new aircraft gas turbine engines are summarized in Table 11.

Typical combustor heat release rates at full power are 5×10^6 Btu/ft³ h atm, the residence time of the gas in the primary zone is 2 to 3 m sec with additional 5 to 6 m sec in the dilution zone. The combustor inlet temperature is up to 1 000°F. The outlet temperature is about 2 300°F, and the primary zone combustion temperatures are in the range of 3 600° to 4 000°F. The difficulty of tailoring combustion conditions to ensure low pollutant emission is increased by the greatly different combustion conditions corresponding to a 4:1 fuel-air ratio variation between full power and idling conditions, with a fuel-flow variation over a range of 50:1 and pressure variation within a range of 10:1.

Fig. 8 illustrates the 'performance envelope' within which combustors must operate satisfactorily at all conditions. The graph relates air-fuel ratio, combustor temperature rise, engine speed and flight altitude as parameters for a gas turbine engine.[34]

Table 12 represents emissions from a typical aircraft engine equipped with low-smoke combustors. It can be seen that CO and hydrocarbon emissions are high at low-power settings while smoke and NO_x emissions increase strongly with engine speed and reach their maxima at take off.

It is generally recognized that the combustor primary zone is the source of most of the smoke and NO_x emission from gas turbine engines. To reduce smoke emission the primary zone fuel-air ratio must not be allowed to run at high values. In recent design practice it is operated at a peak value of 2.0 and preferably limited to a fuel-air ratio of 1·5. (This compares with values of 3·0 at full power operation in previous designs). The reduction in fuel-air ratio in the primary zone is achieved by admitting more air through the swirler vanes into the combustor. The larger mass flow of air, together with good aerodynamic design of the swirl vanes, enables spatially uniform concentration distributions to be obtained in the primary zone. Further improvements can be achieved by finer atomization, or partial prevaporization of the liquid fuel. Premixing the fuel with the combustion air, and with combustion products prior to ignition, aids further the reduction of both smoke and oxides of nitrogen emission.

It is desirable that the primary zone air-fuel ratio be maintained constant over the load range. This requires either a variable-geometry combustor or a multinozzle design where all the fuel nozzles would operate at full load, but only alternate fuel nozzles at low loads. While such modifications would complicate somewhat both design and operation of the combustors they could considerably improve both idle emissions of CO and hydrocarbons, and high load level emissions of smoke and of NO_x. There is promise also in the development of a 'staged' combustor[35] in which an additional fuel

TABLE 12 Emissions from a typical turbofan aircraft engine with low-smoke combustion (after Hazard, H. R.[34])

Throttle position		Idle	Approach	Cruise	Takeoff
Thrust, lb		525	5 100	11 500	13 600
Fuel flow, lb/h		780	3 196	6 740	8 330
Fuel-air ratio		0·0074	0·0109	0·0148	0·0176
Dry carbon:	ppm	—	—	—	—
	lb/h	0·34	1·53	2·38	3·12
	fuel	0·00044	0·00048	0·00035	0·00038
CO:	ppm	590	52	17·2	15
	lb/h	60·26	14·8	7·6	6·9
	lb/lb fuel	0·07726	0·00462	0·00113	0·00083
HC:	ppm as CH_4	200	3·7	1·8	1·3
	lb/h	11·62	0·61	0·39	0·32
	lb/lb fuel	0·01492	0·00019	0·00006	0·00004
NO_x	ppm	7·4	23·2	67·6	98·0
	lb/h	0·75	6·58	29·72	44·98
	lb/lb fuel	0·00097	0·00206	0·00441	0·00540

injector is used downstream of the primary zone. Preliminary theoretical and experimental studies on a simple model combustor of this design have confirmed that fuel staging has a potential for reducing NO_x emission from gas turbine combustors.

3.3. Internal combustion engines

The major difference between the continuous combustion in industrial processing and energy production, and that in internal combustion engines is that combustion in the latter is a transient non-steady state process which takes place in a highly cooled combustion chamber of large surface-to-volume ratio. For purposes of discussion internal combustion engines can be grouped into: (a) spark ignition; and (b) compression ignition engines.

Spark ignition engines take in a homogeneous mixture of fuel and air on the suction stroke, compress this mixture and ignite it by means of a controlled ignition source such as a spark plug. The power output of the engine is controlled by a throttle in the intake port which controls the quantity of mixture taken in. The fuel/air ratio of the mixture remains nearly constant for all loads. Upon ignition, combustion proceeds by means of a flame front travelling through the combustion chamber. In contrast, the intake of a compression ignition or diesel engine is unthrottled and the charge taken in on the intake stroke consists of air only. Fuel is injected near the end of the compression stroke and ignites spontaneously due to the high temperature of the compressed air. No moving flame front exists, and the fuel/air mixture at each point is burned when the conditions of temperature, mixture strength, etc., are suitable. At the time of combustion, the fuel/air mixture is highly heterogeneous (liquid droplets burning in air). The power output of the diesel engine is controlled by varying the fuel input, which means that the fuel/air mixture is extremely lean at low engine loads. These operating characteristics account for the differing behaviour in respect of pollutant emissions.

With spark ignition engines the pollutants of principal concern are CO, hydrocarbons, NO_x polynuclear aromatics (PNA), lead and other particulates. The gaseous hydrocarbons are primarily light paraffins and olefins,

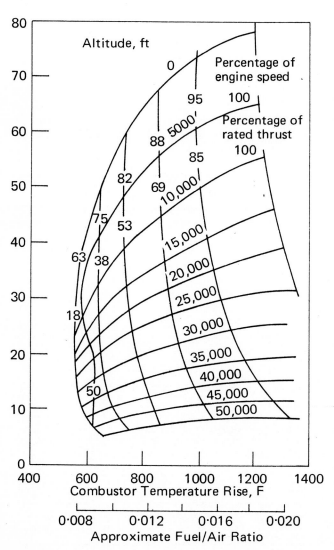

FIG. 8 Typical gas turbine combustor operating characteristics (after Hazard, H. R.[34])

along with aromatics aldehydes and other oxygenates with unmodified fuel molecules being a minor constituent of the hydrocarbon emissions. The emission of smoke and odour are not generally pollutants of great concern at the moment.

The compression ignition or diesel engine pollutants of major concern are CO, HC, PNA, NO_x, odour and smoke. A large proportion of their hydrocarbon emissions are unmodified fuel molecules. Diesel CO emissions are very low compared with petrol engines. The emission of smoke and odour from diesel engines is very noticeable, and is therefore highly objectionable. The range of emission levels characteristic of present diesel engines is shown in Table 13.[36,37,38,39] This can be compared with Table 14 which shows emission levels for a highly loaded petrol engine at constant speed.

TABLE 13 Characteristic emission levels for diesel engines at constant speed.

Pollutant	Idle	Full load
HC ppm	20 to 300	150 to 1000
CO per cent	0·01 to 0·1	0·03 to 0·3
NO_x ppm	200 to 400	750 to 2500

TABLE 14 Emission levels for highly loaded constant-speed spark ignition engine[2]

Pollutant	Concentration	Mass g/hp h
HC ppm	100 to 200	1·7 to 3·4
CO per cent	1·0 to 2·0	30 to 60
NO_x ppm	2 000 to 4 000	10 to 20

The range of values shown in Table 13 is due to the different types of diesel engine, i.e. two-stroke, four-stroke, turbo-charged, naturally-aspirated, direct-injection, pre-combustion chamber-injection.

Of the various spark-ignition engines the two-stroke engines have markedly different emission characteristics for two basic reasons. Firstly, as the products of combustion are scavenged out by the unburned air/fuel mixture, raw fuel finds its way into the exhaust. Secondly since lubricating oil is usually mixed with the petrol some of this passes through the engine unburned, producing a characteristic haze in the exhaust gases. Table 15 shows emission levels observed in a study of a 'driving-cycle' test using a two stroke motor cycle engine.

TABLE 15 Emission levels for a two-stroke engine in a seven-mode driving cycle[40]

Pollutant	Concentration
HC ppm	2 000 to 6 000
CO percent	1·5 to 7·5
NO_x ppm	100 to 500

where it can be seen that CO and hydrocarbons can reach high levels.

Satisfactory predictions of the observed levels of CO and NO_x in petrol engine exhausts have been derived on the basis of analytical investigations of the kinetics of the expanding products of combustion in an engine cylinder.[41,42] These investigations covered stoicheiometric or lean fuel/air ratios. Analytical predictions for fuel-rich mixtures have been less successful; however, CO and NO emissions can be predicted with reasonable

accuracy using the extensive experimental data available.

Unburned hydrocarbons in the exhaust gases have been shown to be largely the result of wall-quenching, or because of crevices or other regions in the combustion chamber where combustion does not take place.[43,44,45]

The characteristics of particulate emission, and of PNA emissions together with their mode of formation have been under investigation for some time.[46,47] However, this work is not complete and is continuing. Lead aerosols in exhaust emissions can obviously be eliminated be removing this antiknock additive from the fuel if the resultant loss in engine performance can be accepted.

The effect of the many design and operating variables on spark-ignition, engine-exhaust emissions is summarized in Table 16.[48] The major direction or change is indicated by an arrow. If the change is relatively small a horizontal line is drawn. Obviously the conclusions one draws about a change depends on the starting point from which the change is made. This table assumes that the starting point is a vehicle which has no special exhaust emission control, such as a 1967 American production car.

At present it is not known whether future improvements in engine and component design will enable manufacturers to meet future USA Federal Regulations. In view of this fact immediate attention is being mainly directed to the fitting of attachments and devices in the exhaust pipes to reduce emissions. Three main methods have had considerable attention by engine manufacturers. These are exhaust gas recirculation devices, thermal reactors or after-burners, and catalytic reactors. At the same time research is being carried out to examine the effects of operational variables on the efficient operation

TABLE 16 Effect of design and operating variables on exhaust emissions and engine air flow (after Patterson, D. J.[46,48])

Variable increased	HC conc.	CO conc.	NO conc.	CH₂O conc.	Intake mass flow constant load
Air-fuel ratio	↓↑	↓↑	∧	↓↑	↑
Load	—	—	↑		↑
Speed	↓	—	↑↓		↑
Spark retard	↓	—	↓	↓	↑
Exhaust back pressure	↓	—	↓		↑
Valve overlap	↓	—	↓		↑
Intake manifold pressure	—	—	↑		↑
Combustion chamber deposits	↑	—	↑		—
Surface to volume ratio*	↑	—	—		—
Combustion chamber area	↑	—	—		—
Stroke to bore ratio	↓	—	—		↑
Displ. per cylinder	↓	—	—		—
Compression ratio	↑	—	↑		↓
Air injection	↓	↓	—↑	↑↓	↑
Fuel injection	↓	↓	↑		—
Coolant temperature	—↓	—↓	↑		—

*Engine changes which decrease surface to volume ratio reduce heat loss to the coolant. As a result NO concentration may increase.

of these attachments. For example, for the efficient operation of catalytic reactors to reduce oxides of nitrogen a predetermined amount of carbon monoxide has to be present in the flue gas. It is interesting to note that the use of lead containing additives results in rapid poisoning of the catalysts in catalytic reactors, which is an additional reason for their elimination from the fuel.

The variation of the exhaust gas emissions from rotary petrol engines of the Wankel type with variation of air/fuel ratio, ignition timing, etc., are qualitatively the same as those from the reciprocating engine. Hydrocarbon emission levels, however, tend to be much higher and NO_x levels significantly lower. This is due to the poorer mixing of the gases in the combustion chamber, and the fact that the surface to volume ratio is higher than in reciprocating engines which results in lower peak gas temperatures.[49]

The two principal examples of combustion process modification that can result in reduced emissions from petrol engines are exhaust gas recirculation, and the use of stratified charge engines, perhaps better known as petrol injection. The latter have been under investigation for many years. Only recently have they possibly been regarded as low emission engines.

The diesel engine combustion process is not well understood compared with existing knowledge of petrol engine combustion. This is because of the heterogeneous nature of the process. The fuel injection, mixing, vaporization, ignition and combustion processes that take place are complex, and difficult to describe mathematically, and much research remains to be done especially in the field of odour measurement, detection and source identification. The results of such research should give an insight into possible modifications to the diesel combustion process to give reduced emissions. Meanwhile there is considerable scope for innovation, and for empirical studies into the way that fuel is introduced and burned.

4. CONCLUSIONS

Research in three major categories, on the effect of air pollution, on new and improved methods of control, and on the laws governing the dispersion of pollutants in the atmosphere, is a necessary requisite for an air pollution control programme.

The largest contributors to air pollution are processes of combustion of fossil fuels.

The major combustion-generated pollutants are non-combustible solids, unburned gases and solids, oxides of sulphur and oxides of nitrogen. Secondary pollutants form as a result of interaction of some of these pollutants with naturally occurring compounds, aerosols and sunlight.

Further research is necessary to provide information

(*a*) for the development of new and improved diagnostic techniques used both in the laboratory, and in the field for determining pollutant concentration;

(*b*) for the better understanding of the mechanism of formation of pollutants in combustion processes, and also in the atmosphere (secondary pollutants).

Based on a discussion of control techniques that can be used in three major areas of combustion-generated pollution: stationary combustion sources, aircraft engines and automobile engines, it is argued that the most prom-

ising methods of control are those aimed at combustion modification.

It is recognized that not all pollutants respond equally to such modifications, and there is need for parallel research and development of processes that remove pollutants from the fuel (e.g., desulphurization), and devices for precipitating or catalytically reducing pollutant emission.

It is recommended that, because of the promise shown by methods of combustion modification for emission control, and also because of the complexity of the research and development problem, a 10-year plan for research and development on air pollution control by combustion modifications be commissioned. This would review the practice of air pollution control methods in the UK and would identify gaps in combustion technology and discuss various routes for research and development, together with estimates of technical and economic feasibility for the application of results. Such a plan would be useful for the formulation and planning of an air pollution control programmes and it could serve also as a source of reference for research organizations, universities and research-sponsoring agencies for co-ordinating research in this important area.

5. REFERENCES

1. BARRATT, R. E. Seminar on New developments for combustion engineering, Penn. State University, July 1971.
2. The Federal R. & D. plan for air-pollution control by combustion modification. Report prepared for EPA under contract CPA 22-69-147 by Battelle Memorial Institute, Columbus, Ohio, 1971.
3. BROADBENT, D. H. Pollution as an international problem. Lecture at Sheffield University, 1973.
4. DERWENT, R. G. and STEWART, H. N. M. *Atmospheric Environment*, 1973, **7**, 385-401.
5. 108th Annual Report on Alkali etc. Works 1971, p.49, HMSO.
6. LEE, K., THRING, M. W., and BEÉR, J. M. *Combustion & Flame*, 1962, **6**, (8), pp. 137-145.
7. APPLETON, J. P. Paper 20, 41st meeting of AGARD (Propulsion and Energetics Panel) on Atmospheric Pollution by Aircraft Engines, April, 1973, London.
8. WHITTINGHAM, G. 3rd Symposium (International) on Combustion, pub. Williams & Wilkins Co., 1949, p. 453.
9. HEDLEY, A. B. The mechanism of corrosion by fuel impurities, CEGB Marchwood Conference, 1963. Butterworths, London, p. 11, p. 204.
10. HEDLEY, A. B. *J. Inst. Fuel*, 1967, **40**, (April), 224.
11. ZELDOVICH, J. *Acta Physiochim*. URSS 1946, **21**, 577.
12. THOMPSON, D., BROWN, T. D., and BEÉR, J. M. *Combustion & Flame*, 1972, **19**, 69-79.
13. GIAMMAR, R. D., and PUTNAM, A. A. Guide for the design of low-noise level combustion systems, A.G.A. Basic Research Project BR-3-5, Battelle Memorial Institute, Columbus, Ohio, USA, Jan. 1971.
14. PUTNAM, A. A. Combustion driven oscillations in industry, Fuel & Energy Science Series, ed. J. M. BEÉR, Elsevier Pub. Co., 1971.
15. GUPTA, A. K., SYRED, N., and BEÉR, J. M. *J. Inst. Fuel*, 1973, **46**, (384), 119-123.
16. TILSON, S. 'Science & Technology,' New York, June, 1965.
17. SMITH, T. B., WEINSTEIN, A. I., and ALKEZWEENEY, A. J. Report No. MR 170-FR-912, Meteorology, Research Inc., Altadena, CA, USA, 1970, 42p.
18. STERN, A. C. Air pollution, 2nd edn., 1968, Vol 1, p. 68. Academic Press.
19. SCHERER, G., and TRANIE, L. A. Paper 83, 14th Symposium (International) on Combustion, Penn. State University, 1972, Combustion Institute, USA.
20. British Patent No. 1 288 824 (UK), 6 May 1971, Improvements in or relating to the control of burners.
21. BREEN, B. P., in Emissions from continuous combustion systems, Ed. W. CORNELIUS & W. G. AGNEW, Plenum Press, New York & London, 1972, pp. 325-340.
22. DIBELIUS, N. R., and MARKS, W. F. *Mechanical Engineering* (ASME), March 1973, p.77.
23. GILLS, B. G. *J. Inst. Fuel.*, 1973, **46**, 71.
24. SHAW, J. T. *J. Inst. Fuel.*, 1973, **46**, 170.
25. BARTOK, W., CRAWFORD, A. R., and PIEGARI, G. S. Paper No. 81, 14th Symposium (International) on Combustion, Penn. State University, 1972, Combustion Institute, USA.
26. HEAP, M. P., LOWES, T. M., and WALMSLEY, R. Paper 82 (ibid.).

27. SALOOJA, K. C. *Journal of Fuel & Heat Technology*, January 1972.
28. Reference 2, page IV-11.
29. PLUMLEY, A. L., JONAKIN, J., MARTIN, J. R., and SINGER, J. G. *Combustion*, 1968, **40,** (1), (July), p.16.
30. THURLOW, G. G. Private Communication.
31. Final Report OAP Contract CPA 70-46, Report No. ER. 1KQA-F. 72, June 1972, Esso Research, Abingdon, UK.
32. US Environmental Protection Agency Office of Air Programmes Report No. AP-87, June, 1971.
33. SAWYER, R. F., *et al.* Paper 22, AGARD Symposium as Reference 7.
34. HAZARD, H. R. Combustors, Chapter 5 in Gas Turbine Engineering Handbook, Gas Turbine Publications Inc., Stamford, Connecticut, 1966 (2nd Ed. 1970).
35. LEFEBVRE, A. H., and FLETCHER, R. S. Paper No. 30, AGARD Symposium as Reference 7.
36. YUMLU, V. S., and CAREY, A. W. SAE Preprint No. 680420 (1968).
37. PEREZ, J. M., and LANDON, E. W. *SAE transactions* 1968, **77,** Sect. 2, p.1516.
38. MERRION, D. F., ibid, p.1534.
39. MARSHALL, W. F., and HUM, R. W. ibid, Sect. 3, p. 2139.

40. WIMETTE, H. J., and VAN DEVVEER, R. T. Report on the determination of mass emissions from two cycle engine operated vehicles prepared for NAPCA/DHEW by Olsen Laboratories Inc., (Jan. 1970), Contract No. CPA 22-69-91.
41. NEWHALL, H. K. 12th Symposium (International) on Combustion, The Combustion Institute, Pittsbrugh, 1969, p. 603.
42. LAVOIE, G. A., HEYWOOD, J. B., and KECK, J. C. *Combustion Science & Technology*, 1970, **1,** p. 313.
43. SCHEFFLER, C. E. *SAE Transactions*, 1967, **75,** Section 1, p. 571.
44. DANIEL, W. A. ibid, 1968, **76,** Section 1, p. 774.
45. DANIEL, W. A. SAE Reprint No. 700108 (1970).
46. HABIBI, K. *Environmental Science & Technology*, 1970, **4,** (March), p. 239.
47. BEGEMAN, C. R., and COLUCCI, J. M. SAE Preprint No. 700469 (1970).
48. PATTERSON, D. J., and HENEIN, N. A. Emissions from combustion engines and their control, Ann Arbor Science Publishers Inc., Michigan, USA 1st Ed. 1972, p. 178.
49. MUROKI, T. Inst. Mech. Eng. London, Conference Proceedings on Air pollution control in transport engines, November 1971 p. 304 (paper C.152/71).